Bioinspired Sensing, Actuation, and Control in Underwater Soft Robotic Systems

Derek A. Paley • Norman M. Wereley
Editors

Bioinspired Sensing, Actuation, and Control in Underwater Soft Robotic Systems

Springer

Editors
Derek A. Paley
Department of Aerospace Engineering and
Institute for Systems Research
University of Maryland
College Park, MD, USA

Norman M. Wereley
Department of Aerospace Engineering
University of Maryland
College Park, MD, USA

ISBN 978-3-030-50475-5   ISBN 978-3-030-50476-2 (eBook)
https://doi.org/10.1007/978-3-030-50476-2

This Springer imprint is published by the registered company Springer Nature Switzerland AG.
The registered company address is: Gewerbestrasse 11, 6330 Cham, Switzerland

Foreword

Aquatic animals occupy a wide range of niches in the ocean and shore, including the ocean depths, and have evolved a wide range of mechanisms for propulsion, manipulation, and bio-sensing to enable them to thrive in environments that are challenging for human technology. If we are able to extract the principles of sensorimotor control, biomechanics, and fluid dynamics of underwater propulsion and control in aquatic organisms and translate these principles of nature into technological capabilities, this would have the potential to exceed current engineering practice and enable animallike agility, maneuverability, and manipulator dexterity in future autonomous underwater systems.

This is a propitious time for engineering such bioinspired systems, because we are on the verge of creating soft matter that can emulate the diverse capabilities of flexible appendages of living organisms and accomplish tasks that are not practical for rigid robots. Robotics systems have the potential to be transformed by the introduction of smart materials, flexible electronics, additive fabrication, 3D printing, and new soft materials, such as elastomeric foams, hydrogels, and dielectric elastomers with both sensing and actuation properties. Indeed, the suite of tools available to engineers, scientists, and biologists now enables the embedding of sensing and control electronics in these new materials. However, many basic research challenges and questions remain for this technology before revolutionary systems exploiting multifunctional materials can be realized. One promising path is to specify and exploit the biological design principles by which living organisms achieve distributed computing, sensing, actuation, and power. Flexible controlled structures are ubiquitous in the natural world, such as the arms or tentacles of squid, octopus, jellyfish, and sea stars, or mantles of cephalopods in the marine environment, and elephant trunks in the terrestrial domain.

The ability to chemically tailor materials with local domains exhibiting intrinsic sensing and actuation properties, variable compliance, and adhesion shows tremendous potential for smart robotic components. However, faced with such a wide state space of possible material and mechanical property distributions for these soft heterogeneous structures, the question becomes this: *What functionality is desired for these novel structures?* One viable path is to study the behavior, physiology, function, and form of aquatic animals in order to characterize, model, and build prototype biologically inspired soft appendages.

From the engineering aspect, soft appendages can create challenging control problems, because they may have essentially infinite degrees of freedom, unlike prior mechanical robots with a small number of joints. Several of the chapters in this book outline approaches to the modeling and control of soft appendages. This book also features several timely reviews covering aquatic animal locomotion mechanisms, soft actuator designs for locomotion, and optimization of shape gait in multiple appendage entities. One chapter analyzes amphibious propulsion mechanisms and introduces a soft amphibious prototype. This volume also provides current overviews of soft materials relevant to robotics. Lastly, this book offers several original research efforts on the modeling, control, and fabrication of octopus arms and prototype jellyfish robots.

Office of Naval Research, Arlington, VA, USA Thomas McKenna
April 8, 2020

Preface

This book summarizes the latest research in the emerging field of bioinspired soft robotics for the underwater domain, primarily drawing from (but not limited to) an ongoing research program in bioinspired autonomous systems sponsored by the Office of Naval Research. The program has stimulated cross-disciplinary research in biology, material science, computational mechanics, and systems and control for the purpose of creating novel robots and robotic appendages for maritime applications.

College Park, MD, USA — Derek A. Paley
College Park, MD, USA — Norman M. Wereley

Contents

Chapter 1
Introduction

Derek A. Paley and Norman M. Wereley

Soft robotics is an emerging field that seeks to replace traditional (rigid) robotic components with flexible sensors and actuators that achieve greater flexibility and degrees of freedom. Bioinspired soft robotics seek to emulate the diverse capabilities of natural systems such as an elephant trunk or octopus arm through novel architectures, materials, and control designs. This book focuses on soft robotic systems that are designed to operate underwater. It includes representative research from the state of the art in the emerging field of soft robotics, with a special focus on bioinspired soft robotics for underwater applications. Topics include novel materials, sensors, actuators, and system design for distributed estimation and control of soft robotic appendages inspired by the octopus and sea star. It contains a comprehensive snapshot of state-of-the-art advances in bioinspired soft robotics with tutorial and original content reinforcing the book's subject. The chapters of the book explore the following themes: mechanisms and motion, materials and muscles, biological analysis and design, and modeling and control.

1.1 Mechanisms and Motion

Biological strategies that have survived through the evolutionary process can be used as inspiration to create robust locomotion of robots using soft materials. Chapter 2, "*Bioinspired Shape-Changing Soft Robots for Underwater Locomotion: Actuation*

D. A. Paley (✉)
Department of Aerospace Engineering, University of Maryland, College Park, MD, USA
e-mail: dpaley@umd.edu

N. M. Wereley
Department of Aerospace Engineering, University of Maryland, College Park, MD, USA
e-mail: wereley@umd.edu

D. A. Paley, N. M. Wereley (eds.), *Bioinspired Sensing, Actuation, and Control in Underwater Soft Robotic Systems*, https://doi.org/10.1007/978-3-030-50476-2_1

and Optimization for Crawling and Swimming," by Hermes et al., reviews how soft underwater organisms change their shape to enhance motion, catalogs how researchers have used soft active materials to mimic these abilities, and discusses how recent work uses optimization to produce more effective underwater robotic locomotion. The authors categorize underwater locomotion into crawling and swimming, where crawling uses contact with a solid substrate for moving through the fluid, and swimming relies purely on hydrodynamic forces for propulsion. They also summarize the main categories of actuators for soft robotics with a focus on how these actuators can be used to create shape-changing mechanisms. The biological inspirations and engineering applications described in this chapter form the basis for the design of soft underwater robots that locomote effectively by creating favorable interactions with fluid environments.

The underwater environment is one of the most hostile environments for both humans and engineered systems. To explore these hard-to-reach regions, engineers have created robust and powerful remotely operated vehicles and autonomous underwater vehicles. Although these existing systems are well suited to moving quickly through open water, they are often noisy, exhibit high energy consumption, and are rigid, thus unable to enter tight confined spaces or interact with fragile objects or organisms. Although traditional robotics systems are capable of fast and precise motions, they are not well suited to mimic or reproduce biologically inspired motions because of their lack of flexibility. Soft robotic systems, on the other hand, can create continuous deformations to produce motions that are highly adaptable to changing and unpredictable environmental situations. Chapter 2 compares and contrasts the principles of soft-material shape-changing locomotion with a focus on actuation and optimization for engineering applications.

Amphibious robots capable of transition from aquatic to terrestrial locomotion face significant challenges associated with propulsive efficacy in each environment. Conventionally, amphibious robots have utilized separate systems for aquatic and terrestrial locomotion, such as rotors and wheels, respectively. Recent approaches have attempted to consolidate the propulsive mechanism footprint and complexity in hopes of creating systems that mirror the performance and adaptability of living organisms. The crux of such a bioinspired design philosophy lies in integrating hydrodynamic profiles and terrestrial mobility into a cohesive robot architecture. Chapter 3, "*Amphibious Robotic Propulsive Mechanisms: Current Technologies and Open Challenges*," by Baines et al., surveys existing amphibious robots and identifies seminal designs. The authors synthesize findings to highlight open avenues of research for the continued development of amphibious robots, including a variable-stiffness morphing limb as a potential next-generation propulsor for amphibious robots.

1.2 Materials and Muscles

Soft actuators, often also referred to as artificial muscles, have been intensively developed in recent decades for constructing novel soft robots and machines. Diverse materials and structures have been designed and fabricated to exhibit various actuating behaviors. Chapter 4, "*Artificial Muscles for Underwater Soft Robotic System*," by Wang et al., reviews several representative soft actuators that have been recently widely explored, including pneumatic/hydraulic actuators, electroactive polymers, liquid crystal elastomers, responsive hydrogels, shape memory polymers, twisted fiber artificial muscles, and magneto-active elastomers. Their fabrication, performance, unique features, and modeling are discussed in Chapter 4, as well as special requirements of soft actuators for underwater robotic systems.

Among the various active soft materials developed for sensors and actuators inspired by biological muscles, stimuli-responsive hydrogels, a class of water-loaded polymers, exhibit large volume change and actuation strain upon environmental cues, enabling them to absorb and release water up to more than 90% of their total weight. These tissue-like, multifunctional, and multi-responsive hydrogels exhibit attractive sensing and actuation capabilities, qualifying them as potential candidates for artificial muscles used in next-generation underwater soft robotics. Chapter 5, "*Bioinspired Sensors and Actuators Based on Stimuli-Responsive Hydrogels for Underwater Soft Robotics*," by Lo et al., introduces a variety of stimuli-responsive hydrogels that can serve as soft sensors for local environment and strain monitoring, and as powerful actuators capable of rapidly generating high force. Recent progress demonstrates the versatility of smart soft materials and their potential for producing autonomous soft robots with self-diagnostic capabilities, built-in feedback control, and increased autonomy.

In the field of soft robotics, ionic polymer-metal composites (IPMCs) have shown great promise in the advancement of bioinspired actuators and sensors. Their affinity for use in aqueous environments, as well as their low-driving voltages, large deformation performance, and capacity for miniaturization, makes them excellent candidates for biomimetic soft robotic systems. Scientists and researchers have explored the many possibilities of using IPMCs in underwater applications as actuators and sensors, including the development of artificial fins and other biocomponents such as cilia arrays, artificial skin, and sensing systems akin to the fish lateral line. Chapter 6, "*Ionic Polymer-Metal Composite (IPMC) Artificial Muscles in Underwater Environments: Review of Actuation, Sensing, Controls, and Applications to Soft Robotics*," by Minaian et al., discusses current advances and implementation of IPMC-based artificial muscles, including an overview of material and fabrication techniques, examples of bioinspired actuator and sensor designs, utilization of shape memory properties and segmented electrodes for more complex actuation, and the application and performance of fabricated devices in underwater environments.

Chapter 7, "*Design and Analysis of Electrohydraulic Systems for Underwater Systems Utilizing Fluidic Artificial Muscle Actuators*," by Chapman and Bryant, explores the coupling between fluidic artificial muscles (FAMs) and the electrohydraulic systems that provide their motive power. FAMs are simple yet high force-density soft robotic actuators capable of operating in both atmospheric and underwater conditions. To explore the application of FAMs in underwater robotics, this chapter considers the case study of a single degree of freedom robotic arm actuated by a single FAM actuator. It presents a fully coupled dynamic model of the constituent subsystems and uses these subsystem models to investigate the effectiveness of using FAMs as actuators for such robotic appendages. This case study can be generalized to various electrohydraulic subsystems in order to consider multiple system design configurations prior to construction.

1.3 Biological Analysis and Design

Fish swim using a combination of active and passive movements. Active swimming is generated by muscles that produce an undulatory wave passing down the body of fish. The undulating body imparts momentum to the water and the fish moves forward. Passive movement, in contrast, occurs when fluid pushes on the body, resulting in undulatory motion without muscular involvement. Animals may simultaneously generate active movement while also experiencing the consequences of morphological characteristics such as fins that generate passive movement. Passive models of fish are easily manufactured, but not sufficiently complex to realistically model fish swimming. Hard robotic models can be active and complex, but they are difficult to design and manufacture. Chapter 8, "*A Soft Robotic Model to Study the Effects of Stiffness on Fish-Like Undulatory Swimming*," by Wolf and Lauder, describes two soft robotic fish models using pneumatic actuators, or PneuNets, that produce active swimming by deforming a central plastic backbone. Using these active pneumatic models, they demonstrate an important interaction between activation frequency and stiffness, and show that variable longitudinal stiffness between anterior and posterior PneuNets can lead to an increase in thrust without simultaneously increasing lateral forces.

Motivated by the swimming mechanisms of jellyfish, Chapter 9, "*A Biomimetic Robotic Jellyfish Based on Shape Memory Alloy Springs*," by Kazemi-Lari et al., describes a novel concept for a soft biomimetic underwater robot that imitates the shape and kinematics of a typical jellyfish. Compared to previous attempts by other researchers to design and fabricate a synthetic jellyfish, this prototype achieves comparable performance in a simpler, compact, solid-state design. The body of the robot is a thin, flat slab of silicone rubber in the shape of a central disk with radially protruding flaps. The central disk of the body is cast with an embedded pre-stretched shape memory alloy (SMA) spring, which is employed as an artificial muscle to create the swimming motion.

Despite recent advances in the field of soft robotics, localized actuation, sensing, and control remain inadequately addressed. While soft robots offer advantages in adaptable configurations and deformation, they still lack precision, speed, and output force. The octopus is an exemplary model for the design of soft robots, offering solutions to such obstacles, with its powerful and agile arms and dexterous, highly precise suckers. Octopus arms are muscular hydrostats, achieving movements through the conservation of volume, yet the forces they can exert through their arms and suckers are impressive. The knowledge obtained about the sensing, actuation, and control mechanisms of octopus arms and suckers will greatly assist in implementing the distributed control of soft robot arms. Chapter 10, "*Control and Functionality of Octopus Arms and Suckers*," by Bagheri et al., highlights the key structure–function relationships of octopus arms and suckers, and how they have inspired the field of soft robotics.

1.4 Modeling and Control

The octopus offers an enticing paradigm for the control of distributed, high-dimensional, underwater systems. Octopus arms are composed almost entirely of muscles, arranged in highly organized patterns, allowing active control of bending, twisting, and stretching. In particular, the octopus can form pointed joints along an arm to quickly fetch objects from a distance, and can use its arms to crawl bipedally on the seafloor. Chapter 11, "*Octopus-Inspired Arm Movements*," by Ling and Kanso, analyzes image data of fetch and crawl motions, then reproduces these behaviors in a three-dimensional elastic filament model of the octopus arm. The authors first constrain the tip of the arm to follow prescribed trajectories consistent with experimental observations and then reverse engineer the active internal forces and moments that produce compatible full arm movements, thus developing blueprints for basic motion primitives. The chapter analyzes the effect of compliance on robustness of these motions to environmental variations and the implications for the motor control of cephalopods and the development of control strategies for soft robotic systems.

Chapter 12, "*Decentralized Estimation and Control of a Soft Robotic Arm Using Linearized Beam Model*," by Shivakumar et al., uses partial differential equations (PDEs) to design decentralized estimation and control laws for a segmented octopus arm. The dynamics of the soft robot arm are formulated as a nonlinear PDE, which is then linearized about setpoints to obtain a linear PDE representation similar to linear Euler–Bernoulli beam equations. The authors use this linearized PDE model to design infinite-dimensional control and estimation laws. The optimal controllers and observers are then discretized during the implementation phase to perform operations such as shape tracking. The chapter shows that the discretized observer or controller can be implemented in a manner that allows decentralized operation in the robot arm.

Soft swimming robots are primarily composed of elastically deformable materials, which typically make up the robot's body, limbs, and/or fins. Such robots can swim by moving their limbs, flapping their fins, or undulating their body in order to control thrust and direction. Chapter 13, "*Modeling Soft Swimming Robots Using*

Discrete Elastic Rod Method," by Huang et al., presents a technique to model these soft swimming robots using a computational framework based on the method of discrete elastic rods (DER). This approach to soft robot simulation draws inspiration from methods to simulate slender structures that are widely used in the computer graphics community. In this framework, the soft robot limbs or fins are treated as flexible rods that deflect in response to internal actuation and surface tractions from contacting bodies and the surrounding fluid. The chapter applies this model to the special case of a sea-star-inspired robot composed of radiating limbs that produce motion through bending and hydrodynamic drag, beginning with an overview of the DER-based framework and followed by simulation results for forward swimming and turning.

Many organisms achieve locomotion through undulatory traveling wave motions, both on land (e.g., snakes, caterpillars, and worms) and underwater (e.g., eels and flagellar single-celled organisms). Recently, caterpillar locomotion has been modeled using the theory of planar discrete elastic rods (PDER). Chapter 14, "*Distributed Control of a Planar Discrete Elastic Rod for Eel-Inspired Underwater Locomotion*," by Scott et al., considers a planar discrete elastic rod model with active control over the intrinsic material parameters of length and curvature distributed along its length. The authors introduce local curvature feedback control laws to drive the shape of the robot to a desired traveling-wave reference trajectory, where the phase of the wave is determined via a central pattern generator or via distributed control with a circulant communication topology. Through numerical simulations utilizing simple models of fluid–body interaction forces, the chapter examines undulatory gaits that give rise to net forward motion. These results show promise for the design of distributed feedback control laws in modular soft robotic systems.

Controlling the motion of soft robotic systems can be a challenging task and researchers have turned to biology to shed light on this complex problem. In particular, researchers have used mathematical models of central pattern generators (CPGs) to control the periodic movement of robotic systems. A useful property of CPGs is that their output can be altered by sensory feedback. However, questions remain about how to acquire sensory feedback from a compliant structure, and how to use sensory feedback to produce desired performance. Chapter 15, "*Bioinspired Neural-Based Control of Flexible Fish-Like Propulsors*," by Carryon and Tangorra, focuses on neural-based control of a flexible fish-like propulsor with spatially varying mechanical properties. The authors show how manipulation of the sensory feedback signal and sensory feedback architecture affect the entrainment state (i.e., the frequency and amplitude) of the fin. These results illustrate how sensory information may be used to modulate the performance of a flapping fin using feedback.

Chapter 2
Bioinspired Shape-Changing Soft Robots for Underwater Locomotion: Actuation and Optimization for Crawling and Swimming

Mark Hermes, Michael Ishida, Mitul Luhar, and Michael T. Tolley

2.1 Introduction

The underwater environment is one of the most hostile environments for both humans and engineered systems. To explore these hard-to-reach regions, engineers have created robust and powerful remotely operated vehicles (ROVs) and autonomous underwater vehicles (AUVs). Although these existing systems are well-suited to moving quickly through open water, they are often noisy, exhibit high energy consumption, and are rigid, thus unable to enter tight confined spaces or interact with fragile objects or organisms [1]. To solve these problems, we take inspiration from nature. Animals are quiet to avoid predation, energy efficient to reduce the amount of food required to survive, and can use their bodies' compliance to squeeze through tight openings. Although traditional robotics systems are capable of fast and precise motions, they are not well-suited to mimic or reproduce biologically inspired motions because of their lack of flexibility. Soft robotics systems, on the other hand, can create continuous deformations to produce motions

Mark Hermes and Michael Ishida authors contributed equally to this work.

M. Hermes · M. Luhar (✉)
Aerospace and Mechanical Engineering, University of Southern California, Los Angeles, CA, USA
e-mail: luhar@usc.edu

M. Ishida · M. T. Tolley
Mechanical and Aerospace Engineering, University of California San Diego, La Jolla, CA, USA

D. A. Paley, N. M. Wereley (eds.), *Bioinspired Sensing, Actuation, and Control in Underwater Soft Robotic Systems*, https://doi.org/10.1007/978-3-030-50476-2_2

that are highly adaptable to changing and unpredictable environmental situations [2]. For this reason, there is a wide overlap between soft robotics and bioinspired robotics. This chapter compares and contrasts the principles of soft material shape-changing locomotion with a focus on actuation and optimization for engineering applications.

We present two definitions that are instrumental in this chapter. First, *soft materials* are characterized by low hardness and high elasticity, which can be used for shape-changing behavior. The material properties of the components differentiate soft robotics from traditional rigid robotics. Soft materials have a bulk elastic moduli in the 10^4–10^9 Pa range, similar to those of soft biological materials (i.e., muscle, cartilage, etc.), whereas materials used in traditional robotics have moduli that are generally at or greater than 10^9 Pa (e.g., aluminum, 10^{10} Pa). We also include in our definition deformable structures and stiffness-changing mechanisms, both of which partially consist of rigid materials exhibiting soft properties in bulk. Thus, soft robots are systems that create autonomous, controllable motions using actuators made of soft materials or with composite structures made of rigid materials that have macroscopic properties similar to those of soft materials [3]. Some underwater organisms such as sea cucumbers consist of only soft materials, which motivates the use of soft materials for creating robots inspired by biology.

Second, *shape-changing* for locomotion in this work is defined as continuous deformation of the soft component(s) interacting with the surrounding media (i.e., the water for swimming systems and the substrate for crawling systems). A robot or organism with a soft interface between the inner body and the fluid environment can change macroscopic shape in one of the following ways: (1) by changing its volume, e.g., intaking water and expanding and (2) by changing the geometry of its body without changing its volume, e.g., undulating the body or rowing with flippers. However, a rigid system can produce similar categories of shape changes: (1) changing its volume by extending an appendage from an internal geometry and (2) changing the geometry of its body by rotating joints between rigid links. However, these examples would not result in continuous deformation in the surfaces interacting with the surrounding fluid, distinguishing it from soft shape changes.

Shape change is a useful mechanism for enhancing or coordinating locomotion through a fluid. We distinguish two primary methods of moving through a fluid: crawling and swimming. The characteristic difference between the two is the presence or absence of interaction with a substrate. Crawling requires a body to be partially attached to semi-solid objects, such as sand or coral. Crawling in nature is accomplished by a synchronous cyclic motion of appendages, exemplified by crustaceans and echinoderms (Fig. 2.1). This strategy has the benefit of continuous attachment to the substrate during locomotion, which reduces the need to control motion perpendicular from the surface, e.g., motion vertically through the water column. In addition, if an organism is moving within a low-velocity boundary layer on the seafloor, assuming the effect from shear stress is minimal compared to drag, crawling can reduce the drag and lift forces from moving in flow. These characteristics aid the stability of the locomotion, but it comes at the cost of speed.

Type of locomotion	Description	Animal	Schematic	Robot
Friction-based crawling (discrete)	• Friction between appendage and substrate allows appendage to push body • Maximum friction force is affected by lift and buoyancy • Discretized actuation	Crustaceans 	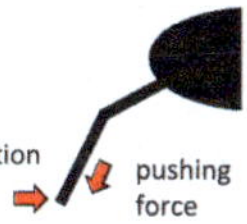	Rigid crab-like hexapod 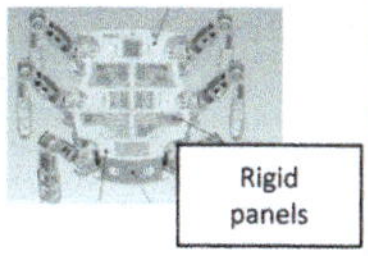
Adhesion-based crawling (distributed)	• Adhesion between appendage and substrate allows appendage to pull body • Maximum adhesive force is not affected by lift and buoyancy • Distributed actuation	Echinoderms 		Echinoderm-inspired tube feet 

Fig. 2.1 Crawling modes observed in animals. Friction-based crawling uses a pushing motion caused by friction between an appendage and the substrate. Adhesion-based crawling creates a separate force, e.g., suction or chemical, to create a pulling motion between the appendage and the substrate. Citations for right-most column: [4, 5]

Natural swimming modes are diverse and make use of a range of different hydrodynamic phenomena, including vortex shedding, jetting, and drag-based propulsion (Fig. 2.2). Swimming is a faster method of locomotion than crawling, but requires control of all six degrees of freedom (three translational and three rotational) instead of three degrees of freedom required for planar crawling motions (two translational and one rotational). However, legged crawling results in a larger configuration space for the robot than swimming does as crawling usually requires multiple legs with multiple degrees of freedom. In addition, swimming often requires energy expenditure to maintain a position in unobstructed flow, rather than using passive friction with a substrate.

The remainder of this chapter is structured as follows. Section 2.2 discusses actuators that can be used to create shape changes in soft robotics. Section. 2.3 details the biological mechanisms that create crawling and swimming motions in nature as well as relevant bioinspired robots. Then, Sect. 2.4 explores techniques to optimize the gait of soft actuators and manipulate hydrodynamic body forces through shape change. Finally, we discuss future research directions for soft underwater mobile robots in Sect. 2.5.

2.2 Actuation in Soft Robotics

Soft robots are distinguished from rigid or traditional robots by the compliance of their constituent materials. These soft materials can perform continuum motions that allow them to form complex shapes, make large deformations, and use their passive material properties to adapt to their environment. Actuators made from

Type of locomotion	Description	Animal	Schematic	Robot
Rear-body undulation	• Observed in carangiform, sub-carangiform, thunniform, ostraciiform • Thrust generated by reverse von Karman street vortex shedding • Oscillating motion of tails, flippers	Cetaceans, most fish 		Soft robotic fish (SoFi) 
Full-body undulation	• Otherwise named anguilliform swimming • Traveling wave produced along entire body accelerates fluid for a net forward thrust • Slower motions produced compared to rear-body undulators	Eels, manta ray pectoral fins 		Eel-inspired soft robot 
Drag-induced swimming	• Oar-like appendage • Thrust generated by drag force opposing direction of motion of appendage • Appendage changes orientation to minimize reverse thrust during the recovery stroke.	Duck, platypus, freshwater turtle 		Turtle-inspired limb 
Jet propulsion	• Water jet out of rear nozzle • Thrust generated by momentum conservation • Changing volume of the body increases acceleration via added mass effect	Jellyfish, octopus 		Scallop-inspired robot 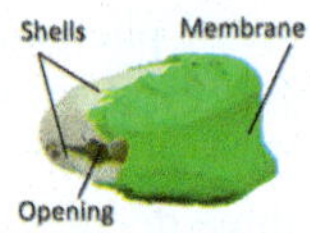

Fig. 2.2 Swimming modes observed in animals. Rear-body undulation and full-body undulation generates thrust through vortex shedding. Drag-induced swimming occurs by exploiting direction-dependent drag control. Jet-based swimming is seen in organisms that eject water through a cavity. Citations for right-most column: [6–9]

soft materials have similar elastic moduli to soft biological tissues such as skin or muscle and are more suitable than rigid structures to mimic biological materials [3]. We previously distinguished soft materials from rigid materials through the use of elastic modulus, wherein general soft materials fall between 10^4 and 10^9 Pa as compared to rigid materials, which have moduli from 10^9 Pa and above. This section details common soft actuators that can be used to create shape changes and shows examples of these actuators applied to soft robots for underwater applications or inspired by aquatic animals.

2.2.1 Soft Materials Actuated by Tendons

Tendon-driven soft robots consist of a soft material with tendon-like cables anchored at various points within the bulk material. The tendons are then connected to an actuator at the base of the soft material. When the actuator at the base applies tension to the cable, it pulls the anchoring point toward the base, causing the desired deformation [15]. The base actuator that pulls the tendon is often a traditional electromechanical component like a motor that can provide large forces [10], which can provide significant velocity or power while the bulk soft material produces continuum deformation or passive compliance. The electromechanical component can alternatively be replaced with a soft linear actuator [16] or smart material [17] to reduce the rigid components required. If a soft surface is attached to tendons at discrete nodes, displacement of the tendons creates a corresponding displacement of the node in a direction determined by the tendon routing. The known displacement of the node can allow control of the surface shape via a model of the soft surface's deformation [18].

2.2.2 *Fluidic Elastomer Actuators*

Fluidic elastomer actuators (FEAs) are a class of soft actuators powered by fluid pressure [19]. These actuators consist of a hyperelastic material with one or more internal cavities into which fluid pressure is applied and can be fabricated through a molding and curing process [20] or by directly 3D printing the soft material [21, 22]. Constraining elements like fibers or sheets are made from flexible but inextensible materials and pattern the deformation of the actuators (Fig. 2.3). When the constraining elements are asymmetric, the actuator creates a bending or twisting motion, e.g., pneumatic networks that bend toward an inextensible layer as the extensible material stretches [23]. Bending actuators can be used to change the shape of a soft robot from a flat contour to a curved surface. When the elements are symmetric, it creates an elongating or contracting motion, e.g., fiber-reinforced actuators that extend axially without expanding radially [24]. Extension or contraction actuators are used as artificial muscles in shape-changing mechanisms as they can change the height of nodes in an array to actively create a predictable surface [25].

Other fluidic actuators forgo an inextensible constraining layer to create unique geometries and motions. Bellowed fluidic actuators have no constraining layer and take advantage of the folding and unfolding of the bellows when pressure is applied to create contraction and extension with minimal straining of the material [26, 27]. Soft tentacles use multiple fluid channels in the bulk cylindrical elastomer cast around a stiffer elastomer core to creating a complex coiling motion when pressurized [28]. Pouch motors take advantage of laminate fabrication methods and can be used to create either a linear motion or a rotational motion around a joint [29]. Inflating a soft pouch can also be used to alter the shape of a body interacting with flow [11].

Type of soft actuator and previous uses	**Description**	**Schematic**	**Robot**
Tendon-driven Carangiform-swimming	• Bulk soft material pulled by a cable • Cable actuated by motors, smart materials, or other soft linear actuators • Good for high power applications	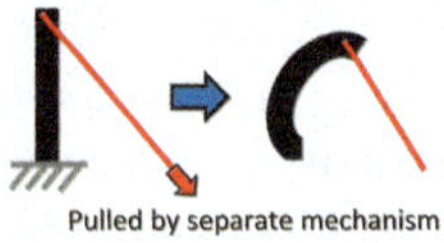	Robotic fish 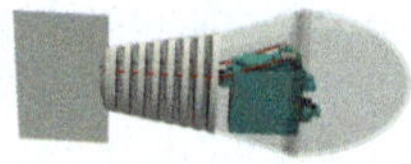
Fluidic elastomer: pneumatic network carangiform-swimming, jetting	• Pressure input causes deformation constrained by inextensible layer • Commonly used to create bending • Requires heavy valves and pumps	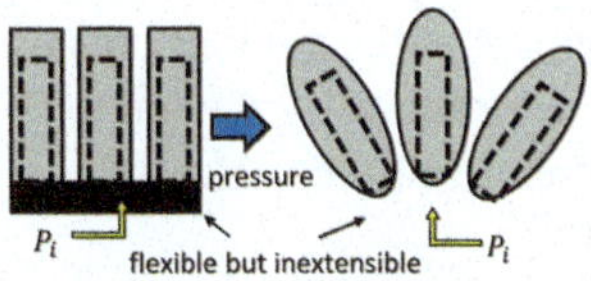	Soft robotic fish (So-Fi) 
Fluidic elastomer: bellows friction-based crawling	• Pressure input causes linear deformation from unfolding material • Bellows can use materials with low strain • Requires heavy valves and pumps	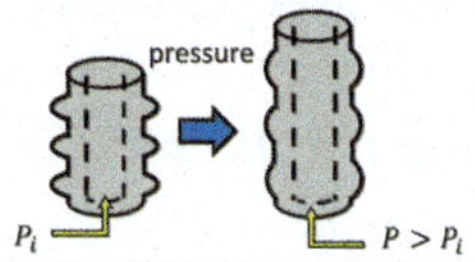	Soft walking robot 
Smart material: dielectric elastomer anguilliform-swimming, carangiform - swimming, jetting	• Maxwell stress causes compression in one plane and extension in the other • Including a strain-limiting layer causes the actuator to bend • High efficiency but requires high voltage and thin composite layers	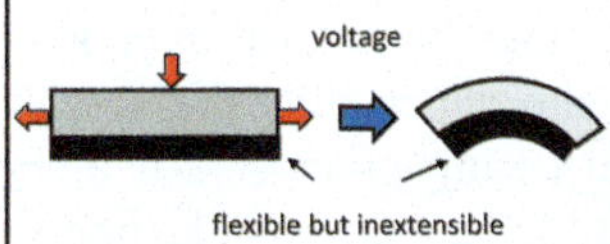	Soft eel-inspired robot 
Smart material: shape memory alloy friction-based crawling	• Deformable at low temperatures, • Restores to a trained shape at high temperatures • Lightweight but has significant time delays due to cooling	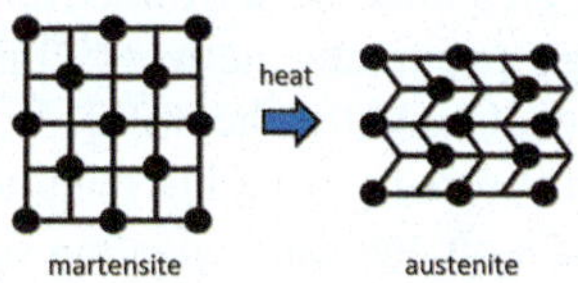	Soft starfish-inspired robot 
Smart material: liquid crystal elastomer Distributed-crawling	• External stimulus changes cross-linking in polymer, creating deformation • Easily reversible • Small form-factor, actuation slow to reverse without heat sink	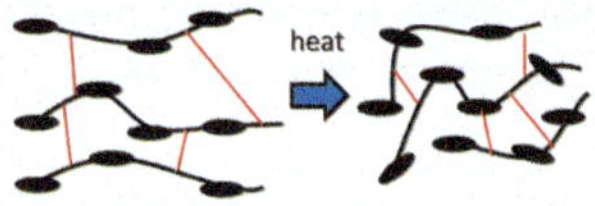	Soft echinoderm-inspired robot 

Fig. 2.3 Soft actuators demonstrated for underwater robots. Tendon-driven actuators use a bulk soft material actuated through tension. Pneumatic networks create bending using inflation constrained by a strain-limiting layer. Bellows produce an extending motion when pressurized as the material unfolds. Dielectric elastomers use Maxwell stress to create deformation in multiple planes. Shape memory alloys and liquid crystal elastomers use heat to trigger deformation. Citations for right column: [6, 10–14]

2.2.3 Smart Materials

Smart materials are a broad class of materials with properties that respond to an external stimulus such as heat or electrical fields. Those that are capable of significant deformation or stiffness change with a controllable and localizable stimulus can be used to create motion. Smart materials can be soft themselves or alternatively used to actuate soft materials. To create a complete actuator, the smart materials are packaged with components that provide the controllable stimulus, such as an embedded heater, a light source, or a high voltage supply [30].

Dielectric elastomer actuators (DEAs) are a type of smart material that uses Maxwell stress to create deformation in a soft material [31]. A soft dielectric polymer is sandwiched between two compliant electrodes, which can be made from conductive metals [32], polymers [31], and fluids [12]. When a voltage is applied to the electrodes, the voltage differential causes the Maxwell stress to compress the elastomer, creating planar extension in both directions perpendicular to the compression direction. The compression force is a function of the dielectric constant, the applied voltage, and thickness of the layer. A DEA can use either the compression force [33] or the stretching motion to create actuation [34]. The planar expansion of DEAs can be used to create shape change in similar ways to FEAs. The inclusion of a flexible but inextensible layer creates a bending motion and can change the shape of a curve interacting with flow [35].

Other smart materials use thermal energy to drive actuation, often created electrically using Joule heating. Shape memory alloys (SMA) are materials that are programmed to deform to a certain shape or pattern at high temperatures. When heat is removed, these materials then restore to the unactuated shape (usually a minimal restoring force is required) [36]. Shape memory polymers (SMP) are a class of polymer-based soft materials that typically deform under a heat stimulus, but are often not able to restore to the unactuated state without additional energy input [30]. Liquid crystal elastomers (LCE) are networks of polymers that organize their structure in predetermined orientations when a temperature threshold is met. This deformation is reversible, as the polymer will return to the original configuration when temperature is decreased [37]. These materials are coupled with resistive heating elements that create the controllable temperature change needed for actuation.

2.2.4 Stiffness Modulation

Active stiffness modulation couples actuation with controllable material properties that can be adjusted on a short time scale. Tunable stiffness can be accomplished by creating a phase change in the actuator materials, e.g., by using embedded heaters to melt a thermoplastic material, thereby softening the actuator [38, 39]. Pressure-induced tunable stiffness, also called jamming, can be accomplished by applying a negative pressure to small granules or high-friction layers within a membrane. This compresses the constituent materials together, creating an effective phase change from a fluid-like to a solid-like bulk material. Previous work has shown the use of jamming actuators for applications such as locomotion [40], haptic feedback [41], and a shape-changing surface [42]. Magnetic field-induced tunable stiffness can be accomplished by subjecting an elastomer with dispersed magnetic particles to a magnetic field, creating a change in stiffness [43]. These stiffness-changing mechanisms cannot create a shape change by themselves, but can be useful alongside the previous actuators. Active control of the stiffness of the surface allows it to be either deformable when the surface must change from one shape to another or rigid when the surface must withstand external forces. The ability of jamming actuators to form an arbitrary shape with a controlled stiffness may be coupled with the shape optimization techniques discussed in Sect. 2.4 to create favorable lift and drag profiles in flow.

2.2.5 Actuator Characteristics and Selection

The wide range of soft robotic actuators have different properties and each type of actuator is suitable for different applications (see Fig. 2.3). Tendon-driven components are often directly actuated by traditional electromechanical actuators like motors and servos, which can give them higher power relative to other soft actuators. However, the electromechanical components are generally bulky and heavy and the tendon routing becomes complex for actuators with many degrees of freedom. In addition, tendons pulling on a soft material often create deformations local to the tendon anchoring point rather than distributed along the continuum, which is not beneficial for operation in fluid. Fluidic elastomer actuators balance efficiency with speed and power, but similarly require valves and pumps, which are typically rigid, heavy, and expensive. Jamming actuators are well-suited for using stiffness to hold a position, but often require another actuator to create the motion to transfer between different states. Both fluidic elastomer actuators and jamming actuators require changing the amount of fluid in the actuator; for operation underwater, this generally necessitates using oil or water to maintain a constant buoyancy.

Smart materials have different strengths and challenges since they do not require the same electromechanical actuators like motors or pumps. Dielectric elastomers are capable of rapid actuation and can create large strains. To create high efficiency with a DEA, the actuator requires high voltage and a thin membrane that is susceptible to puncture. DEAs require soft electrodes that can be challenging to fabricate, so robots designed for underwater operation have capitalized on the surrounding fluid to create electrodes [7]. Shape memory alloys are often challenging to fabricate and train. Although shape memory polymers can be relatively easy to fabricate, they usually do not create reversible deformations, requiring energy input for restoration to their original positions. Liquid crystal elastomer actuators create reversible motions but can exhibit hysteresis and low actuator bandwidth due to the thermal actuation. When operated underwater, both SMA and LCE actuators benefit from increased cooling rates (shorter relaxation time) due to the heat transfer with the water, at the cost of slower heating rates (longer actuation time). The success of a soft robot is greatly dependent on the matching of the actuation type to the application and the corresponding advantages or disadvantages as discussed here.

2.3 Crawling and Swimming in Soft Underwater Robots

We characterize underwater locomotion broadly into swimming and crawling. As defined in Sect. 2.1, crawling occurs when a body remains in contact with a substrate (see Fig. 2.1), whereas swimming relies on hydrodynamic forces for propulsion. We further classify swimming based on the mechanism used for thrust generation: jetting, appendage-based propulsion, and undulatory swimming, which generally relies on the generation of vortices (Fig. 2.2). Many propulsive strategies in nature are characterized by deforming a soft continuum surface that interacts with the fluid environment. Using these soft materials often leverages the elasticity of the material to reduce energy consumption during the recovery portions of the robot's or animal's gait [44]. In addition, the soft actuators produce continuum surfaces that are more desirable for interacting with flow than sharp edges. Thus, application of new soft technologies can enhance and optimize motion for underwater travel. Animals have inspired many underwater soft robots, which have reproduced characteristic locomotion methods with varying degrees of success (see Fig. 2.4).

Here, we plot locomotion efficiency against overall system velocity. Efficiency is defined as the reciprocal of cost of transport (COT), a commonly used metric for quantifying the effectiveness of biological locomotion techniques:

$$efficiency = \frac{1}{COT} = \frac{mv}{P}, \tag{2.1}$$

where m, v, and P are the animal or robot mass, velocity, and power input, respectively. In general, animals are both faster and more efficient than their robotic counterparts with the exception of commercially available propeller-driven crafts. These large-scale systems such as submarines, AUVs, and ROVs can reach high speeds and efficiencies through large power usage. The penalty for high power usage in the cost of transport metric can be offset if the robot is moving a large amount of mass or at a high velocity. Furthermore, these propeller-driven systems are established technologies that include powertrain and energy efficiency optimizations. However, the high efficiency and speed exhibited by animals shows the effectiveness of soft materials and bioinspired gaits at meso or small scales.

The robots that use established rigid, electromechanical actuators are in general both faster and more efficient than their soft robot counterparts. While rigid swimming robots currently outperform soft robots, this is not necessarily an indictment

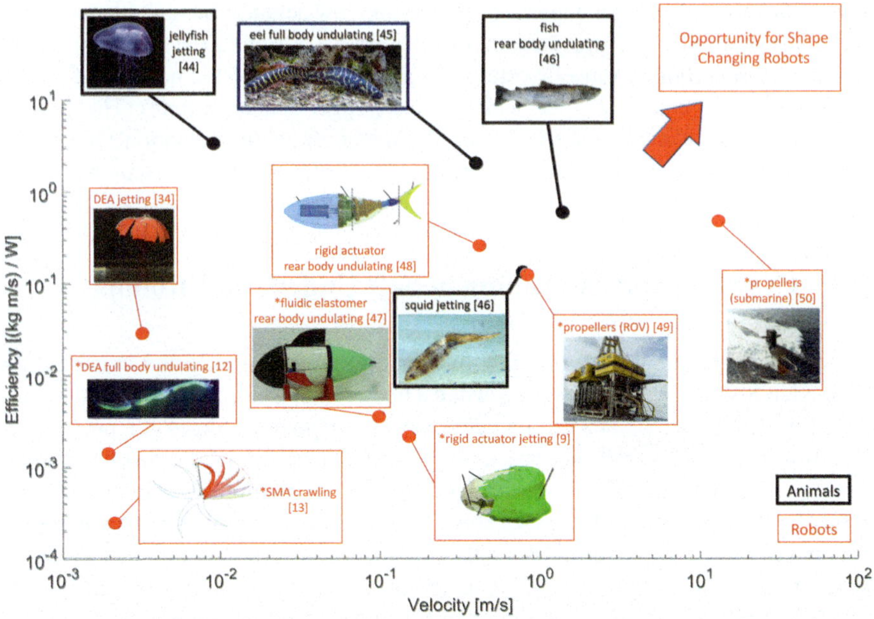

Fig. 2.4 Plot of the relationship between system velocity and locomotion efficiency defined as 1/COT for animals (black text and outline) and underwater robots (red text and outline). All velocity values taken directly from papers cited. Cost of transport values taken directly from papers when given; otherwise, values were calculated from other listed values or representative public data and are denoted with asterisk. Citations for animal metrics: jellyfish [44], eel [45], squid and fish [46], all images open source. Citations for robots: DEA jetting [34], SMA crawling [13], DEA full-body undulating [12], fluidic elastomer rear-body undulating [47], rigid actuator jetting [9], rigid actuator rear-body undulating [48], propellers (ROV) [49], propellers (submarine) [50]

of soft robots or the locomotion modes enabled by soft actuators. Soft robotics is a relatively new field and soft actuators are not as developed or optimized as rigid actuators like motors. We think that with more development, soft materials can be used to create high-efficiency propulsion, especially at small and low-power scales. One example of this can be seen in the cases of jetting using soft DEAs [34] and a DC motor [9]. The jellyfish robot that uses soft actuators is 13 times more efficient than the scallop robot that uses a rigid actuator, whereas the scallop robot is 50 times faster than the jellyfish robot. This illustrates the future potential usage of soft actuators for creating efficient locomotion.

Note that the examples given are what we consider to be representative of both locomotion strategies and actuators. There are many animals and robots that employ each type of locomotion, so each example lies within a range of demonstrated efficiencies and velocities for that category. When cost of transport calculations were not included in the publications, we calculated COT using other information such electrical power or flow rate and pressure. We also acknowledge that there are other metrics that measure efficiency, such as Froude efficiency (fraction of thrust creating motion over total thrust produced) and we chose cost of transport to incorporate total input (measured by power) normalized by system size (measured by mass) and performance (measured by speed).

2.3.1 *Crawling*

Crawling is a common method of locomotion for both terrestrial and aquatic animals. Although there are parallels between crawling in air and water, the difference in density between the two media creates significant differences in the crawling motion. Because fluid forces such as lift and buoyancy have a lower magnitude in air than in water, terrestrial animals rely mainly on friction with a substrate to move. In contrast, underwater crawlers such as echinoderms with hydrostatic skeletons that are close to being neutrally buoyant cannot primarily rely on friction forces. In general, neutrally buoyant crawlers use adhesive actuators arranged in a distributed pattern, whereas aquatic crawlers that are not neutrally buoyant use discrete appendages and rely on friction to create a crawling motion similar to terrestrial crawlers.

2.3.1.1 Friction-Based Crawling

Underwater friction-based crawling is similar to terrestrial friction-based crawling. The organism uses friction between its appendages and the substrate to push forward against the substrate. As a first approximation, the friction between the appendage and the substrate can be modeled as $F_f = \mu N$, where μ is the coefficient of friction and N is the normal force. For simplicity, this formulation does not consider static friction, which can vary significantly depending on the local rupture dynamics at the appendage-substrate interface [51]. In terrestrial locomotion where extreme air currents are absent, the normal force is equivalent to the weight W of the animal or robot. However, in aquatic locomotion, there are additional fluid forces to consider. The normal force includes both buoyancy $\rho V g$ and lift $\frac{1}{2}\rho C_L A U^2$, where ρ is the density of water; g is the acceleration due to gravity; V, C_L, and A are the volume, lift coefficient, and planform area of the animal or robot; and U is the velocity of the surrounding flow. So for underwater crawlers, friction is defined as

$$F_f = \mu \left(W - \rho V g - \frac{1}{2}\rho C_L A U^2 \right). \tag{2.2}$$

Friction is the force that generates locomotion. If other lateral forces opposing motion such as hydrodynamic drag or gravity (when the crawler is on a slope) are greater than the friction force, the crawler will not be able to move. Thus, to improve locomotion performance, a crawler can reduce the upward lift or buoyancy on its body or it can reduce the drag opposing its motion [11].

Crustaceans such as lobsters and crabs represent the majority of benthic animals that use friction-based crawling [52], though brittle stars and other echinoderms can produce this type of locomotion for land-based motion [53]. Since these animals rely on friction to move, they have adaptations or behaviors that influence the lift and drag on their bodies (see Eq. (2.2)). In high-flow environments, crabs improve their locomotion by changing their orientations with respect to flow, thereby decreasing their drag profiles [54], and crayfish change their body posture to create downward lift (i.e., downforce) to enhance the normal force and therefore the friction at the substrate [55]. On a longer timescale, sea stars displaced from a low-flow to a high-flow environment change the aspect ratio of their arms, perhaps because this reduces drag and lift, thus increasing the force needed to dislodge their bodies [56]. A case study presented in Sect. 2.4 builds on this observation to explore the possibility of real-time shape optimization of a soft sea star body.

Friction-based locomotion is commonly used in underwater crawling robots. Rigid hexapods *Little Crabster* (pictured in Fig. 2.1) [4] and *AQUAROBOT* [57] were both developed for crawling along the sea floor and exploring benthic environments. Jin et al. used SMA wires to actuate soft arm actuators around a central disc to create a robot inspired by the arm motion of sea stars [17]. Cianchetti et al. created an octopus-inspired robot using cable-driven actuators that deformed soft tentacles to create a walking motion robust to varying terrains [58]. Most recently, the concept of changing body shape in flow to aid friction-based locomotion was incorporated into a fluidically actuated soft quadruped robot using a morphing body capable of adjusting lift and drag [11]. The continuous deformation created by soft actuators allows the robot to change its buoyancy (by changing its volume), its lift (changing its area and lift coefficient), and its drag (changing its area and drag coefficient). Such shape optimization is further discussed in Sect. 2.4.

2.3.1.2 Adhesion-Based Crawling

Adhesion-based crawling is an alternative to friction-based crawling where a separate active mechanism is used to create traction with the substrate. Because active adhesion generates contact forces independent of friction, buoyancy and lift considerations, (Eq. 2.2) become less important. Thus, adhesion creates much more versatile locomotion patterns capable of traversing steep slopes as well as moving in an inverted orientation. To maximize these advantages, adhesion-based crawlers have many adhesive elements spread over a large area. Unlike friction-based crawlers that push off a substrate with their appendages, adhesion-based crawlers reach forward with their appendages, secure themselves to the substrate, and pull their bodies forward [59].

Animals use two primary methods of distributed adhesion. Echinoderms such as sea stars and sea urchins use many appendages called tube feet to perform adhesion-based crawling. Each tube foot has a single adhesive disc that secretes a chemical adhesive to secure the tube foot to the substrate; it subsequently secretes a second substance to counteract the adhesive as needed during the crawling gait [60]. In contrast, the octopus uses suction generated by many suckers distributed along a few appendages to perform a crawling-like motion [61]. These suckers attach to a substrate by using muscles that create a region of very low pressure inside the sucker (up to 800 kPa lower than the ambient pressure) to produce an adhesive suction force [62].

Due to the technical challenges involved with chemical adhesives, underwater bioinspired robots have focused on the use of suction mechanisms to produce reversible adhesion [63]. For example, dielectric elastomers have been used to create an octopus-inspired sucker where the bending motion of the DEA creates the region of low pressure [64]. A similar adhesive mechanism inspired by the sea urchin used a fluidic piston to create the deformation of the sucker [65]. Another adhesive disc inspired by the clingfish was developed with additional soft geometries specifically to improve performance on rough and non-planar surfaces [66]. However, none of these adhesive systems have been integrated into a crawling robot.

Limited work has been done to replicate adhesive crawling in soft robotics from either a distributed actuation or an underwater adhesion standpoint. He et al. created LCE tubular actuators similar to the tube feet of echinoderms for use in a quadrupedal crawling robot [14]. However, these tubular actuators were not incorporated into a large, distributed array and did not integrate adhesive mechanisms. Previous work also includes an urchin-inspired robot comprising an array of rigid spines and tube feet made from soft fluidically actuated bellows [67] and a sea star-inspired robot with bi-stable buckling actuators [5]. Although these echinoderm-inspired robots used biologically inspired appendages, using magnetic tips for adhesion restricts their crawling to ferrous surfaces. Further development of distributed actuation coupled with reversible adhesion is an open area of research for mobile underwater robots.

2.3.2 Swimming

As noted earlier, we classify swimming locomotion into three broad categories: jetting takes advantage of momentum conservation as expelling water backwards causes forward motion of the body; appendage-based swimming relies on the net drag and lift produced by the cyclic motion of an oar-like appendage (e.g., pectoral fins) to push the body; and undulatory swimming uses a traveling wave passing through the body to interact with the surrounding fluid to create motion [68]. Steady swimming is usually accomplished by cyclic actuation. However, the transient hydrodynamic effects associated with shape change are often critical for short-term bursting or maneuvering. The incorporation of soft actuators or bodies has the potential to improve efficiency, speed, or maneuverability for each of these swimming modes.

2.3.2.1 Swimming via Jetting

Jet propulsion is a locomotion technique in which water is slowly drawn into a chamber (intake stroke) and quickly expelled from the chamber (thrust stroke). During the intake stroke, the body increases in volume as water enters. During the thrust stroke, the chamber contracts, reducing the volume and pushing the water out in a jet. The body thus moves forward in accordance with the conservation of linear momentum. The amount of thrust produced and the energetic efficiency depend both on the velocity of the jet produced by the contraction and the structure of the vortices formed by the ejected water [69, 70]. The sudden decrease in body volume also creates acceleration due to the added-mass effect as the cross-sectional area of the body decreases [71, 72]. In nature, this method of swimming is used prominently by cephalopods, bivalves, and jellyfish.

Jet propulsion is often modeled by the following equation, which takes into account the thrust created by the jet, the drag acting on the body, as well as any inertial or gravity effects:

$$F_g + m_{eff}\frac{dU}{dt} + \frac{1}{2}C_D\rho AU^2 + q\frac{dm}{dt} = 0, \tag{2.3}$$

where F_g is any net force caused by gravity in the direction of motion, $m_{eff}\frac{dU}{dt}$ is the inertial force caused by acceleration of the body taking into account the added-mass effects, $\frac{1}{2}C_D\rho AU^2$ is the drag force on the body, and $q\frac{dm}{dt}$ is the jet thrust [73]. Here, $\frac{dU}{dt}$ is the time rate of change of the body's velocity, m is the mass of the body, m_{eff} is the effective mass including contributions from the added-mass effect, q is the velocity of the jet with respect to the body, and $\frac{dm}{dt}$ is the time rate of change of the mass of the body plus enclosed water. From this equation, we can see that during horizontal swimming, the motion of the swimmer at steady state is balanced by the drag force on the body and the jetting force caused by the outflow of fluid.

Jet-propulsion mechanisms among animal species are similar in method but very different in performance. Both squids and octopuses have antagonistic muscles around their mantles that expand the mantle to draw in water and contract the mantle to rapidly expel water [74]. This mechanism creates large bursts of speed that can exceed the speed of fish at the expense of locomotion efficiency [75]. Similarly, a jellyfish uses the muscles around its bell to force water out, forming the jet. However, once the thrust stroke is over, it releases its muscles and the bell returns to its large-volume configuration through the elasticity in the bell itself rather than additional musculature. Thus the jellyfish only expends energy during the thrust stroke and not during the intake stroke, greatly increasing its swimming efficiency [44].

Soft robots use various mechanisms to create locomotion via jet propulsion. RoboScallop, a bivalve-inspired robot, used a motor-powered hinge to open and close rigid shells to expel water [9]. A Hoberman sphere covered in a flexible skin has also been used to create jet-based thrust, taking advantage of the mechanism's large capability for volume change [76]. Another octopus-inspired soft robot used a hyperelastic skin pressurized with water in which the elastic restoring force of the skin forces the jet of water out of the body [77]. Finally, Serchi [78] developed a cephalopod-inspired pulsed propulsion robot that first demonstrated the advantages of combining soft actuation with pulsative jet thrust.

Several different approaches have been used to create jellyfish-inspired robots, in an attempt to reproduce the lightweight and efficient mechanisms found in nature. Hydraulic actuators [79] and SMA actuators [80] have been used to create jellyfish-like robots, but neither have been able to reproduce the high locomotion efficiencies characteristic of biological jellyfish. Recently, DEA actuators were used to create the most efficient untethered jellyfish robot to date [81].

2.3.2.2 Appendage-Driven Swimming

For the purposes of our analysis, we describe the primary hydrodynamic forces associated with appendage-based propulsion as lift and drag. Though both forces may be present on a body as it moves in the water, animals will generally rely on one of the two forces for the purpose of generating motion. A discussion of how lift-based swimming emerged in terrestrial animals to increase speed and efficiency can be found in Fish [82]. This section provides a brief discussion of lift-based propulsion using appendages. Lift forces are also important for swimmers that rely on body undulations for locomotion, which is discussed in Sect. 3.2.3.

In simple terms, the key difference between lift-based and drag-based swimming is in the direction of the fluid forces generated relative to the direction of appendage movement. In either case, the hydrodynamic thrust is in the direction of net animal motion. If the appendage is primarily moving perpendicular to the direction of the animal's net motion, it is considered lift-based swimming; if the appendage is moving parallel to the direction of the animal's net motion, it is considered drag-based swimming. By changing the angle of attack of the appendage, the animal can change the shape of its appendage interacting with flow and thus its hydrodynamic profile throughout the strokes of the gait. Lift and drag forces are often modeled by the quasi-steady drag and lift equations

$$F_D = \frac{1}{2} C_D \rho A U^2 \tag{2.4}$$

$$F_L = \frac{1}{2} C_L \rho A U^2, \tag{2.5}$$

where C_D and C_L are the coefficients of drag and lift, respectively, which depend on the shape of the appendage, ρ is the density of the fluid, A is the characteristic area of the appendage, and U is the fluid speed with respect to the appendage. A swimmer can change the forces generated by its appendage by changing the angle of attack with respect to its body motion, thus changing the shape of the appendage interacting with the surrounding fluid. This changes both the coefficients of lift and drag, as well as the reference area (depending on its formulation). The quasi-steady empirical formulations (see Eqs. (2.4) and (2.5)) are often used with time-varying velocities, areas, and lift or drag coefficients to estimate the net forces generated over a cycle. For example, Vogel [83] used the quasi-steady lift and drag equations (see Eqs. (2.4) and (2.5)) to estimate thrust by assuming constant speed, hydrodynamic coefficients, area, and a specific duty cycle of 50% (thrust is produced during half of the gait cycle) for drag-based and 100% (thrust is produced during the entire gait cycle) for lift-based (no reversal delay) motion.

Drag-based swimming operates on a similar principle to oar mechanics for boat travel. The two phases that characterize drag-based swimming are the recovery stroke and the power stroke. The animal or robot moves an appendage through the water and the drag force opposing that motion pushes the body forward. For net travel, the thrust produced by the power stroke must be greater than the drag produced by the recovery stroke. There are various ways to accomplish this. For example, freshwater softshell turtles twist their fore-flippers so that the cross-sectional area of the flipper interacting with flow is smaller during the recovery stroke than the power stroke, leading to lower drag on the recovery stroke [84]. Platypus propel themselves using directionally compliant fins. The fins flex on the upstroke and, once at maximum forward extension, expand and push backward. Frogs, on the other hand, actively pull their legs in after the power stroke and glide until an optimal velocity is reached before beginning the cycle again.

Drag-based propulsion has been used for several swimming robots. A whirligig beetle-inspired robot made by Jia et al. that used the principle of directional compliance [85] was found to be very energy efficient at low velocities. A robot developed by Tang et al. used dielectric elastomer actuators [86] to create a frog-like burst-coast motion. This locomotion strategy can often generate high accelerations but lacks maneuverability. Kramer et al. developed variable-stiffness limbs to showcase a morphing design that creates both a flipper for paddling and a leg for crawling [8]. Compared to other swimming strategies, drag-based locomotion is demonstrated more by amphibious or terrestrial animals because land-based appendages can be readily adapted to this method [82]. This type of propulsion can be very efficient, as demonstrated by the whirligig beetle [87], and rapid, as seen in platypus and frogs. However, according to Fish [82], drag-based propulsion is a less efficient strategy for fully aquatic mammals, partly due to the drag generated by the non-streamlined body arrangements associated with such propulsion. Despite the disadvantage of potential drag increase by having these appendages, drag-based swimming is a widely used method for soft robot swimming and can be an energy-efficient design solution, particularly for amphibious locomotion.

An inexhaustive list of animals that use more lift than drag to propel themselves with appendages include species in the Pinniped clade (walrus, sea lions, fur seals), humans (kicking in front crawl), and Spheniscidae (penguins). The mode of propulsion is similar to flapping for flying animals; however, in aquatic animals, gravitational forces are offset by buoyancy. Lift-based swimming typically relies on the shedding of a vortex ring in the direction opposite to the direction of motion.

Further discussion of the vortex production and roll-off are provided in Sect. 2.3.2.3. Sea turtle-inspired limbs are an example of biological lift-based propulsion, and have also been used to showcase soft actuators with multi-modal actuation. For example, a limb made from SMA wire was designed to be capable of both bending and twisting motions, allowing different cross-sections of the limb to interact with flow during the power and recovery phases of the stroke [88].

2.3.2.3 Undulatory (Body-Driven) Swimming

Many swimming organisms propel themselves through the water by generating undulatory traveling waves along their bodies. Dominant underwater undulation motion strategies can be broken into anguilliform (i.e., eel-like swimming), carangiform (i.e., mackerel-like swimming), and thunniform (i.e., tuna-like swimming). Though there is a continuum of behavior between these three categories, they serve as useful boundaries for classification. The differences between the three categories can be explained based on the percentage length of the body involved in the traveling wave: fish that primarily generate thrust with their caudal segments are considered thunniform swimmers; fish that generate thrust with their entire bodies are considered anguilliform swimmers; and fish between the two are considered carangiform swimmers [83].

Most soft robots operate in the high Reynolds number regime, where inertial rather than viscous fluid forces are dominant. For this inertial regime, two main explanations have been proposed for how traveling waves interact with the fluid to create propulsion, i.e., how body waves impart directional momentum to the fluid. The first theory suggests that propulsive thrust is generated by reactive forces acting on the body [89, 90]. Fluid parcels next to the body are accelerated toward the rear of the body because of the pressure variations generated by the traveling wave. This imparts backwards momentum to the fluid, thereby generating a forward propulsive force on the body. Though this explanation provides some insight into the fluid forces responsible for propulsion, the fluid accelerations are tightly coupled to vortex shedding at the tail. This vortex shedding also provides a propulsive force. Specifically, evidence suggests that vortex rings are generated twice per cycle, which, through their interactions, create a reverse von Kármán vortex street (RvKVS), where rearward momentum is imparted to the fluid [91, 92]. According to von Kármán [93, 94], the thrust generated by a RvKVS can be predicted using

$$T = -\frac{\Gamma^2}{2\pi a} + \frac{\Gamma b}{a}(1 + 2\nu), \tag{2.6}$$

where T is thrust, Γ is the vortex intensity, a is wavelength, b is wake width, and ν is the vortex traveling velocity. The first component is an induced drag element, whereas the second term is the propulsive element. Thrust can be enhanced by increasing the coefficient $\frac{b}{a}(1 + 2\nu)$: increasing the undulation frequency will decrease wavelength a; increasing the amplitude of oscillation will increase the wake width b; increasing oscillation frequency and amplitude will also increase ν and, correspondingly, thrust T.

Previous work also suggests that there exists an optimal oscillation frequency for swimming that can be expressed in terms of the dimensionless Strouhal number, $St = fL/U$, where f is the oscillation frequency, L is the characteristic length (often the amplitude of the caudal fin oscillation for fish), and U is the velocity. Fish typically travel with Strouhal numbers of 0.25–0.35 [95] and, similarly, robotic undulating fish are also designed to operate in this Strouhal number range [96].

The physical principles that relate vortex shedding to propulsion are complex. Not only does the body oscillation generate thrust, it can also serve as a drag-reduction mechanism by ensuring optimal vortex arrangements in the wake behind the body and potentially re-laminarizing the turbulent boundary layer formed on the body [97]. Further, because of the complex and interdependent nature of the fluid-structure interactions involved in undulatory swimming, researchers often have difficulty distinguishing between the various mechanisms involved, e.g., reactive forces, added-mass effects, vortex-based propulsion, and drag reduction. However, despite the challenges associated with modeling these phenomena, many robots have been developed that successfully exploit undulatory swimming motions.

Undulatory swimming is successful in part because it can be accomplished using streamlined body shapes with few external appendages. This characteristic can be replicated in robots using soft materials to actuate and connect segments. For instance, a recent study used DEA segments to create a robot capable of anguilliform propulsion [12]. Similarly, electro-active polymer actuators have been used to generate undulations in a ray-inspired robot [98]. For faster propulsion, researchers have developed carangiform-type swimming robots by discretizing the traveling wave into a limited number of actuated body segments. These segments are wrapped in a deformable skin, generally made from a synthetic rubber such as latex or silicone. *RoboTuna* [99], the first successful robotic fish, used foam and Lycra to create a soft deformable interface. Since this seminal work was published, many more robotic fish have been developed with various actuation components, such as servo motors encased in corrugated rubber tubing [100], a silicone bidirectional PneuNet actuator powered by a miniature hydraulic pump [6], as well as a motor driving antagonistic tendons around a free-to-rotate tail joint capable of high actuation frequency [48]. These examples demonstrate that soft, deforming materials are advantageous for developing bioinspired robots and implementing the motion strategies that make animals successful.

2.4 Optimizing Shape and Kinematics

The examples presented thus far show that soft actuators can help design robotic systems more similar in morphology to biological systems than rigid structures. Next, we consider how shape change enabled by soft actuators can be used to improve locomotive efficiency with gait-pattern optimization and hydrodynamic force manipulation. There are two aims for applying optimization techniques to soft robots. First, the designer hopes to identify parameters that create the *most favorable* dynamic sequence or shape change with respect to a goal metric. Second, the designer would like to find this *most favorable* parameter set *most quickly*. Striving toward these two objectives is what motivates this section.

Gait-pattern optimization has been used to generate efficient, segmented, and oscillating robots using multiple servo motors as position-control units, shown in Fig. 2.5a [100]. Shape optimization for drag reduction and lift enhancement is also a well-studied area, particularly for aerodynamic aircraft applications [101] and biomimetic robotics [102] (Fig. 2.5c). Recent work also considers hydrodynamic applications such as the design of ship hulls [103] and autonomous underwater vehicles [104] (Fig. 2.5e), as well as the development of a benthic crawling robot [11] (Fig. 2.5b). However, the vast majority of these optimization studies consider the *a priori* design of rigid body shapes. As far as the authors are aware, no studies have considered real-time shape optimization of soft underwater structures. To inspire further applications in the field, Sect. 2.4.3 offers a sample study of shape optimization inspired by the morphological plasticity exhibited by sea stars in response to hydrodynamic conditions [56].

2.4.1 Gait-Pattern Optimization

Several prior studies have addressed optimal swimming kinematics to explain biological observations and to understand the underlying fluid-dynamic phenomena [70, 105, 106]. This subsection considers gait optimization for the control of soft robot locomotion in water. The problem of gait optimization for locomotion begins with defining a performance objective and parameterizing actuator sequences. This definition could be understood in a discrete sense, by using individual state positions as parameters for a given time interval, i.e., finite state transitions. Another approach is to assume a specific continuous function is optimal with the correct coefficients. In general, this approach is implemented using sinusoidal motor inputs, as steady-state animal underwater locomotion is commonly performed using periodic actuation. Our intention for describing techniques used in gait optimization is to address the need for determining control sequences of actuators interfacing with a continuum soft structure.

Inspired by neuro-muscular control, where the brain regulates complex harmonic muscular activity with simplified commands, researchers have studied the use of Central Pattern Generators (CPG) for discovering optimal gait sequences for robotics [107, 108]. CPG gait optimization reduces the high dimensionality associated with finite state machines in time and space. Generally for CPG implementation, a periodic input is assumed to generate the optimal gait. The parameters associated with a sinusoidal input, i.e., amplitude, phase, and frequency, can then be optimized autonomously by using an onboard processor to solve a system of first-order ODEs with limit-cycle behavior. For instance, a stochastic population-based

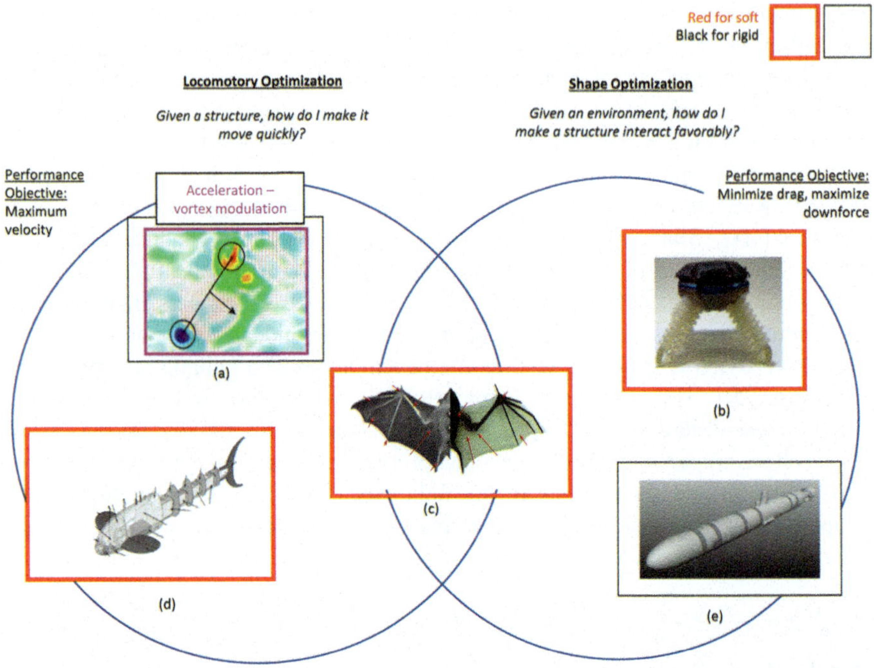

Fig. 2.5 Diagram showing examples of gait optimization: (**a**) changing amplitude of oscillating caudal fin controls vortex intensity, which regulates acceleration [92]; (**b**) shape optimization is used by Ishida et al. [11] for minimizing drag and maximizing downforce on a walking quadrupedal robot; (**c**) Colorado et al. [102] implemented shape morphing using SMA actuators on a bat robot where the goal of speed optimization can be achieved through proper gait sequence selection and aerodynamic force control through shape change; (**d**) learning optimal CPG parameters for multi-actuator fish robot [100]; (**e**) structural optimization is observed by Joung et al. [104] for shape optimization of an AUV

evolutionary algorithm, such as a genetic algorithm or particle-swarm algorithm, can be used to select and optimize the parameter combinations to be tested. Because fish undulation is well synchronized and approximately sinusoidal, CPG techniques are effective for optimizing gait sequences of soft robots demonstrating undulatory locomotion [109].

An advantage of using CPG gait generation is that the output transition of position commands resulting from discontinuous parameter changes is continuous, because the patterns are generated by a set of differential equations. These equations can be solved online using microcontrollers [107]. Thus, in a large testing environment, the robot can operate for long periods of time, given sufficient power resources [109]. A disadvantage of this method is that optimal gait patterns are assumed to be sinusoidal. For complex, nonlinear functions, where actuators do not necessarily have independence, this method may not converge to an optimal solution.

Despite the potential disadvantages, many robots still use CPG methods to avoid the complexity of choosing dynamic gait sequences for optimization. Alternatively, there are other studies that have used state transitions instead of CPG methods. For example, Lal et al. [110] used physical model simulations in conjunction with genetic algorithm-based rule selection to optimize gait sequences for a five-legged brittle star-inspired robot and then implemented the pattern on a physical system for testing. Such decentralized control approaches may be useful for investigating more complex gait-function spaces.

Another component of developing online optimal shape-changing strategies for locomotion is the performance-evaluation component. For performance goals to be adaptive, the robot must be able to distinguish changes in objective value. Thus, sensing is very important for system feedback. If a system is tethered and operator interaction is involved, a Eulerian frame with object tracking using image feeds may be used to estimate state information relevant to performance evaluation. However, for unmanned systems, the sensing must be in a Lagrangian framework, and thus housed onboard. Several researchers have used inertial measurement units, which yield accelerometer, gyroscope, and compass data to estimate position and velocity [109, 111]. Researchers have also used infrared sensing and video imaging to avoid collision with walls and identify targets [109]. To estimate dynamic fluid interactions, Bernoulli's principle (or other simplified models) have been used in conjunction with pressure measurements to obtain estimates of flow speed and body forces [112].

2.4.2 Fluid-Structure Interaction Optimization

The efficacy of underwater locomotion depends to a large extent on drag, as motion can be greatly restricted if a body is not streamlined. Additionally, in moving water, it can be advantageous for a robot to be able to change its shape to minimize drag across different flow conditions (e.g., Reynolds number ranges, changes in direction). Rigid, inflexible robots are often incapable of optimizing for such dynamic environmental conditions. Thus, researchers have increasingly been turning to soft materials to perform shape optimization for drag reduction. In addition to steady flow forcing, transient hydrodynamic effects associated with shape change can also result in added lift, drag, or thrust. Such transient effects have not been studied thoroughly yet and could be exploited using soft materials for control or optimization purposes.

The field of shape optimization has seen tremendous growth following the development of Computational Fluid Dynamics (CFD) technologies. Adjoint methods for gradient-based functional optimization have been used successfully in numerous applications. These methods define a surface using grid points and a fitting curve or surface. CFD simulation is then used to evaluate and optimize surface parameters. As computing resources improve, so does the quality and speed of these simulations. Despite the great success of shape optimization through simulation, unmodeled interactions are still substantial enough to mandate additional wind and water tunnel testing. Furthermore, optimization through simulation can also be very computationally intensive particularly when considering flexible structures interacting with high Reynolds number turbulent flows. Thus, for a robot adapting quickly to changing flow conditions, shape optimization would either require applying a known mapping of shape to hydrodynamic force or using simplified models for estimating hydrodynamic force.

The idea of morphing structures being used to tune body shape based on the environment is not new when considering aircraft [113]. However, recent examples demonstrate that some of the actuation technologies discussed earlier can inspire variations in design and create opportunities to generate new capabilities. For example, Han et al. used SMA actuators to control wingtip vortices for adapting to flight conditions [114]. Rodrigue et al. similarly used SMA wires to create a wing-twisting effect [115]. Comparatively fewer studies have been performed for morphing watercraft. Some ideas for morphing torpedo hulls are provided by Rufino et al. [116]. However, because the Reynolds numbers and fluid forces can be very large for traditional watercraft applications, little attention has been given to optimizing flexible structures for hydrodynamic performance. Below, we provide an example of a soft underwater shape-changing robot as inspiration for further research.

2.4.3 Case Study: Real-Time Shape Optimization for a Soft Sea Star

We present a case study illustrating how shape change capabilities enabled by soft materials can be used for engineering benefit. In particular, we show that hydrodynamic body forces on a robot in a steady flow environment can be tuned for drag minimization or downforce maximization (negative lift). For this study, we aim only to provide an example of how shape change may be used for hydrodynamic optimization. Rather than focus on the method, we encourage the reader to view the study as a platform for connecting optimization with soft actuation and physical systems.

Biological observations show that intertidal sea stars exhibit morphological plasticity in response to changing flow conditions [56]. Specifically, sea stars in wave-exposed regions have narrower arms and smaller frontal area compared to sea stars from sheltered sites. Further, sea stars transplanted from sheltered sites to exposed regions develop narrower arms. One explanation for this morphological plasticity is that the changes in body shape may enable sea stars to resist dislodgment by reducing drag or maximizing downforce, which can enhance friction as per Eq. (2.2). Though such shape changes occur on a time scale of several weeks for sea stars, soft robots are only limited by the speed of the actuators. This section

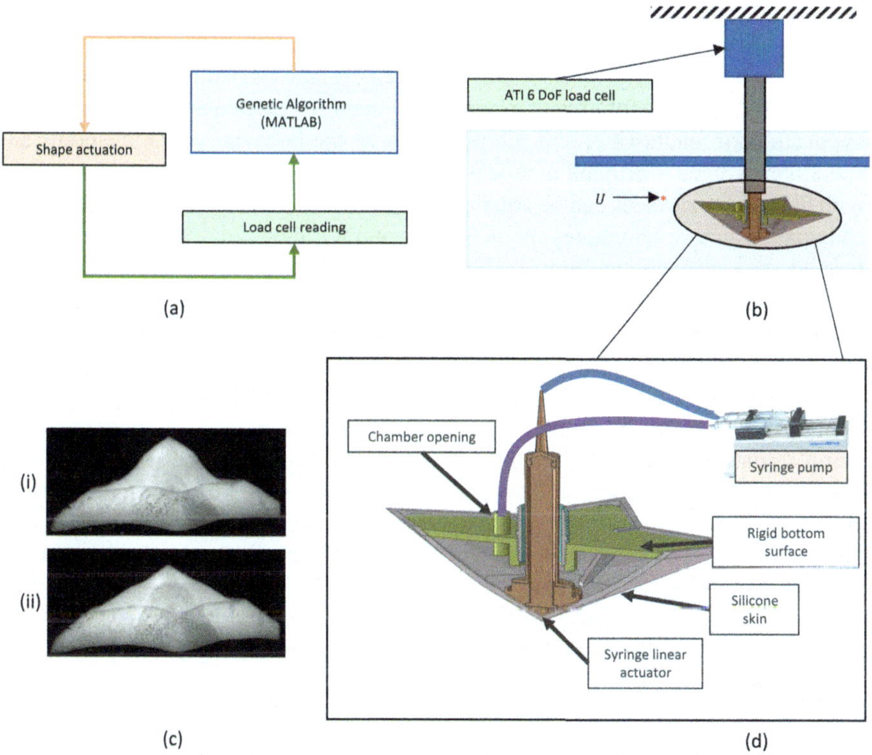

Fig. 2.6 Experimental shape optimization. (**a**) Schematic logic flow of the optimizer. Sensor readings evaluate performance and a genetic algorithm computes the next iteration of test parameter combinations. (**b**, **d**) A sea star-inspired morphing body with silicone skin is attached to a load cell in a water tunnel. (**c**) Images showing the actuation range: (i) volume is minimal and height is maximal; (ii) volume is maximal and height is minimal

presents a short case study on experimental shape optimization that is inspired by the morphological plasticity observed in sea stars (Fig. 2.6).

We cast a silicone body similar in size to the 5-arm sea star (*Pisaster ochraceous*) that exhibits morphological plasticity in response to flow conditions [56]. The body shape was controlled by pressurizing a hollow chamber with water using a syringe pump (100 mL syringe, pump uncertainty ±0.01 mL) and extending a linear actuator in the center (2 cm stroke, ±0.1 cm). In other words, the control parameters were body volume and height. As shown in Fig. 2.6b, this shape-morphing body was attached to a load cell and subjected to hydrodynamic forcing. A genetic algorithm was used to identify drag-minimizing and downforce-maximizing optima based on the load cell measurements.

The experimental approach was inspired by early studies on evolutionary algorithms by Rechenberg in the 1960s [117]. This prior work demonstrated that evolution-based design of both a segmented plate and a 180° pipe bend for drag minimization can reduce the number of experiments required compared to using a fine grid search. Furthermore, for highly nonlinear function spaces such as a shape-to-drag mapping, a grid- or gradient-based search may only find a local optimum. Evolutionary shape optimization via experiment is particularly well-suited for applications in underwater soft robotics because the behavior of the materials and actuators involved is difficult to model and the fluid-structure interactions associated with deforming surfaces can be equally complex. Development of a physical shape-changing structure eliminates the need for complex modeling or CFD simulation. Instead, the physical system is allowed to interact with the environment and embedded sensor measurements are used to guide shape optimization. Because the method is completely autonomous, it is a practical design tool for reducing test facility and operator time.

The shape-changing robot was created by casting silicone rubber (Smooth-On, Ecoflex 00-30) in 3D printed negative molds (Fig. 2.6b, d). Two hydraulic syringe-pump actuators were developed using NEMA23 stepper motors to (1) accurately control fluid injection into a deforming cavity and (2) drive a secondary 50 mL syringe, which was used as a linear actuator to change height. Thus, volume and height change served as a surrogate for the shape changes observed in nature. The actuation range is illustrated in Fig. 2.6c. The syringe–pump actuators were controlled using an Arduino Uno PWM with stepper motor drivers. The dimensions of the base for the morphing body and the height range were guided by biological observations [56].

A water-channel facility with cross-section dimensions 0.9×0.6 m, capable of generating flows at speeds up to 0.6 m/s, was used for the experiments. The morphing structure was mounted to an ATI Gamma load cell (Fig. 2.6b). The load cell has 1/160 N resolution in the streamwise (drag) direction and 1/80 N resolution in the wall-normal (lift) direction. The load cell was connected to a National Instruments data acquisition device that interfaced with MATLAB to generate force readings. Minimum measured drag and lift forces were 0.1 N.

We used MATLAB as the logic control to interface with the load-cell measurements and actuation system. This logic control is lumped into the genetic algorithm in Fig. 2.6a. We used an 8-bit binary encoding method for each actuator state, and performed a uniform crossover and mutation procedure to evolve populations. We used high probability for mutation in the system (10%) so that the parameter space (height and volume) could be explored sufficiently and quickly. The load cell measurements were used to construct a probability density function (PDF) for the population as a function of height and volume. The location of the maximum for this PDF was interpreted as the parameter set that maximized fitness. Though we chose to use a genetic algorithm as a search technique, any stochastic parameter search would be equally applicable. The experiments were run only once for 300 samples. We subsampled with 100 samples (after 100 sample seed-time for GA) to ensure that the probability distribution of the optimized parameter set is mean-stationary, and we found the means to be $\text{Mean}_{N\in[100:200]} = [33.9, 1.8]$ mL, $\text{Mean}_{N\in[200:300]} = [30.7, 1.7]$ mL, where N is the sample index. Thus, because the means vary by only

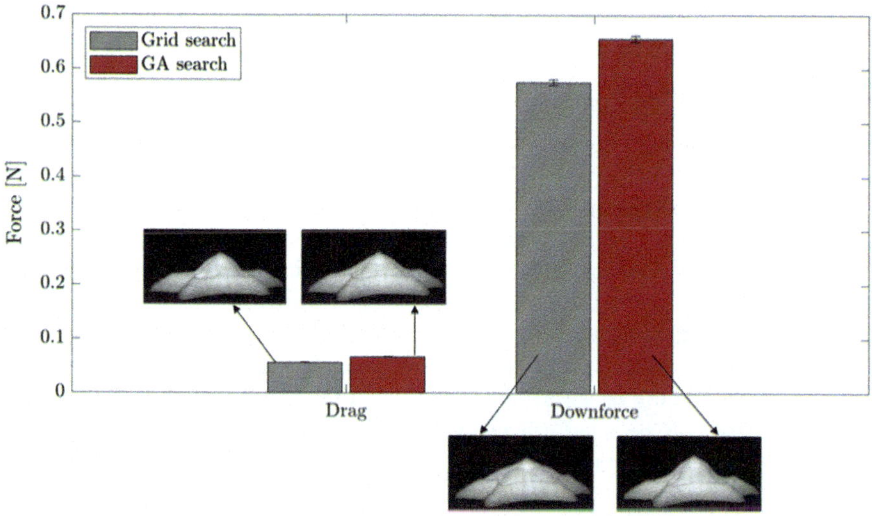

Fig. 2.7 Minimum drag and maximum downforce (negative lift) obtained using a grid search and the genetic algorithm (GA). The uncertainty reported is the standard error associated with the force measurement

3/100 and 1/100 of the respective actuation ranges, we conclude that our process is stationary.

Results obtained from the genetic algorithm are compared with those obtained from a grid search in Fig. 2.7. For drag minimization, the optimal shape identified by the grid search yielded better performance (i.e., lower drag) than the shape identified by genetic algorithm. However, both shapes are qualitatively similar. Moreover, we observed the genetic algorithm still populating the minimal parameter set space with samples when the experiment completed, indicating that additional run-time may result in a more optimal parameter set. On the other hand, for downforce maximization, the genetic algorithm identified a better configuration than the grid search.

Importantly, this short case study demonstrates that a morphing body can be used to tune hydrodynamic performance. This tuning can happen autonomously and in near real-time with appropriate controllers. Therefore, in addition to guiding the design of underwater soft robots, similar methods might also be used for active shape control in untethered systems. For example, an underwater robot could stabilize itself by expanding like a pufferfish in response to a measured disturbance.

2.5 Conclusions and Future Directions

This work discusses the ways in which the advantages of soft robotics may be used for underwater bioinspired robots. Soft materials support complex motions that closely mimic the crawling or swimming behaviors of underwater animals and create continuous curvatures well-suited for interactions with underwater flow. These soft actuators may be actuated with tendons, fluid pressure, heat, or high voltage, producing various levels of power and efficiency.

Although friction-based crawling is a common approach to underwater robots, this method of locomotion is greatly influenced by fluid forces such as lift and buoyancy, requiring underwater crawlers to adjust shape or posture for optimal performance. Limited work on distributed adhesion-based crawling has been completed to date and we consider this a promising area of research that has the potential to create significantly more robust robots. In the swimming domain, we see that jellyfish jetting produces the highest efficiency and slowest velocity, whereas squid jetting produces the highest velocity, albeit with lower efficiency. These variations of jet propulsion leverage the elastic restoring force of the soft structures more than drag-based swimming or vortex-shedding swimming and we think that working to refine the trade-off between efficiency and speed using soft materials is another promising avenue for future development.

2.5.1 Future Actuation Directions

One open area of research for underwater locomotion is the issue of reversible underwater adhesion. Adhesion can be used to aid locomotion by creating greater traction with a substrate to apply a propulsive force. While many soft underwater organisms like the octopus and the sea star are capable of suction or chemical adhesion, existing robotics techniques have not been very successful in replicating these aspects of underwater animals. Modern techniques for increasing traction against a surface such as gecko adhesive do not function underwater and carrying and excreting an onboard adhesive and de-adhesive for each foot is not practical for a mobile robot. Commercial suction cups are unsuccessful at adhering to real-life surfaces that are not smooth, so exploring suction geometries used by animals could produce new insights into engineering solutions. Developing an underwater reversible adhesion method will benefit robots by increasing their locomotion efficiency as well as their ability to manipulate objects.

Another promising area of research is to progress toward a greater combination of locomotion efficiency and total velocity. Although nature displays a trade-off between efficiency and overall velocity, current techniques generally lag behind their biological counterparts. Since fluid forces are more significant underwater than in air due to the higher density of water, soft actuators that are well-suited for operation in air are not necessarily optimal for operation in water. Soft fluidic actuators can take advantage of lift and buoyancy to modify upward forces on the robot and to reduce the effective weight. Rather than incrementally improving and optimizing existing soft actuators, new soft actuators can be developed specifically for the underwater environment to create advantageous interactions with the surrounding fluid.

2.5.2 *Future Optimization Directions*

Developing optimal segment actuation for fish-inspired robots began by analyzing actual fish bodies and directly mapping the curves to motor positions. This method works well, but it does not account for differences between the robotic implementation and the biological analogy. Because of this, researchers have moved to implementing controllers that optimize actuators online using CPG algorithms. This generally works well for oscillation-driven bodies, but because parameter combinations can span a large search set, these implementations can take many iterations to learn an optimal configuration. Thus, a clear future direction is to minimize the time needed to learn optimal parameters and investigate additional limit-cycle generators to reproduce nonlinear sequences. Further, these methods select parameters for steady flow environments; adapting to a time-dependent environment (a periodic flow field) may require a more sophisticated method to learn time-varying parameters. A future direction is also developing algorithms suited to controlled motion in oscillatory flow environments.

In addition to control sequence optimization, a promising field for soft material underwater robotics is dynamic shape control. Numerical topology optimization has been well-explored using CFD analysis and parametric CAD models of rigid structures. However, few physical experiments have been developed for online topology optimization. We describe a case study demonstrating the potential of using soft deforming materials for design optimization, and envision this idea to be applied to an autonomous, untethered system. In this setting, the distinction between gait optimization and topology optimization becomes obscured. A future direction for topology optimization is to develop a system capable of autonomously evaluating its hydrodynamic performance and adapting to flow conditions in complex environments. This system could be coupled to a swimming propulsor, thereby connecting hydrodynamic and gait optimizations to achieve maximal performance.

Building off the advances detailed in this chapter is a step toward creating more effective soft underwater mobile robots. By carefully considering the shape and function of soft bodies and actuators, we aim to design robots that create favorable interactions with flow in variable fluid environments. These soft robots can be used for exploration of underwater environments not suitable for human or ROV traversal and for monitoring and measuring of vulnerable species and habitats.

Acknowledgments This chapter is based on work supported by the US Office of Naval Research under grant number N00014-17-1-2062 (Program Manager: Dr. Thomas McKenna).

References

1. T.E. Laidig, L.M. Krigsman, M.M. Yoklavich, Reactions of fishes to two underwater survey tools, a manned submersible and a remotely operated vehicle. Fish. Bull. **111**(1), 54–67 (2013)

2. G.M. Whitesides, Soft robotics. Angew. Chem. Int. Ed. **57**(16), 4258–4273 (2018)
3. D. Rus, M.T. Tolley, Design, fabrication and control of soft robots. Nature **521**(7553), 467 (2015)
4. J.-Y. Kim, B.-H. Jun, Design of six-legged walking robot, Little Crabster for underwater walking and operation. Adv. Robot. **28**(2), 77–89 (2014)
5. M.A. Bell, I. Pestovski, W. Scott, K. Kumar, M.K. Jawed, D.A. Paley, C. Majidi, J.C. Weaver, R.J. Wood, Echinoderm-inspired tube feet for robust robot locomotion and adhesion. IEEE Robot. Autom. Lett. **3**(3), 2222–2228 (2018)
6. R.K. Katzschmann, J. DelPreto, R. MacCurdy, D. Rus, Exploration of underwater life with an acoustically controlled soft robotic fish. Sci. Robot. **3**(16), eaar3449 (2018)
7. C. Christianson, N.N. Goldberg, D.D. Deheyn, S. Cai, M.T. Tolley, Translucent soft robots driven by frameless fluid electrode dielectric elastomer actuators. Sci. Robot. **3**(17), eaat1893 (2018)
8. R.L. Baines, J.W. Booth, F.E. Fish, R. Kramer-Bottiglio, Toward a bio-inspired variable-stiffness morphing limb for amphibious robot locomotion, in *2019 Second IEEE International Conference on Soft Robotics (RoboSoft)*, pp. 704–710 (IEEE, Piscataway, 2019)
9. M.A. Robertson, F. Efremov, J. Paik, Roboscallop: a bivalve inspired swimming robot. IEEE Robot. Autom. Lett. **4**(2), 2078–2085 (2019)
10. F. Gibouin, C. Raufaste, Y. Bouret, M. Argentina, Study of the thrust–drag balance with a swimming robotic fish. Phys. Fluids **30**(9), 091901 (2018)
11. M. Ishida, D. Drotman, B. Shih, M. Hermes, M. Luhar, M.T. Tolley, Morphing structure for changing hydrodynamic characteristics of a soft underwater walking robot. IEEE Robot. Autom. Lett. **4**(4), 4163–4169 (2019)
12. C. Christianson, N. Goldberg, S. Cai, M.T. Tolley, Fluid electrodes for submersible robotics based on dielectric elastomer actuators, in *Proceedings of the Electroactive Polymer Actuators and Devices (EAPAD) 2017*, vol. 10163 (International Society for Optics and Photonics, 2017), p. 10163
13. S. Mao, E. Dong, H. Jin, M. Xu, S. Zhang, J. Yang, K.H. Low, Gait study and pattern generation of a starfish-like soft robot with flexible rays actuated by SMAs. J. Bionic Eng. **11**(3), 400–411 (2014)
14. Q. He, Z. Wang, Y. Wang, A. Minori, M.T. Tolley, S. Cai, Electrically controlled liquid crystal elastomer-based soft tubular actuator with multimodal actuation. Sci. Adv. **5**(10), eaax5746 (2019)
15. M. Calisti, M. Giorelli, G. Levy, B. Mazzolai, B. Hochner, C. Laschi, P. Dario, An octopus-bioinspired solution to movement and manipulation for soft robots. Bioinspiration Biomimetics **6**(3), 036002 (2011)
16. P. Ohta, L. Valle, J. King, K. Low, J. Yi, C.G. Atkeson, Y.-L. Park, Design of a lightweight soft robotic arm using pneumatic artificial muscles and inflatable sleeves. Soft Robot. **5**(2), 204–215 (2018)
17. H. Jin, E. Dong, G. Alici, S. Mao, X. Min, C. Liu, K. Low, J. Yang, A starfish robot based on soft and smart modular structure (SMS) actuated by SMA wires. Bioinspiration Biomimetics **11**(5), 056012 (2016)
18. A.A. Stanley, K. Hata, A.M. Okamura, Closed-loop shape control of a haptic jamming deformable surface, in *2016 IEEE International Conference on Robotics and Automation (ICRA)* (IEEE, Piscataway, 2016), pp. 2718–2724
19. P. Polygerinos, N. Correll, S.A. Morin, B. Mosadegh, C.D. Onal, K. Petersen, M. Cianchetti, M.T. Tolley, R.F. Shepherd, Soft robotics: review of fluid-driven intrinsically soft devices; manufacturing, sensing, control, and applications in human-robot interaction. Adv. Eng. Mater. **19**(12), 1700016 (2017)
20. F. Ilievski, A.D. Mazzeo, R.F. Shepherd, X. Chen, G.M. Whitesides, Soft robotics for chemists. Angew. Chem. Int. Ed. **50**(8), 1890–1895 (2011)
21. T. Wallin, J. Pikul, R. Shepherd, 3D printing of soft robotic systems. Nat. Rev. Mater. **3**(6), 84 (2018)

22. D. Drotman, S. Jadhav, M. Karimi, P. Dezonia, M.T. Tolley, 3D printed soft actuators for a legged robot capable of navigating unstructured terrain, in *2017 IEEE International Conference on Robotics and Automation (ICRA)* (IEEE, Piscataway, 2017), pp. 5532–5538
23. B. Mosadegh, P. Polygerinos, C. Keplinger, S. Wennstedt, R.F. Shepherd, U. Gupta, J. Shim, K. Bertoldi, C.J. Walsh, G.M. Whitesides, Pneumatic networks for soft robotics that actuate rapidly. Adv. Funct. Mater. **24**(15), 2163–2170 (2014)
24. F. Connolly, P. Polygerinos, C.J. Walsh, K. Bertoldi, Mechanical programming of soft actuators by varying fiber angle. Soft Robot. **2**(1), 26–32 (2015)
25. M.A. Robertson, M. Murakami, W. Felt, J. Paik, A compact modular soft surface with reconfigurable shape and stiffness. IEEE/ASME Trans. Mechatron. **24**(1), 16–24, (2018)
26. D. Drotman, M. Ishida, S. Jadhav, M.T. Tolley, Application-driven design of soft, 3-D printed, pneumatic actuators with bellows. IEEE/ASME Trans. Mechatron. **24**(1), 78–87 (2018)
27. B.N. Peele, T.J. Wallin, H. Zhao, R.F. Shepherd, 3D printing antagonistic systems of artificial muscle using projection stereolithography. Bioinspiration Biomimetics **10**(5), 055003 (2015)
28. R.V. Martinez, J.L. Branch, C.R. Fish, L. Jin, R.F. Shepherd, R.M. Nunes, Z. Suo, G.M. Whitesides, Robotic tentacles with three-dimensional mobility based on flexible elastomers. Adv. Mater. **25**(2), 205–212 (2013)
29. R. Niiyama, X. Sun, C. Sung, B. An, D. Rus, S. Kim, Pouch motors: printable soft actuators integrated with computational design. Soft Robot. **2**(2), 59–70 (2015)
30. C. Majidi, Soft-matter engineering for soft robotics. Adv. Mater. Technol. **4**(2), 1800477 (2019)
31. R. Pelrine, R. Kornbluh, G. Kofod, High-strain actuator materials based on dielectric elastomers. Adv. Mater. **12**(16), 1223–1225 (2000)
32. S. Hsien Low, L. Lynn Shiau, G.-K. Lau, Large actuation and high dielectric strength in metallized dielectric elastomer actuators. Appl. Phys. Lett. **100**(18), 182901 (2012)
33. C. Christianson, N.N. Goldberg, M.T. Tolley, Elastomeric diaphragm pump driven by fluid electrode dielectric elastomer actuators (FEDEAs), in *Electroactive Polymer Actuators and Devices (EAPAD) XX*, vol. 10594 (International Society for Optics and Photonics, 2018), p. 10594
34. C. Christianson, C. Bayag, G. Li, S. Jadhav, A. Giri, C. Agba, T. Li, M.T. Tolley, Jellyfish-inspired soft robot driven by fluid electrode dielectric organic robotic actuators. Front. Robot. AI **6**, 126 (2019)
35. D. Chen, Q. Pei, Electronic muscles and skins: a review of soft sensors and actuators. Chem. Rev. **117**(17), 11239–11268 (2017)
36. M. Sreekumar, T. Nagarajan, M. Singaperumal, M. Zoppi, R. Molfino, Critical review of current trends in shape memory alloy actuators for intelligent robots. Ind. Robot. Int. J. **34**(4), 285–294 (2007)
37. C. Ohm, M. Brehmer, R. Zentel, Liquid crystalline elastomers as actuators and sensors. Adv. Mater. **22**(31), 3366–3387 (2010)
38. J. Sun, J. Zhao, An adaptive walking robot with reconfigurable mechanisms using shape morphing joints. IEEE Robot. Autom. Lett. **4**(2), 724–731 (2019)
39. M.A. McEvoy, N. Correll, Shape-changing materials using variable stiffness and distributed control. Soft Robot. **5**(6), 737–747 (2018)
40. E. Steltz, A. Mozeika, N. Rodenberg, E. Brown, H.M. Jaeger, JSEL: jamming skin enabled locomotion, in *2009 IEEE/RSJ International Conference on Intelligent Robots and Systems* (IEEE, Piscataway, 2009), pp. 5672–5677
41. T.M. Simon, R.T. Smith, B.H. Thomas, Wearable jamming mitten for virtual environment haptics, in *Proceedings of the 2014 ACM International Symposium on Wearable Computers* (ACM, New York, 2014), pp. 67–70
42. A.A. Stanley, A.M. Okamura, Controllable surface haptics via particle jamming and pneumatics. IEEE Trans. Haptics **8**(1), 20–30 (2015)
43. Z. Varga, G. Filipcsei, M. Zrínyi, Magnetic field sensitive functional elastomers with tuneable elastic modulus. Polymer **47**(1), 227–233 (2006)

44. B.J. Gemmell, J.H. Costello, S.P. Colin, C.J. Stewart, J.O. Dabiri, D. Tafti, S. Priya, Passive energy recapture in jellyfish contributes to propulsive advantage over other metazoans. Proc. Nat. Acad. Sci. **110**(44), 17904–17909 (2013)
45. V. van Ginneken, E. Antonissen, U.K. Müller, R. Booms, E. Eding, J. Verreth, G. van den Thillart, Eel migration to the Sargasso: remarkably high swimming efficiency and low energy costs. J. Exp. Biol. **208**(7), 1329–1335 (2005)
46. D. Webber, R. O'dor, Monitoring the metabolic rate and activity of free-swimming squid with telemetered jet pressure. J. Exp. Biol. **126**(1), 205–224 (1986)
47. R.K. Katzschmann, A.D. Marchese, D. Rus, Hydraulic autonomous soft robotic fish for 3d swimming, in *Experimental Robotics* (Springer, Berlin, 2016), pp. 405–420
48. J. Zhu, C. White, D. Wainwright, V. Di Santo, G. Lauder, H. Bart-Smith, Tuna robotics: a high-frequency experimental platform exploring the performance space of swimming fishes. Sci. Robot. **4**(34), eaax4615 (2019)
49. Nautilus, ROV Hercules (2003). https://nautiluslive.org/tech/rov-hercules. Accessed 10 Feb 2020
50. US Navy, Attack submarines—SSN 2019. https://www.navy.mil/navydata/fact_display.asp?cid=4100&tid=100&ct=4. Accessed 10 Feb 2020
51. O. Ben-David, J. Fineberg, Static friction coefficient is not a material constant. Phys. Rev. Lett. **106**(25), 254301 (2011)
52. J. Ayers, Underwater walking. Arthropod Struct. Dev. **33**(3), 347–360 (2004)
53. H.C. Astley, Getting around when you're round: quantitative analysis of the locomotion of the blunt-spined brittle star, *Ophiocoma echinata*. J. Exp. Biol. **215**(11), 1923–1929 (2012)
54. M. Weissburg, C. James, D. Smee, D. Webster, Fluid mechanics produces conflicting, constraints during olfactory navigation of blue crabs, *Callinectes sapidus*. J. Exp. Biol. **206**(1), 171–180 (2003)
55. S.H. Maude, D.D. Williams, Behavior of crayfish in water currents: hydrodynamics of eight species with reference to their distribution patterns in southern Ontario. Can. J. Fish. Aquat. Sci. **40**(1), 68–77 (1983)
56. K.J. Hayne, A.R. Palmer, Intertidal sea stars (*Pisaster ochraceus*) alter body shape in response to wave action. J. Exp. Biol. **216**(9), 1717–1725 (2013)
57. J. Akizono, T. Tanaka, K. Nakagawa, T. Tsuji, M. Iwasaki, Seabottom roughness measurement by aquatic walking robot, in *Oceans' 97. MTS/IEEE Conference Proceedings*, vol. 2 (IEEE, Piscataway, 1997), pp. 1395–1398
58. M. Cianchetti, M. Calisti, L. Margheri, M. Kuba, C. Laschi, Bioinspired locomotion and grasping in water: the soft eight-arm OCTOPUS robot. Bioinspiration Biomimetics **10**(3), 035003 (2015)
59. D. Dodou, P. Breedveld, J. De Winter, J. Dankelman, J. van Leeuwen, Mechanisms of temporary adhesion in benthic animals. Biol. Rev. **86**(1), 15–32 (2011)
60. R. Santos, S. Gorb, V. Jamar, P. Flammang, Adhesion of echinoderm tube feet to rough surfaces. J. Exp. Biol. **208**(13), 2555–2567 (2005)
61. J. Von Byern, W. Klepal, Adhesive mechanisms in cephalopods: a review. Biofouling **22**(5), 329–338 (2006)
62. W.M. Kier, A.M. Smith, The structure and adhesive mechanism of octopus suckers. Integr. Comp. Biol. **42**(6), 1146–1153 (2002)
63. H. Albitar, K. Dandan, A. Ananiev, I. Kalaykov, Underwater robotics: surface cleaning technics, adhesion and locomotion systems. Int. J. Adv. Robot. Syst. **13**(1), 7 (2016)
64. M. Follador, F. Tramacere, B. Mazzolai, Dielectric elastomer actuators for octopus inspired suction cups. Bioinspiration Biomimetics **9**(4), 046002 (2014)
65. A. Sadeghi, L. Beccai, B. Mazzolai, Design and development of innovative adhesive suckers inspired by the tube feet of sea urchins, in *2012 Fourth IEEE RAS and EMBS International Conference on Biomedical Robotics and Biomechatronics (BioRob)* (IEEE, Piscataway, 2012), pp. 617–622
66. J.A. Sandoval, S. Jadhav, H. Quan, D.D. Deheyn, M.T. Tolley, Reversible adhesion to rough surfaces both in and out of water, inspired by the clingfish suction disc. Bioinspiration & Biomimetics. IOP Publishing, **14**(6), 066016 (2019)

67. T. Paschal, M.A. Bell, J. Sperry, S. Sieniewicz, R.J. Wood, J.C. Weaver, Design, fabrication, and characterization of an untethered amphibious sea urchin-inspired robot. IEEE Robot. Autom. Lett. **4**(4), 3348–3354 (2019)
68. M. Calisti, G. Picardi, C. Laschi, Fundamentals of soft robot locomotion. J. R. Soc. Interface **14**(130), 20170101 (2017)
69. J.O. Dabiri, S.P. Colin, J.H. Costello, Fast-swimming hydromedusae exploit velar kinematics to form an optimal vortex wake. J. Exp. Biol. **209**(11), 2025–2033 (2006)
70. J.O. Dabiri, Optimal vortex formation as a unifying principle in biological propulsion. Ann. Rev. Fluid Mech. **41**, 17–33 (2009)
71. S. Steele, G. Weymouth, M. Triantafyllou, Added mass energy recovery of octopus-inspired shape change. J. Fluid Mech. **810**, 155–174 (2017)
72. F. Giorgio-Serchi, G.D. Weymouth, Drag cancellation by added-mass pumping. J. Fluid Mech. **798**, R3 (2016)
73. W. Johnson, P. Soden, E. Trueman, A study in jet propulsion: an analysis of the motion of the squid, *Loligo vulgaris*. J. Exp. Biol. **56**(1), 155–165 (1972)
74. J.M. Gosline, M.E. DeMont, Jet-propelled swimming in squids. Sci. Am **252**(1), 96–103 (1985)
75. C. Wardle, Limit of fish swimming speed. Nature **255**(5511), 725 (1975)
76. S. Iacoponi, G. Picardi, M. Chellapurath, M. Calisti, L. Cecilia, Underwater soft jet propulsion based on a Hoberman mechanism, in *2018 IEEE International Conference on Soft Robotics (RoboSoft)* (IEEE, Piscataway, 2018), pp. 449–454
77. G. Weymouth, V. Subramaniam, M. Triantafyllou, Ultra-fast escape maneuver of an octopus-inspired robot. Bioinspiration Biomimetics **10**(1), 016016 (2015)
78. F.G. Serchi, A. Arienti, C. Laschi, Biomimetic vortex propulsion: toward the new paradigm of soft unmanned underwater vehicles. IEEE/ASME Trans. Mechatron. **18**(2), 484–493 (2012)
79. J. Frame, N. Lopez, O. Curet, E.D. Engeberg, Thrust force characterization of free-swimming soft robotic jellyfish. Bioinspiration Biomimetics **13**(6), 064001 (2018)
80. A. Villanueva, C. Smith, S. Priya, A biomimetic robotic jellyfish (Robojelly) actuated by shape memory alloy composite actuators. Bioinspiration Biomimetics **6**(3), 036004 (2011)
81. T. Cheng, G. Li, Y. Liang, M. Zhang, B. Liu, T.-W. Wong, J. Forman, M. Chen, G. Wang, Y. Tao, et al., Untethered soft robotic jellyfish. Smart Mater. Struct. **28**(1), 015019 (2018)
82. F.E. Fish, Transitions from drag-based to lift-based propulsion in mammalian swimming. Am. Zool. **36**(6), 628–641 (1996)
83. S. Vogel, *Life in Moving Fluids: The Physical Biology of Flow* (Princeton University Press, Princeton, 1994)
84. C.M. Pace, R.W. Blob, M.W. Westneat, Comparative kinematics of the forelimb during swimming in red-eared slider (*Trachemys scripta*) and spiny softshell (*Apalone spinifera*) turtles. J. Exp. Biol. **204**(19), 3261–3271 (2001)
85. X. Jia, Z. Chen, A. Riedel, W.R. Hamel, M. Zhang, Energy-efficient propulsion inspired by whirligig beetles, in *2014 IEEE International Conference on Robotics and Automation (ICRA)* (IEEE, Piscataway, 2014), pp. 2463–2468
86. Y. Tang, L. Qin, X. Li, C.-M. Chew, J. Zhu, A frog-inspired swimming robot based on dielectric elastomer actuators, in *2017 IEEE/RSJ International Conference on Intelligent Robots and Systems (IROS)* (IEEE, Piscataway, 2017), pp. 2403–2408
87. X. Jia, Z. Chen, A. Riedel, T. Si, W.R. Hamel, M. Zhang, Energy-efficient surface propulsion inspired by whirligig beetles. IEEE Trans. Robot. **31**(6), 1432–1443 (2015)
88. H.-J. Kim, S.-H. Song, S.-H. Ahn, A turtle-like swimming robot using a smart soft composite (SSC) structure. Smart Mater. Struct. **22**(1), 014007 (2012)
89. P.W. Webb, Form and function in fish swimming. Sci. Am. **251**(1), 58–68 (1984)
90. M. Sfakiotakis, D.M. Lane, J.B.C. Davies, Review of fish swimming modes for aquatic locomotion. IEEE J. Oceanic Eng. **24**(2), 237–252 (1999)
91. R. Godoy-Diana, J.-L. Aider, J.E. Wesfreid, Transitions in the wake of a flapping foil. Phys. Rev. E **77**(1), 016308 (2008)

92. O. Akanyeti, J. Putney, Y.R. Yanagitsuru, G.V. Lauder, W.J. Stewart, J.C. Liao, Accelerating fishes increase propulsive efficiency by modulating vortex ring geometry. Proc. Nat. Acad. Sci. **114**(52), 13828–13833 (2017)
93. P.G. Saffman, *Vortex Dynamics* (Cambridge University Press, Cambridge, 1992)
94. N.E. Kochin, *Theoretical Hydromechanics* (Interscience Publishers, Geneva, 1964)
95. C. Eloy, Optimal Strouhal number for swimming animals. J. Fluids Struct. **30**, 205–218 (2012)
96. D. Tzeranis, E. Papadopoulos, G. Triantafyllou, On the design of an autonomous robot fish, in *Proceedings of the 11th IEEE Mediterranean Conference on Control and Automation* (2003), pp. 17–20
97. D. Barrett, M. Triantafyllou, D. Yue, M. Grosenbaugh, M. Wolfgang, Drag reduction in fish-like locomotion. J. Fluid Mech. **392**, 183–212 (1999)
98. A. Punning, M. Anton, M. Kruusmaa, A. Aabloo, A biologically inspired ray-like underwater robot with electroactive polymer pectoral fins, in *International IEEE Conference on Mechatronics and Robotics*, vol. 2004 (2004), pp. 241–245
99. M.S. Triantafyllou, G.S. Triantafyllou, An efficient swimming machine. Sci. Am. **272**(3), 64–70 (1995)
100. W. Zhao, J. Yu, Y. Fang, L. Wang, Development of multi-mode biomimetic robotic fish based on central pattern generator, in *2006 IEEE/RSJ International Conference on Intelligent Robots and Systems* (IEEE, Piscataway, 2006), pp. 3891–3896
101. L. Leifsson, S. Koziel, Aerodynamic shape optimization by variable-fidelity computational fluid dynamics models: a review of recent progress. J. Comput. Sci. **10**, 45–54 (2015)
102. J. Colorado, A. Barrientos, C. Rossi, K.S. Breuer, Biomechanics of smart wings in a bat robot: morphing wings using SMA actuators. Bioinspiration Biomimetics **7**(3), 036006 (2012)
103. K. Kostas, A. Ginnis, C. Politis, P. Kaklis, Ship-hull shape optimization with a T-spline based BEM-isogeometric solver. Comput. Methods Appl. Mech. Eng. **284**, 611–622 (2015)
104. T.-H. Joung, K. Sammut, F. He, S.-K. Lee, Shape optimization of an autonomous underwater vehicle with a ducted propeller using computational fluid dynamics analysis. Int. J. Naval Archit. Ocean Eng. **4**(1), 45–57 (2012)
105. S. Kern, P. Koumoutsakos, Simulations of optimized anguilliform swimming. J. Exp. Biol. **209**(24), 4841–4857 (2006)
106. A.J. Smits, Undulatory and oscillatory swimming. J. Fluid Mech. **874**, P1 (2019)
107. A.J. Ijspeert, Central pattern generators for locomotion control in animals and robots: a review. Neural Netw. **21**(4), 642–653 (2008)
108. C. Stefanini, S. Orofino, L. Manfredi, S. Mintchev, S. Marrazza, T. Assaf, L. Capantini, E. Sinibaldi, S. Grillner, P. Wallén, et al., A novel autonomous, bioinspired swimming robot developed by neuroscientists and bioengineers. Bioinspiration Biomimetics **7**(2), 025001 (2012)
109. D. Korkmaz, G. Ozmen Koca, G. Li, C. Bal, M. Ay, Z.H. Akpolat, Locomotion control of a biomimetic robotic fish based on closed loop sensory feedback CPG model. J. Marine Eng. Technol., 1–13 (2019)
110. S.P. Lal, K. Yamada, S. Endo, Studies on motion control of a modular robot using cellular automata, in *Australasian Joint Conference on Artificial Intelligence* (Springer, Berlin, 2006), pp. 689–698
111. Z. Shen, J. Na, Z. Wang, A biomimetic underwater soft robot inspired by cephalopod mollusc. IEEE Robot. Autom. Lett. **2**(4), 2217–2223 (2017)
112. M. Kruusmaa, P. Fiorini, W. Megill, M. de Vittorio, O. Akanyeti, F. Visentin, L. Chambers, H. El Daou, M.-C. Fiazza, J. Ježov, et al., FILOSE for svenning: a flow sensing bioinspired robot. IEEE Robot. Autom. Mag. **21**(3), 51–62 (2014)
113. S. Barbarino, O. Bilgen, R.M. Ajaj, M.I. Friswell, D.J. Inman, A review of morphing aircraft. J. Intell. Mater. Syst. Struct. **22**(9), 823–877 (2011)
114. M.-W. Han, H. Rodrigue, H.-I. Kim, S.-H. Song, S.-H. Ahn, Shape memory alloy/glass fiber woven composite for soft morphing winglets of unmanned aerial vehicles. Compos. Struct. **140**, 202–212 (2016)

115. H. Rodrigue, S. Cho, M.-W. Han, B. Bhandari, J.-E. Shim, S.-H. Ahn, Effect of twist morphing wing segment on aerodynamic performance of UAV. J. Mech. Sci. Technol. **30**(1), 229–236 (2016)
116. R.J. Rufino, T.F. Miller, F.S. Gandhi, Morphing hull concepts for unmanned underwater vehicles, in *OCEANS 2008* (IEEE, Piscataway, 2008), pp. 1–11
117. I. Rechenberg, Case studies in evolutionary experimentation and computation. Comput. Methods Appl. Mech. Eng. **186**(2–4), 125–140 (2000)

Chapter 3
Amphibious Robotic Propulsive Mechanisms: Current Technologies and Open Challenges

Robert Baines, Frank Fish, and Rebecca Kramer-Bottiglio

3.1 Introduction

Much research effort has been dedicated to underwater robots, as evidenced by numerous papers and the contents of this book [1]. An even larger body of literature concerns terrestrial robots. But what about amphibious robots that can operate both in water and on land? From the literature, we glean that amphibious robots are fantastic vehicles for studying autonomous navigation strategies in unstructured, complex environments [2]. Furthermore, they have proven useful as physical models for gaining deeper insight into the gait patterns and mechanics of animal locomotion [3–5], analyzing the health of ecosystems [6, 7], and understanding physical principles underlying propulsion in various media [8, 9]. Beyond the academic space, advances in amphibious robotics are projected to be a significant boon to industries such as reconnaissance, surveying, offshore mine detection, and water quality monitoring, where seamless transitioning between locomotion modes is critical to success [10–13].

In spite of their versatility, relatively few amphibious robots have been reported in the literature. Significant challenges remain for designing, building, and implementing amphibious systems outside of the laboratory setting. Central to realizing an amphibious robot is designing propulsive mechanisms for effective water- and land-based locomotion. In doing so, the engineer must strive to balance conflicting features. As we shall investigate in this chapter, both in natural and physical systems,

R. Baines · R. Kramer-Bottiglio (✉)
Department of Mechanical Engineering and Materials Science, Yale University, New Haven, CT, USA
e-mail: rebecca.kramer@yale.edu

F. Fish
Department of Biology, West Chester University, West Chester, PA, USA
e-mail: ffish@wcupa.edu

D. A. Paley, N. M. Wereley (eds.), *Bioinspired Sensing, Actuation, and Control in Underwater Soft Robotic Systems*, https://doi.org/10.1007/978-3-030-50476-2_3

functional shapes conducive to load bearing and terrestrial maneuverability often detract from hydrodynamic efficiency and water compatibility.

The transition between water and land, called the littoral zone, also presents challenges. The littoral zone is characterized as a turbulent environment due to wave action, intermixing of heterogeneous substrates, suction forces, and abrasive flows imposed by fluidized sediment. Rocks, shoals, uneven slopes, dense algal beds, and reefs are all obstacles that an amphibious robot might encounter and have to negotiate in a transition zone [14, 15]. The dynamic onslaught of physical phenomena and obstacles in the littoral zone constitute a multi-faceted problem for which there are no obvious robot design solutions.

To provide biological inspiration in the design of amphibious propulsors, this chapter first analyzes animals' body plans and their locomotor adaptations in Sect. 3.2. Animal morphology and physiology is, in fact, chiefly influenced by evolutionary pressures for effective movement in an environment [16]. For the sake of brevity, we hone in on key examples from semi-aquatic, semi-terrestrial, and highly specialized species that typify the range of propulsive modes exhibited by animals.

With biological context, we transition to a survey of existing amphibious robotic platforms in Sect. 3.3. Designs striving to address the slew of environmental challenges amphibious robots face can be broadly classified into systems that locomote using separate or united propulsive mechanisms. We define separate to mean that distinct subsystems move a robot on land and though water, whereas movement with a united mechanism is achieved in both media via the same subsystem. While separating propulsive mechanisms is a more traditional approach and may allow robots to locomote with specialized modes of transit in each environment (i.e., using wheels to move on land and jets to move through water), uniting propulsive mechanisms has gained popularity as a means of reducing system complexity and exploring propulsive architectures inspired by amphibious animals (i.e., using snake-like undulations to move in water and on land) [17]. Beyond separate and united propulsive mechanisms, we further sort robots into (1) wheeled, (2) legged, (3) undulatory, or (4) soft categories, based on their body plans and primary means of propulsion. Section 3.3 further expounds on what distinguishes these classifications from each other.

After a synthesis of existing work, Sect. 3.4 identifies promising avenues for future research on amphibious robotics. Lastly, Sect. 3.5 presents a case study: our research on a variable stiffness morphing limb, a design that seeks to unite various propulsive functionalities into a single cohesive mechanism.

3.2 Biological Perspectives on Amphibious Locomotion

3.2.1 Movement through Different Media

There are physical differences between water and air that determine the different mechanisms of animal locomotor modes in either environment. Density and viscosity are the most important physical properties [18]; water is 800 times denser than air and 55 times more viscous. The ratio of inertial to viscous effects is also of paramount importance to aquatic locomotion [19], and can be expressed quantitatively via the Reynolds number:

$$Re = \frac{\rho U L}{\mu}, \tag{3.1}$$

where ρ is the fluid density, U speed, L characteristic length of the body in the fluid, and μ dynamic viscosity. We will touch on the ramifications of scale with regard to specific propulsive mechanisms later. Regardless of scale, swimming animals tend toward a density close to that of water to support their body weight via buoyancy. They use the high density and viscosity of the medium to generate hydrodynamic forces for propulsion. Swimming is accomplished by propulsors that can be broadly classified as undulatory, lift-based oscillatory, drag-based oscillatory, or jetting [20–22].

On land, an animal moves through air, so gravitational forces predominate and the weight of an animal has to be supported by rigid or hydrostatic skeletons. Animals apply frictional forces from contact of a body or limbs on the solid ground for propulsion. Broadly speaking, terrestrial locomotion is enabled by propulsors that induce undulatory motions, limbed locomotion, a combination of both undulating and limbed locomotion, or rolling [23–26].

Many animals are capable of moving between water and land [22]. The reasons why animals adapted for multi-modal locomotion between water and land stem from survival adaptations, including catching prey, escaping from predators, mating, and searching for food [27]. For vertebrate animals, the shift from finned swimming to legged terrestrial locomotion in the transition from fish to amphibians is considered one of the watershed events in evolution. Aside from amphibians, mammals, reptiles, and birds have amphibious species exhibiting varying degrees of terrestrial and aquatic locomotor adaptations. Amphibious behaviors are also exhibited by invertebrate lineages, most notably mollusks and arthropods. All amphibious animals, regardless of classification, must strike a balance between being both semi-aquatic and semi-terrestrial, and utilize united or separate propulsive mechanisms for locomotion in either media.

3.2.2 Amphibious Animals with United Propulsive Mechanisms

Numerous amphibious animals exist in nature that utilize the same propulsive mechanism in both mediums [28]. Amphibians, notably frogs, hop and swim using a similar movement pattern: maximum extension of hind limbs, followed by a sweeping recovery phase [29]. Mammalian limbs can relatively easily engage in drag-based propulsion in water. Drag-based swimming and quadrupedal walking locomotion modes are embodied by mammals such as muskrats, elephants, and opossums [30]. When swimming in a drag-based regime, propulsive drag force is produced only through half of the stroke cycle by the rearward movement of the appendage, since forward motion is a non-thrust generating recovery phase. Consequently, limbed amphibious mammals demonstrate high locomotor costs due to their inability to specialize for water or land [22].

Amphibious reptiles, like freshwater turtles, also utilize united propulsive mechanisms—swinging of limbs—for drag-based swimming and quadrupedal walking. Amphibious snakes, on the other hand, leverage bodily undulations to move in the water and on land. Evolutionary adaptations that enhance aquatic locomotion at the expense of terrestrial locomotion, such as diminished ventral plates and a flattened tail, can be observed between some species of snakes. [31]. Mollusks like the octopus represent a rather unique case in that they use unsupported limbed locomotion to crawl along the ocean bottom and on land [32].

Propulsive mechanisms for amphibious animals that walk both on the surface of water and on land are very much governed by size. Small animals with hydrophobic surfaces, such as water striders, can exploit surface tension to support the body [33]. As the size of an animal increases, surface tension becomes insufficient to support its weight. The basilisk lizard can run atop water by simultaneously generating surface-level drag and expanding an air cavity underwater [34]. As the size of a basilisk lizard increases, the kinematics of the foot stroke change to keep it atop the surface [35]. For even larger organisms, like aquatic birds, movement atop water requires the addition of broad wings impacting the surface [36].

3.2.3 Amphibious Animals with Separate Propulsive Mechanisms

A less common model in nature than united propulsive mechanisms is to utilize separate propulsive mechanisms for either medium. Among mammals, otters engage in quadrupedal walking and undulatory swimming [37]. Reptiles of the order Crocodilia, including alligators and crocodiles, use tail undulations in the water but rely on quadrupedal gaits on land. Similarly, newts and salamanders undulate their bodies in a standing wave while using their limbs to walk [28]. Some at high speeds will undulate on land with their legs tucked in to their sides.

3.2.4 Specialization for Water or Land

Generally, semi-aquatic or semi-terrestrial animals—like those mentioned up to this point—are less energy efficient and slower (in terms of body lengths per second) than animals that are specialized for one environment [22]. Yet in engineering, an amphibious robot is not limited by the same physiological factors that limit animals. Animals must move by oscillations of appendages or body undulations powered by muscles, must respire with gills or lungs for oxygen to fuel cells constituting their muscles, are composed of biomaterials such as bone, cartilage, and chitin that have lower strengths compared to metals used in engineered systems, and have a large portion of the body devoted to reproductive functions. An amphibious robot can take advantage of specialized, rapid, and efficient aquatic and terrestrial propulsive mechanisms, or those that are dangerous for animals [5]. It is thus worth enumerating some of the more optimal propulsive mechanisms and body plans for water and land exhibited by highly derived (specialized for certain environments as a result of evolution) species.

Inhabiting an exclusively aquatic environment, tuna and dolphins have streamlined, hydrodynamic bodies and rely on oscillation of their caudal fins to generate thrust as a vector component of lift. Other fast and efficient aquatic animals with lateral fins and flippers can produce thrust by undulatory (bluegill, sunfish, stingray) swimming, as well as oscillatory wing-like movements (sea lion, sea turtle). Though not sustainable over long periods, jetting can enable rapid accelerations and is seen in jellyfish, squid, and octopus. Larger aquatic animals, like those mentioned above, are considered to be nekton, that is, capable of swimming long distances independent of water currents. Nektonic animals swim at high Reynolds number ($Re > 10^3$ up to 10^8) [38]. At high ranges of Re, swimming is performed by accelerating a mass of water for propulsion. Viscous forces are small, whereas inertial forces are large [38]. Yet the overwhelming number of animals that exist in the oceans are small, like plankton, and use ocean currents pushing on their bodies as a propulsive mechanism to move long distances. Most planktonic animals (e.g., copepods) lie within an intermediate range of Re ($1 < Re < 10^3$), where viscous and inertial forces are both important [39]. At even smaller scales, bacteria cilia and flagella operate between 10^{-5} and 10^{-6} Re, where viscous effects dominate [40].

Terrestrial vertebrates are generally not streamlined. They have a defined neck, no blubber to contour the body shape, and if they have limbs, the limbs are generally cylindrical with an approximately circular cross section. As size (and thereby mass) of a terrestrial organism increases, gravity becomes a more dominant force, unlike in water where weight is supported by buoyancy. Among limbed terrestrial animals, peak limb stresses can increase with increasing body size, so posture of the skeletal elements of the legs tends toward a more columnar (upright) stance [41, 42]. Such a change in posture maintains a safety factor independent of size, but at the expense of accelerative capability and maneuverability. Horses and cheetahs are exemplar animals with skeletal components disposed to upright walking and high-speed linear galloping. Highly derived terrestrial animals, like the horse and cheetah, exploit spring-like interactions of their body with the ground to enhance thrust produced with each stride [43]. The storage and release of elastic energy as a propulsive mode actually increases in efficiency at higher speeds in some animals [44]. In addition to limbed propulsion aided by elastic potential energy storage, it is instructive to mention another (perhaps more unusual) specialized terrestrial propulsive mechanism that harnesses gravity: rolling. Passive rolling is demonstrated by arachnids, while species of caterpillar actively build up angular momentum during rapid escape maneuvers [26]. Top speeds of rolling animals can be an astounding tens of body lengths per second.

3.2.5 Biological Inspiration for Design of Amphibious Robotic Propulsive Mechanisms

Both environmental medium and scale influence an organism's propulsive mechanism adaptations. The intermediate status of amphibious animals compromises their locomotive performance in either environment, but the mechanics of these intermediate species can potentially serve as a template to develop a new generation of amphibious robots. Furthermore, amphibious robots can be designed to incorporate locomotor mechanics that are specialized for either environment or any scale, expanding upon what nature has been capable of producing through evolution.

It is thus the charge of the robotic designer to synthesize various propulsive mechanisms and corresponding body plans found in nature—even to look beyond these natural examples to synthetic solutions—to innovate and produce an effective amphibious robotic system. Such a methodology to realizing future amphibious robots is depicted in Fig. 3.1. The next sections review current work in amphibious robotics. An exposé of novel designs, biologically inspired or of purely synthetic origin, as well as the advantages and disadvantages of these designs, supplies additional foreknowledge to synthesize next-generation amphibious robots.

3.3 Classification of Amphibious Robots

A strict template for classification does not necessarily encompass any given robot's propulsive mechanisms. While robots with united propulsive mechanisms may be much easier to classify, robots with separate propulsive mechanisms defy any strict classification. For the sake of review, we sort amphibious robots into wheeled, legged, and undulating based on the most salient aspect of their morphologies. Wheeled, legged, and undulating amphibious robots utilize significantly different mechanisms to move in water and on land. Wheeled systems include rounded entities concentric to axles that, when engaged in rolling, provide leverage based on the radius. Wheels can be driven passively (by gravity) or actively (by motors).

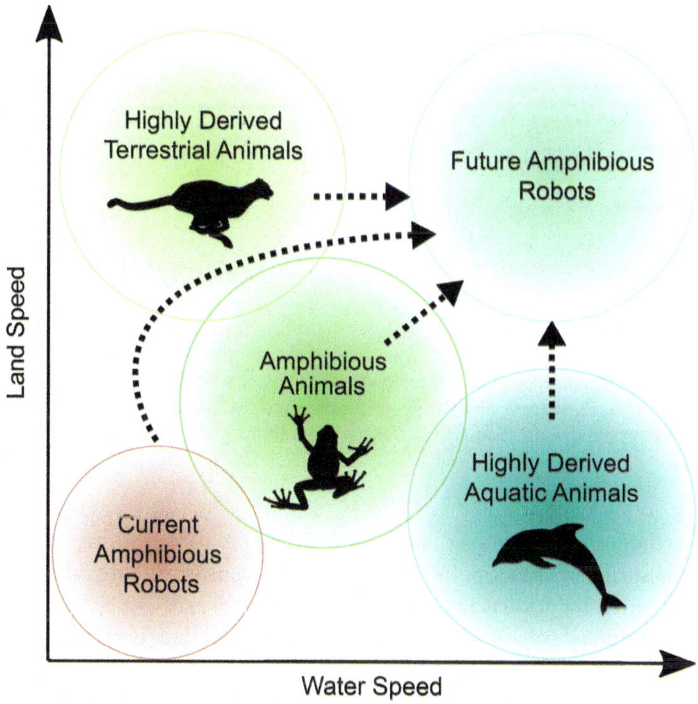

Fig. 3.1 Combining knowledge of the propulsive mechanisms of amphibious animals, the specialized mechanisms exhibited by highly derived species, and existing amphibious robot designs is key to developing the next generation of amphibious robots. Note that the relative positioning of groups is based on the authors' qualitative assessment of performance

In water, movement of treads through the fluid medium can serve as discretized paddles for drag-based thrust. On land, wheels rest on the ground at all times and rely on friction contact force on the substrate at a point and repetitive revolutions to generate a thrust vector.

Legs are a more generic morphology. They can be articulated, multi-degree-of-freedom, or single degree-of-freedom links of various shapes and kinematic configurations. In water, legs engage in oscillations or power strokes to generate thrust. On land, legged systems utilize discrete footholds to move. The body is supported off the ground with either upright or sprawled legs. Having a lower center of mass, sprawled posture is generally more stable than the full upright orientation of legs. A robot's legs must leave the ground in periodic increments to achieve bodily displacement.

Undulating robots come in elongated forms, often consisting of a series of modules connected together. In water, undulatory swimmers display a gradient of propulsive modes for thrust production—from undulatory waves encompassing the total length of the body (anguilliform), through progressively expanded posterior regions of the body (carangiform), culminating in thrust production confined to the

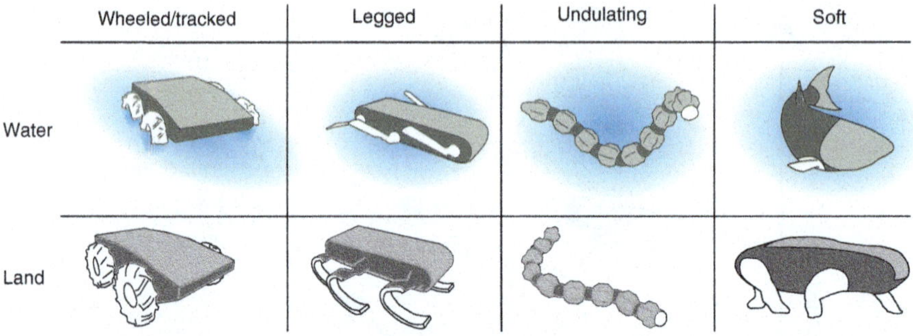

Fig. 3.2 Amphibious robots sorted into four distinct categories based on their primary propulsive mechanisms and structural features: wheeled/tracked (note, wheels are half-submerged in water in the depicted image), legged, undulating, and soft. Here, we illustrate how each category might unite the same propulsive mechanism, or transform its limbs or body plan to accommodate water and land

caudal fin (thunniform). On land, undulating robots move in a fashion characterized by multiple surface contact points. Waves passing posteriorly down the body push on solid substrate for forward movement.

Soft amphibious robots may locomote in a way that is consistent with the previous three delineated categories, but are distinguished by their composition of continuously deformable materials, typically having a Young's modulus on the order of, or less than, one MPa. Consequently, soft robots do not commonly use traditional rigid motors. Reliance on soft actuators [45] is thus another factor that separates them from the other three categories.

Figure 3.2 depicts how each category might unite the same propulsive mechanism, albeit with morphological transformations present in the legged and soft categories. Each category is associated with a unique set of advantages and disadvantages that motivate discussion on what constitutes an effective amphibious robot design. The following subsections address each of the classes of robots in the listed order. For each, we open with a discussion of its respective advantages. We then highlight seminal amphibious robots belonging to that class. Our intention is to not to detail every amphibious robot reported in literature, but to summarize key innovations as embodied by the seminal designs. Aspects we focus on when appraising an amphibious robot are its capacity to bear payloads (i.e., sensors, camera, equipment; this criteria is of course dependent on size), ease of control, efficiency, maneuverability, and speeds in water and on land. Aside from absolute speeds in water and on land, which we have plotted against each other for a number of designs and comparably sized animals in Fig. 3.3, the other metrics are not consistently reported for amphibious robots throughout literature. We instead provide a qualitative assessment of these metrics. Each section closes with a synopsis of a particular class' drawbacks, and therefore opportunities for future research.

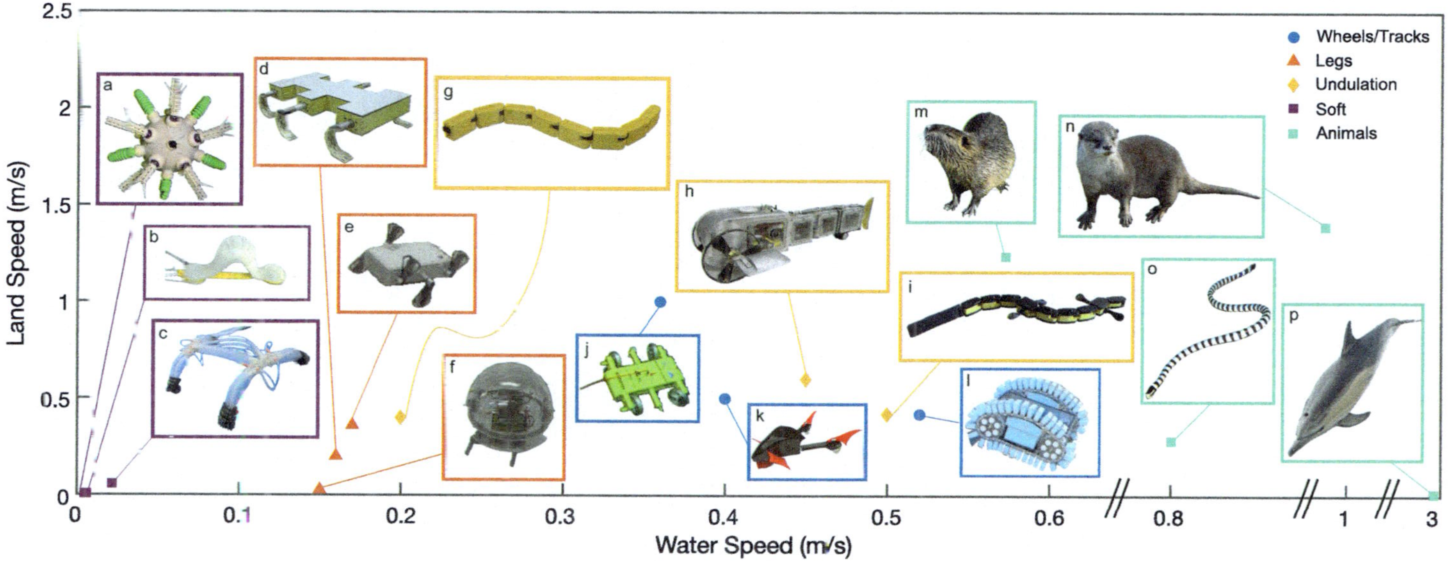

Fig. 3.3 A comparison of absolute sustained speeds between representative amphibious robots from each of the delineated categories, as well as semi-aquatic, semi-terrestrial, and aquatic animals. Note that many of the increases in robots' speed have naturally come hand-in-hand with improved research hardware, such as higher torque motors and lower footprint microcontrollers. We try and account for this factor by featuring a variety of older and newer published work from each category. Citations for (**a–l**) are [4, 6, 9, 46–54]. We include comparably sized animals operating in a similar Reynolds regime (thus excluding whales, bacteria, etc.) for comparison to the robots. The animals are from (**m–p**): muskrat, river otter, yellow-lipped sea krait, and bottlenose dolphin. Citations for speeds (**m–p**): [37, 55–59]

3.3.1 Wheeled and Tracked Amphibious Robots

Wheels represent a unique propulsive mechanism for amphibious robots since, compared to legs and undulations, wheel-like forms are less readily used in nature [26]. On land, wheels pose an energetic benefit because when traveling at a constant speed, the kinetic energy remains constant [40]. Compare this energy profile to most other natural propulsive mechanisms that necessitate trajectories of cyclic acceleration and deceleration that may incur a sizable energy cost. It is not surprising then, that a common—and perhaps the simplest—propulsive mechanism for amphibious robots is wheels or tracks.

A ubiquitous amphibious robot design, similar to many amphibious vehicles, uses wheels to move on land and on top of water. Such robots are far along in the development pipeline and are commercially available [60]. Though rotating wheels to move in water and on land unite the same propulsive mechanism, simplifying control, it comes at the expense of speed in water. Speed is limited because wheels increase drag due to the heightened relative velocity from the rotation of the wheel, which can induce flow separation [61].

Yamada et al. built what appears to be a quintessential wheeled amphibious robot, but with a clever twist for improved aquatic locomotion. Their four-wheeled platform, R-Crank, incorporates a ribbed crank link between its front and back wheels on each of its sides. The crank link generates thrust for surface-based aquatic locomotion as the tires spin [62].

While the previously mentioned robots include wheels as a part of their hardware, other robots are, by no stretch of definition, wheels. A design perk of wheel-bodied robots over robots with wheels is that the need for a bulky chassis is eliminated, reducing weight, and potentially conferring hydrodynamic benefits. Consider that the drag force on a body is

$$F_D = C_D A \frac{\rho V^2}{2}, \tag{3.2}$$

where C_D is the drag coefficient, ρ is the density of the fluid, V is the flow velocity relative to the body, and A is the reference area. Note removing a chassis will substantially decrease A. Also, the C_D of a wheel, typically shaped like a sphere or disk can be much lower than typically angular, rectangular chassis.

One wheel-bodied system, Groundbot, was developed by Rotundus AB in Sweden as a multi-terrain surveillance robot. Spherical, treaded, made of rubber, and fully resembling a tire, Groundbot can roll and steer itself using an offset internal weight actuated by motors. It boasts up to 3 m/s (5 bl/s) sustained speeds on level ground. It employs the same rolling mechanism to traverse water as it does for land, floating atop the surface and generating thrust with its specifically engineered treads that essentially serve as a series of paddles [63].

Another robot with a wheel-like body is the triphibious MUWA. MUWA consists of a ring of polystyrene foam surrounding a multicopter. The polystyrene ring gives

MUWA enough buoyancy to float atop water, but also provides geometry conducive to rolling. When rolling on land, MUWA controls its trajectory by adjusting the pitch of its rotating propellers [64].

Transition areas rife with loose or fluidized sediment threaten to ensnare wheels. Rolling resistance of wheels increases in proportion with soil compliance [65]; in fact, one study found that rolling resistance on concrete can be 10–15 times less than that of sand [66]. Terramechanics research models physical interactions with the substrate and provides an estimate of the extent to which sinking into the substrate impedes motion [67]. Concentrated weight is a primary cause of local soil fracture, causing ensnarement. Tracks distribute the weight of a robot over a larger area to mitigate sinking and improve locomotion across muddy terrain, but may sacrifice some speed and maneuverability.

Tracked amphibious robots have been built that are intended to sink below the surface to crawl on lake beds [11]. Sinking to the bottom is not always practical, though. Motivated by the need to monitor an estuary system composed of multiple rapid transitions from shallow water to soggy land, one robot was designed with buoyant tracks so it can engage in surface swimming while retaining the ability to navigate muddy sections [6]. This surface-swimming robot has the fastest speed on water reported in Fig. 3.3.

The aforementioned wheeled or tracked designs have not demonstrated ability to swim in 3D. For certain applications, amphibious robots need to be able to engage in 3D swimming so that they can transition between the surface and underwater and explore the water column in between. In order to enable 3D swimming while retaining the merits of wheels on land, one group introduced a class of robot with hybrid wheels/propellers and separate fins [53]. The hybrid wheels/propellers have spokes emanating radially from the termination of their axle, effectively providing the hydrodynamic thrust of propellers and generating sustained speed of up to 0.36 m/s (0.375 bl/s) in water (Fig. 3.3j). The fins on the robot are used to steer when swimming. Owing to its wheeled design, this robot also exhibits the top speed on land out of all designs in Fig. 3.3, at 1 m/s (1.04 bl/s). Transitions from aquatic to terrestrial locomotion are accomplished by orienting the wheels/propellers so their revolution will provide forward thrust in water or serve as a wheel on the land.

Another hybrid wheel/propeller mechanism that allows for 3D swimming, dubbed the eccentric paddle, consists of a shaft embedded in a wheel on which paddles are radially distributed. The paddles move in and out of the wheel to adjust the extent of their interaction with the environment. Equipped with eccentric paddles, the robot demonstrates an unusual capability: a quasi-walking mode of locomotion by cyclically protruding the paddles from the wheel [68–70]. Offering the speed of wheels as well as aquatic mobility, hybrid approaches, like combined wheels/propellers and separate fins or the eccentric paddle, nonetheless rely on complex agglomerations of mechanisms that are difficult to control. Such intricate hybrid mechanisms with many moving components may also be susceptible to debris in the littoral zone compromising their function.

Overall, although they offer efficiency and high top speeds on flat land and the requisite structural integrity to support large payloads, a major drawback of wheeled and tracked robots is their lack of ground clearance and consequently diminished ability to traverse uneven terrain [71, 72]. Heavy systems with wheels, specifically, have a propensity to become trapped in shallow, fluidized sediment [73]. Moreover, wheels do not scale well to small sizes because they become sensitive to substrate compliance and uneven terrain impacting forward motion [40]. A promising design strategy, creating a wheel-bodied robot or a hybridized paddle-wheel mechanism, can reduce mass, enable 3D swimming, and address minor challenges posed by uneven terrain [53, 70]. Yet, hybrid mechanisms can be plagued by their host of moving parts, especially in the littoral zone rife with obstacles. If creating hybrid mechanisms, designers should err on the side of simplicity to provide resilience against environmental detritus. Lastly, robots with wheels or tracks can leave destructive trails behind them—a heavy wake or dislodged soil. While observing fragile ecosystems or trying to maintain stealth, disturbances to the environment can undermine mission success. More work on wheeled systems that minimize environmental impact thus represents an open area of research. When navigation of uneven terrain or stealth are prerequisites to a successful mission, other types of amphibious robots might be better options than current designs with wheels or tracks.

3.3.2 Legged Amphibious Robots

Legged amphibious robots tend to be more complex in design and control architecture than wheeled robots due to the multiple controlled degrees of freedom associated with each leg. Primary advantages of legged robots include their capability to traverse obstacles wheels or tracks cannot, and need for only discrete footholds to locomote, as opposed to a continuous supporting surface. On especially soft ground, legged robots deform terrain less than wheeled or tracked systems and thereby can diminish the energy required for traversal [74]. Due to their multiple degrees of freedom, legged robots can also change direction without slippage. Like wheeled robots, legged robots can sustain concentrated payloads undulating robots cannot, since their motion is not dependent on dynamic oscillations of interconnected bodily modules.

Spurred by calls to develop systems to locate and destroy mines in the surf zone, one of the first examples of a legged amphibious robot—let alone one of the first amphibious robots—was published in 1996. The report detailed a hexapod robot, Ariel, inspired by the crab that walks at a shallow depth along the seabed [12]. Since Ariel, similar work has been published focusing on the creation of lobster-inspired robots with the same purpose [75]. A major concern for benthic walking robots is the large hydrostatic pressures associated with great depths—not only for water-proofing but also for feasibility of walking. Large moments would be created about the leg joints at depths, and it is unlikely compact actuators could overcome such moments.

In contrast to robots that crawl along the seabed are surface walkers: amphibious robots inspired by insects and reptiles (mentioned in Sect. 3.2) that are able to walk on the surface of water by exploiting the physical properties of their feet, and/or their small body mass. Park et al. presented a platform inspired by the basilisk lizard. The crux of their design revolves around a light-weight robot body and a compliant foot pad, which transfers elastic energy to propulsive momentum [76–78]. Yet such a robot cannot bear payloads that would compromise its light weight and would therefore be limited to minimal integrated circuits. Another surface-walker-inspired robot is bulkier and able to bear payloads, compensating for its increased weight via Styrofoam spherical feet that provide buoyant forces [79]. A cockroach-inspired microrobot fabricated by Chen et al. uses partially submerged foot pads to paddle on the surface of the water, exploiting surface tension. Unlike the previous two mentioned platforms, it is able to dive from the surface to the bottom by emitting high voltage from its padded feet to temporarily break surface tension. It can then walk on underwater surfaces as it does on land [80]. Nevertheless, the robot is unable to replicate the speed and dexterity of an actual water strider due to constraints on the force density of such small actuators. In addition to those mentioned, there have been a variety of other surface-walking robots [81].

Though bottom and surface walking may be sufficient in some scenarios, 3D swimming offers greater surveying capability. To this end, a new chapter in amphibious legged robotics started with AQUA [82, 83]. AQUA, based on Boston Dynamics' R-hex platform [84], utilizes six independently controlled paddles on single degree-of-freedom (DOF) joints as control surfaces during swimming. It also has interchangeable, curved cockroach-style legs for walking that act as springs, efficiently storing and releasing potential energy with each stride. AQUA represents the first robot of its kind—legged, able to proficiently move on land and traverse obstacles, but also able to engage in 3D swimming to fair depths. There are many subsequent amphibious robots that drew design inspiration from AQUA's body plan and leg design.

One series of robots inspired by AQUA are hexapods equipped with Whegs™ (a portmanteau for wheel-legs). Whegs™ integrate swimming and walking mechanisms, combining the simplistic control of a wheel with the articulated cadence and ability to traverse some obstacles that legs typically can [85]. Whegs™ are similar to the cockroach-style legs initially equipped to AQUA, but are built with three protruding paddles equally distributed radially about the center shaft, as opposed to AQUA's single paddle. Whegs™ have been implemented on amphibious robotic platforms as a combined propeller/wheel with great success [86–88]. Despite their critical appeal, it should be noted that the drag-based propulsion supplied by Whegs™ is inefficient relative to lift-based locomotion [30]. Also, like propellers, Whegs™ induce turbulent vortices that may pose too much of an environmental disturbance for some applications. Lastly, the geometry of Whegs™ does not allow backwards locomotion on land (the paddles contact the ground at a point, acting as a rigid rod and stunting motion), which detracts from their terrestrial mobility.

Bearing strong resemblance to Whegs™-style robots but with modifications to address the fact that Whegs™ are not able to back-drive, Ninja Legs incorporate a thin circular wire enclosure around flippers. The enclosure protects the compliant flippers and simultaneously serves as an offset wheel for hybrid rolling/walking [89]. In a similar vein as Ninja Legs, one robot named RoboTerp unites flippers and legs into a single hybrid propulsion mechanism. The structure of one limb consists of a flap passively hinged to a grate. The grate serves as a rigid load-bearing leg to support the robot's weight on land. While swimming, during the robot's power stroke, the flap flattens against the grate thereby increasing paddle surface area and producing forward thrust. On the upstroke, the flap freely swings back, reducing drag [90].

RoboTerp's passive paddle mechanism is not conducive to high aquatic mobility. This reinforces the observation that walking legs, in addition to producing less thrust compared to more hydrodynamic, high aspect ratio surfaces, are far from optimal forms for maneuvering in the water. Indeed, the original AQUA platform had separate flipper and leg modules for water and land, respectively, optimized independently of one another. However, the need for human intervention to manually exchange limb designs undermines a system's ability to autonomously transition between environments.

One group introduced AmphiHex-I (Fig. 3.3d) and in doing so initiated a new amphibious legged robot design paradigm. AmphiHex-I [48, 91, 92] features a transformable leg-flipper propulsion mechanism. The leg-flipper consists of interlocking rigid segments connected via a cable. When the cable is pulled, the interlocking segments are compressed to create a curled cockroach-style leg. When released from tension, the limb becomes a compliant flipper. Although the robot's speeds of 0.16 m/s (0.18 bl/s) underwater and 0.2 m/s (0.23 bl/s) on land rank lower middle-tier in Fig. 3.3, the AmphiHex-I design philosophy seems promising in terms of efficiency. Namely, as opposed to relying on separate propulsive mechanisms that may impede each other's performance, or a united propulsive mechanism that sacrifices specialization for average performance, transforming a propulsive mechanism's shape offers the ability to greatly hone locomotive performance in both environments.

After AmphiHex-I came AmphiHex-II (Fig. 3.3e). This robot introduced an entirely different mechanism than AmphiHex-I: manually adjustable variable stiffness legs. Its semi-circular legs are rigid, fan-shaped frames, and protect flexible flippers within. By adjusting a pin joint along the leg's length and the robot's chassis, one can set the leg to five different stiffnesses. Experiments with the robot underscore that modulating the stiffness improves locomotive performance based on fluid content in a soil-like terrain. It was found that higher stiffness limbs allow the robot to locomote fastest in sandy substrates and soft soils. An intermediate stiffness was found to be more efficient for locomotion in fluidized (25%) soil reminiscent of the littoral zone. Lastly, for swimming, the researchers found the highest stiffness elicited maximum achievable, sustained velocity [9].

Though legs give a robot the ability to skillfully traverse a wide swath of terrain, legged robots are generally slower and less energy efficient on flat land, especially when compared to wheeled robots [5]. Moreover, control of legged robots is much more complex than that of wheeled robots. Particularly in the aquatic environment, controlling legs as hydrodynamic surfaces for efficient propulsion represents an open area of research. Whegs™-style legs simplify the control problem by replacing articulated joints with just 1-DOF. However, Whegs™, as well as a majority of the legged amphibious robot designs mentioned, employ drag-based paddling to move through water. As mentioned, paddling is not an efficient mode of swimming, because thrust is only generated in half of the stroke [30]. Thus, alternative modes of aquatic locomotion must be explored and mechanisms developed to achieve those modes. One example of an alternative locomotion mode, incorporating separate jet nozzles on the bottom of a robot's legs [49, 93], seems like a promising strategy but requires highly complex modeling of jet orientation to optimize locomotion (moreover, this specific robot is extremely slow, as seen by its relative position in Fig. 3.3f). If land speed, hydrodynamics, or control complexity are precursors for a successful mission, legged amphibious robots may not fare well.

3.3.3 *Undulating Amphibious Robots*

The third class of amphibious robots is undulating robots, often those with a serpentine body. Key advantages of undulating robots are their high maneuverability in water, small turning radii on land, and multiple DOF that lend themselves to novel locomotion strategies and negotiation of restricted spaces. Arguably one of the most significant jumps in amphibious robotics coincides with the advent of the snake-inspired robot, AmphiBot. Unlike other land-based snake robots of the time, AmphiBot is capable of swimming and crawling, all with the same mechanism: undulation [17, 94]. AmphiBot's body is composed of interconnected, independently actuated modules. Selective actuation of each of the modules via a central pattern generator allows AmphiBot to undulate to move smoothly through water with a speed of 0.2 m/s (0.26 bl/s). However, it has wheels to overcome friction and attain higher speeds on land. In terms of relative speed to other robots, AmphiBot falls into the middle of the pack (Fig. 3.3g). Subsequent to AmphiBot, there have been a fair number of snake-like robots capable of amphibious locomotion. These follow-ups make slight variations to hardware in efforts to generate more productive thrust with each oscillation, such as including radially distributed wheels around the body, and placing continuous ridges along the body [95–97].

One group sought to integrate undulatory caudal fin swimming, propeller-generated thrust, and flipper-like control surfaces for locomotion in water with wheels for locomotion on land: the results were Amphirobots-I and II [98, 99]. Amphirobots are composed of a multi-link serial chain whose units possess passive wheels and has axles that can swivel both laterally and dorsoventrally, unlike previous amphibious undulating robots. This clever mechanism enables both dolphin- and fish-style swimming up to 0.45 m/s (0.64 bl/s) as well as serpentine crawling on land of 0.6 m/s (0.86 bl/s), positioning Amphirobot-II near the top of those surveyed in both categories (Fig. 3.3h). Caudal oscillation in the fashion of some fish happens to yield one of the lowest costs of transport among swimming modes as mass increases [22]. Yet, Amphirobot's agglomeration of components increases the projected area in Eq. (3.2), proportionally increasing drag, countering to some extent the merits of the efficient propulsive mode. Additionally, as mentioned, having multiple separate propulsive mechanisms obfuscates control, which the authors concede reduces the efficacy of their system [98].

The most recent amphibious undulating robot at the time of this writing originated at Pliant Energy Systems, an energy firm based out of Brooklyn, NY. Their robot, Velox, consists of a static, rigid body lined with two sets of dynamically undulating, elongate fins [100]. These fins are based on Pliant's blade-less energy-harvesting turbines that passively harness fluid flow. A geared transmission system internal to the robot pulls parts of the flexible fins up in a wave. The resulting coordinated undulation endows the robot with high mobility over a variety of terrain. Videos of the robot traversing snow, flat ground, and swimming freely in three dimensions testify to its effective united propulsive mechanism design [101].

Despite its superiority in water, undulating propulsion over land has several drawbacks. First, ongoing challenges with snake-like robots are to get them to traverse uneven terrain, and developing appropriate contact models for such traversal [102]. Second, undulation relies on high surface area contact between a robot's body and the underlying substrate. Friction or smoothness of a substrate therefore governs cost of transport much more than a legged robot experiencing less surface contact. Although friction issues have been partly addressed via the incorporation of passive wheels onto multi-segment robots, wheels are bulky protrusions to a robot's body plan that may further reduce ability to clear obstacles. From a hydrodynamic perspective, wheels can incur undesired drag forces. Thus it stands as an open challenge to devise undulating robots that modulate their substrate friction coefficient without negatively impacting other aspects of locomotion. Third, serpentine undulating robots do not generally have the capacity to bear large, centralized payloads due to size limitations of the modules composing their bodies. More work on centralized chassis-based undulating systems, like Pliant Energy Systems Velox, represents a promising future direction in this regard. In brief, if desiring a simplistic platform for transporting larger payloads or navigating uneven ground, current undulating robot designs may fall short.

3.3.4 Soft Amphibious Robots

The fourth and final class of amphibious robot, ones made primarily of soft materials, constitute a subset of the already small body of literature pertaining to amphibious robots. It is well-established that robots can benefit from new research in soft, responsive materials [45, 103]. The properties of soft materials—compliance, continuous deformation, stretchability, incompressibility, conference of hydrodynamic benefits [104, 105], and resilience to damage and harsh conditions [106]—make them apt candidates for incorporation into amphibious robotics.

To the best of our knowledge, the first explicitly declared soft amphibious robot was developed by Faudzi et al. Inspired by the salamander, the robot has an elongated body with short legs. The body and legs consist of McKibben pneumatic actuators. Though quite slow, selective contraction of McKibbens on various parts of its body enables the robot to traverse solid ground, sand, and patches of shallow water [47]. To enable underwater crawling of an amphibious soft robot, Tang et al. integrated two switchable adhesion actuators on distal ends of a bending actuator. Their robot can execute inchworm-style gaits while adhered to both submerged and dry substrates [46].

The salamander-inspired robot and the underwater inchworm-like crawler, being composed of entirely soft actuators, likely cannot bear significant payloads. Another amphibious crawling robot that demonstrated payload capacity and separated its soft actuators from its body is the un-tethered sea urchin-inspired robot developed by Paschal et al. [107]. The sea urchin bot consists of a rigid body, rigid spines, and soft bending actuators (analogs to tubercle feet on a real sea urchin). A unique aspect of the sea urchin robot is that it uses actuation of rigid spines in tandem with bending actuators to engage in bio-mimetic, in-place turning motions. The robot is also able to drag itself on land or under the surface of water. In spite of its high mobility, the robot is confined to ferrous surfaces because it relies on embedded magnets in the tips of its bending actuators for anchor points while dragging [107].

In another instance of utilizing separate soft actuators on a rigid robot body, soft pneumatic bending actuators were equipped as legs to an amphibious dog-inspired robot. The soft actuators were not implemented in a traditional sense, as with the previous two examples. Instead, they are pre-inflated and routed with cables. By pulling on the pre-inflated actuator with the cable, it straightens. Upon release of the tension in the cable, the actuator's stored energy snaps it back into the bent configuration. The researchers leveraged this mechanism to show the dog-inspired robot trotting on land and dog-paddling in water [4]. Though they provide sufficient force to propel the robot to land speeds of 0.18 m/s (0.34 bl/s)—a rate much faster than other soft amphibious robots—the pre-inflated actuators necessitate additional motors to pull the attached cables.

The same material properties that make soft amphibious robots flexible and resilient impede their locomotive performance. Soft amphibious robots substantially lag behind their rigid counterparts in terms of their reported maximum sustained speeds, as indicated by Fig. 3.3a–c. Lack of speed can be attributed to the dissipative effects of soft materials and the low force density of soft actuators relative to motors used on rigid amphibious robots. An open design challenge is thus to enhance speed of soft robots, both in water and on land, while retaining their desirable rheological properties. A wealth of soft actuation schemes that still have yet to be applied to amphibious soft robots, including shape-memory materials, chemically induced volumetric expansion, dielectric elastomers, and combustion [108–110], might facilitate higher speeds. A second drawback is soft robots' lack of payload-bearing capabilities. Even if they could be designed to move faster, entirely soft structures would buckle under external loads. Fortunately, recent work regarding granular and layer jamming and variable stiffness polymers offers novel ways to endow a compliant robot with rapid stiffness-changing capacity [111–115] and might prove useful if applied to amphibious soft robots.

3.4 Overarching Challenges

As evidenced in the literature, the exact design specifications for an amphibious robot traditionally depend on the task it is intended to complete. In some applications, like shoreline monitoring, capability to traverse muddy terrain and surface-level immersion in water with a small camera will suffice to complete all mission objectives. Other applications, however, like routine inspections of offshore rigs, may mandate climbing uneven terrain and deeper dives with specialized sensor suite payloads.

The various presented propulsive mechanism architectures—wheeled, legged, undulating, and soft—bring their own unique sets of advantages and disadvantages to the design table. As research strives to realize increasingly autonomous systems that are not specifically designed for a task, but rather are multi-functional entities, the following question stands: how does an engineer synthesize current amphibious robot propulsive mechanism designs to create new, highly effective ones? To help guide answers to this question, let us observe facts about the current state of robotic technology and review areas for improvement in each category.

Amphibious robots of all categories are outclassed by animals in terms of absolute speed, as shown in Fig. 3.3. Even the state-of-the-art robots that operate exclusively terrestrially or underwater fall short of animals in terms of max sustained speed (both in body lengths per second and absolute speed), operational duration capacity, acceleration, turning ratio, ability to maneuver in compact spaces, cost of transport, and stealth [116, 117]. Though metrics other than speed are not well reported in the literature for amphibious robots, one can infer that amphibious robots suffer from the same shortcomings (maybe more so) as their exclusively terrestrial or aquatic counterparts. With these observations in mind, improvements in the force density and compactness of actuators would be a significant leap for the field.

Wheeled amphibious robots could become much more effective if they are augmented to accommodate uneven terrain, swim in 3D, mitigate flow separation from wheel surfaces in water, and reduce environmental impact. Many of these improvements could be made by creating hybrid mechanisms that augment wheel morphology.

Legged platforms would benefit from simpler control policies (and correspondingly, actuation systems), as their lack of deployment in field missions to-date is attributed to complex control [118]. Existing force models for legged locomotion on granular media rely on assumptions of size and uniformity of particulates [119] that diminish their accuracy in the highly unstructured littoral zone. Thus, more detailed experiments should be conducted focusing on legged locomotion in fluidized sediment to converge on robust physical models that can inform actuation policies. Second, legged systems can be improved by hydrodynamic form factors, or more generally, ways to augment morphology between functional terrestrial and aquatic streamlined shapes.

Undulating robots could be modified for enhanced ground traversal. In contrast to limbless animals that can navigate almost any surface [120], undulating robots struggle with uneven surfaces or low friction coefficients [102]. Snake locomotion strategies of undulation, including concertina motion (used to move through restricted spaces like tunnels), sidewinding (limited to sandy terrain), and rectilinear movement (uses bottom scales to move without undulations of the body; useful for moving along tree branches), have been applied to robotics, but are far from mirroring the fluidity and efficiency of natural systems [102]. Simplified controllers that can replicate the diversity of complex undulating gaits would thus expand the utility of undulating amphibious robots. As with legged platforms, a fundamental understanding of undulating propulsion over fluidized sediment needs comprehensive experimental analysis and could help inform locomotion policies for littoral zones. Lastly, the lack of large centralized payload capability seems to be an inherent shortcoming of undulating. One solution could be towing a payload behind the robot. Another could be creating systems with a centralized chassis like Pliant Energy's Velox [100].

A consistent disadvantage among most wheeled, legged, and undulating robots presented in the preceding sections is that they are composed of immutable structures with high stiffness, precluding the capability to substantially adapt a propulsive mechanism or body structure for more effective locomotion in a particular environment. Zhong et al. took a step in this direction and showed how a transformable robot limb geometry could be used to enhance performance in water and on land [121]. Zhong et al. also showed that modulating stiffness of a robot's limbs can benefit its locomotion across fluidized sediment [9]. The results of studies such as these, in tandem with studies of terrestrial and aquatic animal locomotor adaptions [22, 30, 116], motivate two pillars to strive for with future amphibious robot designs: (1) endow shape change and (2) devise mechanisms that can switch between soft and rigid states. Soft amphibious robots are well-suited to adapting their shape and stiffness to the environment and therefore represent an untapped reservoir of potential for addressing the multi-faceted challenge of transitions between aquatic and terrestrial locomotion. However, they are not a panacea; as mentioned, soft robots have the prominent disadvantages of low speeds and reduced ability to bear payloads. A promising research avenue is thus developing systems that marry the advantages of soft and rigid materials into a cohesive robotic platform.

3.5 Case Study: Example Propulsive Mechanism for Efficient Amphibious Locomotion

Our group is currently building an amphibious turtle/tortoise-inspired robot with variable stiffness limbs that change between the shape of a flipper and a leg [115, 122]. We drew inspiration from the specialized flipper propulsor of the green sea turtle and the legs of the Galapagos tortoise (Fig. 3.4a). Aside from the propulsors, we noted that sea turtles and tortoises demonstrate similar body plans. In addition to being quadrupedal, they have protective shells occupying a large portion of their bodies. A hybrid robotic platform based on the turtle/tortoise body plan should permit highly efficient movement in both water and on land, moderate speeds and maneuverability in water, as well as stability to negotiate obstacles in the littoral zone. Furthermore, a rigid central shell lends itself to storing control system hardware and a heavy payload.

The limb of the robot is pictured alone in Fig. 3.4b and in the context of a rendered quadruped robot in Fig. 3.4c. The limb consists of an antagonistic pneumatic actuator pair whose strain limiting layers are thermoset polymers that are rigid below 60 °C. Heating up the thermoset polymer past its glass transition temperature softens it, and it is subsequently deformed into a round geometry using the pneumatic actuators. The actuators are held at the inflated state until the material cools, at which point it retains the leg shape. Heating up the material again induces relaxation and it morphs back to the flipper shape.

Placing the morphing limb in a flow tank, we varied its angle of attack and monitored lift and drag forces (inset of Fig. 3.4d shows components). Figure 3.4d illustrates how the airfoil geometry of the flipper markedly increased its glide ratio (ratio of lift to drag forces on an object) in water compared to the leg. Like a traditional airfoil, the peak of the flipper state's glide ratio occurred around an 8° angle of attack. The leg state, on the other hand, resembles a thick hydrofoil, delaying onset of stall to 30° and exhibiting much lower glide ratio [123].

We subjected the limb to compression tests in either its flipper or leg phase (Fig. 3.4e). The leg's circular cross section (and correspondingly increased moment of inertia) enhanced its capability to bear compressive loads relative to the flipper. Consider that the critical buckling load for a beam (leg) under compression is

$$P_{cr} = \frac{\pi^2 EI}{(kL)^2}, \tag{3.3}$$

where E is the elastic modulus of the variable stiffness material, I is the cross-sectional area moment of inertia, k is the length factor (1 for pin boundaries on either side), and L is the unsupported length of the limb. Based on the fact that

$$I_{circle} = \frac{\pi}{4}\left(r_2^4 - r_1^4\right) \tag{3.4}$$

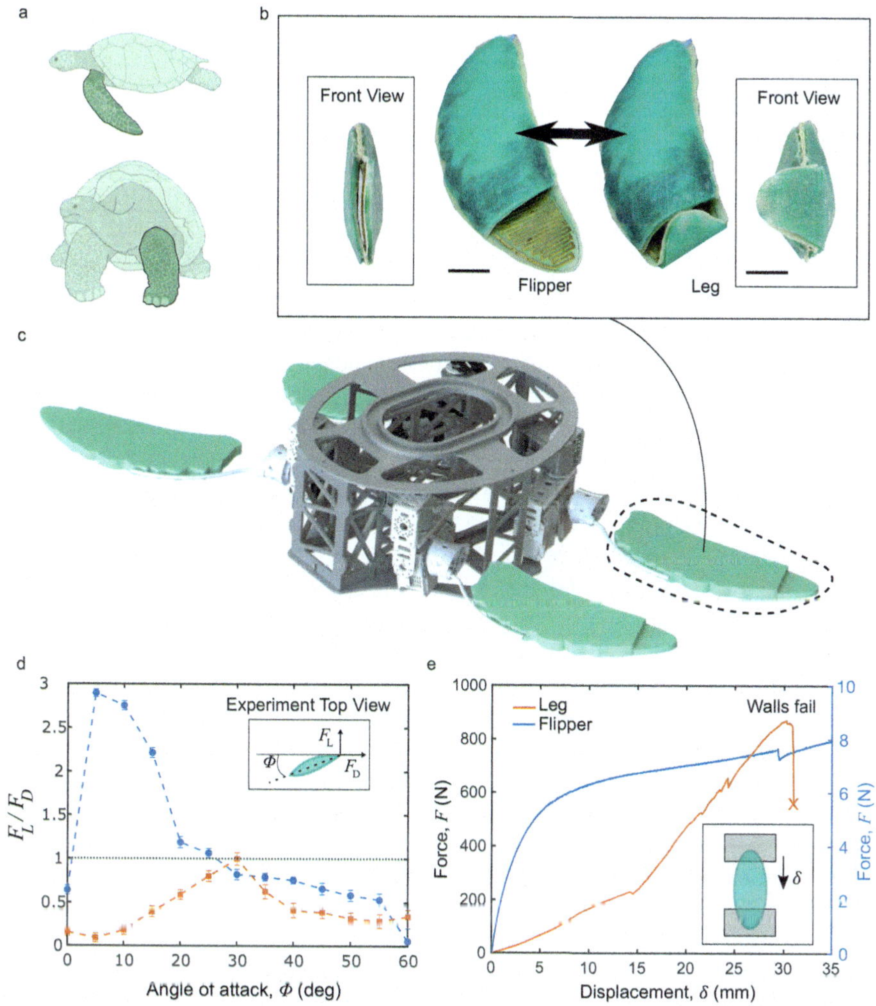

Fig. 3.4 (**a**) Sea turtles and tortoises demonstrate specialized propulsors for aquatic and terrestrial environments, respectively (images adapted from [115]). (**b**) Turtle-inspired limb: morphing between streamlined and load-bearing geometries allows it to perform well in water and serve as a strong leg for land. Scale bar: 40 mm. Inset scale bar: 30 mm (images adapted from [122]). (**c**) Rendering of morphing limbs equipped to quadruped amphibious robot. (**d**) Flow-tank test results at 0.6 m/s water speed. Ratio of lift to drag on the flipper and leg. (**e**) Compress-to-failure test for the flipper and leg

$$I_{rectangle} = \frac{bh^3 - b_1 h_1^3}{12}, \tag{3.5}$$

having as close to a circular cross section as possible enhances the limb's critical Euler buckling load [122].

Moreover, modulating the material's stiffness by heating it past T_g gives us the ability to tune the limb's mechanical response to external forces [115]. This ability may prove useful if bumping into obstacles, or further tuning propulsor performance for a given media [9]. By leveraging on-demand shape and stiffness changes and modulating the gait of the quadruped robot it is attached to, we hypothesize the limb will enable a legged amphibious robot to traverse land and water with high speed, efficiency, maneuverability, and payload capacity.

Beyond our case study, there are myriad soft, responsive, and rigid materials that could be incorporated into a single robot design to facilitate shape change and variable stiffness. Studies should seek to leverage shape change not just of a single propulsor, but of the entire robot body, to improve amphibious locomotion. For instance, changing its body from a fish to a legged quadruped form might grant a robot high speeds in water and on land (Fig. 3.2–soft). We also see work on varying stiffness in amphibious locomotion as a promising domain for future research. Combining variable stiffness materials with wheels, for instance, may enable a robot to selectively distribute its weight over a higher area to easily traverse patches of fluidized sediment; variable stiffness in combination with undulation may allow a robot to passively harness flow of the water to generate productive thrust; toggling between stiff states during the power stroke and compliant states during the retraction stroke can enhance swimming of legged robots; and so on. Lastly, new complementary control strategies must be developed in parallel with hardware that changes shape and stiffness. Controllers must account for an internal representation of the shape and stiffness states of a robot at all times to adapt to underwater currents, terrain grade, and viscosity of substrate, among other factors. For instance, central pattern generators commonly employed to control amphibious robots could be altered with terms that scale amplitude and phase offset as a robot changes shape or stiffness.

3.6 Conclusion

A brief discussion of biological propulsors provided insight into solutions that animals use to transition between aquatic and terrestrial locomotion. Amphibious animals strive to balance performance in various media, which results in mediocre performance. Contrastingly, highly derived species are well-adapted to a specific environment, often at the expense of locomotion in the other. The diversity of aquatic and terrestrial locomotor strategies and body plans invites researchers to combine mechanisms in clever ways to converge on some measure of optimal performance (Fig. 3.1). Our discussion on biology segued into a survey of amphibious robots, which we classify into those that utilize separate or united propulsive mechanisms. Further, we break down amphibious robots into (1) wheeled, (2) legged, (3) undulating, and (4) soft categories. We sort them by analyzing salient aspects of their body plans and locomotion strategies. The advantages and disadvantages of specific robot propulsive mechanism designs from each category highlight areas where future research effort is needed. In particular, it seems that soft, stiffness-changing materials offer significant opportunities to enhance mechanical resilience, hydrodynamic efficiency, and shape-morphing capability of amphibious robots. Our preliminary results toward a turtle/tortoise-inspired quadruped robot with variable stiffness morphing limbs provide a case study of this design philosophy. We believe there is promising future research oriented around amphibious robots that follow a similar design paradigm, one in which shape-morphing, variable-stiffness materials serve to balance the locomotive merits of rigid and soft robots in water and on land.

References

1. F. Fish, Advantages of aquatic animals as models for bio-inspired drones over present AUV technology. Bioinspir. Biomim. **15**, 025001 (2020)
2. A.J. Ijspeert, A. Crespi, D. Ryczko, J.-M. Cabelguen, From swimming to walking with a salamander robot driven by a spinal cord model. Science **315**(5817), 1416–1420 (2007)
3. J.A. Nyakatura, K. Melo, T. Horvat, K. Karakasiliotis, V.R. Allen, A. Andikfar, E. Andrada, P. Arnold, J. Lauströer, J.R. Hutchinson, M.S. Fischer, A.J. Ijspeert, Reverse-engineering the locomotion of a stem amniote. Nature **565**(7739), 351–355 (2019)
4. Y. Li, F. Fish, Y. Chen, T. Ren, J. Zhou, Bio-inspired robotic dog paddling: kinematic and hydro-dynamic analysis. Bioinspir. Biomim. **14**(6), 066008 (2019)
5. A.J. Ijspeert, Biorobotics: using robots to emulate and investigate agile locomotion. Science **346**, 196–203 (2014)
6. L. Cui, P. Cheong, R. Adams, T. Johnson, AmBot: a bio-inspired amphibious robot for monitoring the swan-canning estuary system. J. Mech. Des. **136**(11), 115001 (2014)
7. S. Dhull, D. Canelon, A. Kottas, J. Dancs, A. Carlson, N. Papanikolopoulos, Aquapod: a small amphibious robot with sampling capabilities, in *2012 IEEE/RSJ International Conference on Intelligent Robots and Systems* (2012), pp. 100–105
8. K. Tadakuma, R. Tadakuma, M. Aigo, M. Shimojo, M. Higashimori, M. Kaneko, Omni-Paddle amphibious spherical rotary paddle mechanism, in *2011 IEEE International Conference on Robotics and Automation* (2011), pp. 5056–5062
9. B. Zhong, S. Zhang, M. Xu, Y. Zhou, T. Fang, W. Li, On a CPG-based hexapod robot: amphiHex-II with variable stiffness legs. IEEE/ASME Trans. Mechatron. **23**(2), 542–551 (2018)
10. H. Zhang (ed.), *Climbing and Walking Robots: Towards New Applications* (I-Tech Education and Publisher, Vienna 2007), OCLC: 254375799
11. T.R. Consi, B.R. Ardaugh, T.R. Erdmann, M. Matsen, M. Peterson, J. Ringstad, A. Vechart, C. Verink, An amphibious robot for surf zone science and environmental monitoring, in *OCEANS 2005 MTS/IEEE* (2005), p. 7
12. H. Greiner, A. Shectman, C. Won, R. Elsley, P. Beith, Autonomous legged underwater vehicles for near land warfare, in *Proceedings of Symposium on Autonomous Underwater Vehicle Technology* (1996), pp. 41–48
13. M. Dunbabin, L. Marques, Robots for environmental monitoring: significant advancements and applications. IEEE Rob. Autom. Mag. **19**(1), 24–39 (2012)

14. R.H. Harkins, T. Dunbar, A.S. Boxerbaum, R.J. Bachmann, R.D. Quinn, R. Vaidyanathan, S.C. Burgess, Confluence of active and passive control mechanisms enabling autonomy and terrain adaptability for robots in variable environments, in *Advances in Electrical and Electronics Engineering—IAENG Special Edition of the World Congress on Engineering and Computer Science 2008* (2008), pp. 138–149
15. J. Ayers, J. Witting, C. Wilbur, P. Zavracky, N. McGruer, D. Massa, Biomimetic robots for shallow water mine countermeasures, in *Autonomous Vehicles Mine Countermeasures Symposium* (2000), p. 16
16. M.H. Dickinson, How animals move: an integrative view. Science **288**(5463), 100–106 (2000)
17. A. Crespi, A. Badertscher, A. Guignard, A. Ijspeert, AmphiBot I: an amphibious snake-like robot. Robot. Auton. Syst. **50**(4), 163–175 (2005)
18. A.A. Biewener, S.N. Patek, *Animal Locomotion* (Oxford University, Oxford, 2018)
19. C. Li, T. Zhang, D.I. Goldman, Locomotion: energy cost of swimming, flying, and running. Science **177**(4045), 222–228 (1972)
20. P. Webb, Hydrodynamics and energetics of fish propulsion. Bull. Fish. Res. Board Can. **190**, 1–158 (1975)
21. P. Webb, R. Blake, Swimming, in *Functional Vertebrate Morphology*, ed. by M. Hildebrand, D.M. Bramble, K.F. Liem, D.B. Wake (Harvard University Press, Cambridge, 1985), pp. 110–128
22. F.E. Fish, Biomechanics and energetics in aquatic and semiaquatic mammals: platypus to whale. Physiol. Biochem. Zool. **73**(6), 683–698 (2000)
23. C. Gans, *Biomechanics: An Approach to Vertebrate Biology* (Lippincott, Philadelphia, 1974)
24. M. Hildebrand, The adaptive significance of tetrapod gait selection. Am. Zool. **20**, 255–267 (1980)
25. R.J. Full, Mechanics and energetic of terrestrial locomotion: bipeds to polypeds, in *Energy Transformations in Cells and Organisms*, ed. by W. Wieser, E. Gnaiger (Thieme, Stuttgart, 1989), pp. 175–182
26. H.-T. Lin, G.G. Leisk, B. Trimmer, GoQBot: a caterpillar-inspired soft-bodied rolling robot. Bioinspir. Biomim. **6**(2), 026007 (2011)
27. K. Low, T. Hu, S. Mohammed, J. Tangorra, M. Kovac, Perspectives on biologically inspired hybrid and multi-modal locomotion. Bioinspir. Biomim. **10**, 020301 (2015)
28. R. Lock, S. Burgess, R. Vaidyanathan, Multi-modal locomotion: from animal to application. Bioinspir. Biomim. **9**, 011001 (2014)
29. S.E. Peters, L.T. Kamel, D.P. Bashor, Hopping and swimming in the leopard frog, *Rana pipiens*: I. Step cycles and kinematics. J. Morphol. **230**(1), 1–16 (1996)
30. F.E. Fish, Transitions from drag-based to lift-based propulsion in mammalian swimming. Am. Zool. **36**(6), 628–641 (1996)
31. L.A. Isaac, P.T. Gregory, Aquatic versus terrestrial locomotion: comparative performance of two ecologically contrasting species of European natricine snakes. J. Zool. **273**(1), 56–62 (2007)
32. M. Calisti, A. Arienti, F. Renda, G. Levy, B. Hochner, B. Mazzolai, P. Dario, C. Laschi, Design and development of a soft robot with crawling and grasping capabilities, in *Proceedings of the 2012 IEEE International Conference on Robotics and Automation* (2012), pp. 4950–4955
33. D.L. Hu, B. Cean, J.W.M. Bush, The hydrodynamics of water strider locomotion. Nature **424**, 663–666 (2003)
34. J.W. Glasheen, T.A. Mcmahon, A hydrodynamic model of the locomotion in the Basilisk Lizard. Nature **380**, 340–342 (1996)
35. S.T. Hsieh, Three-dimensional hindlimb kinematics of water running in the plumed Basilisk Lizard (*Basiliscus plumifrons*). J. Exp. Biol. **206**, 4363–4377 (2003)
36. W.T. Gough, S.C. Farina, F.E. Fish, Aquatic burst locomotion by hydroplaning and running in common eiders *(Somateria mollissima)*. J. Exp. Biol. **218**, 1632–1638 (2015)
37. T. Williams, M. Ben-David, S. Noren, M. Rutishauser, K. McDonald, W. Heyward, Running energetics of the North American river otter: do short legs necessarily reduce efficiency

on land? in *Comparative Biochemistry and Physiology Part A: Molecular and Integrative Physiology*, vol. 133 (2002), pp. 203–212
38. S. Vogel, *Life in Moving Fluids* (Princeton University, Princeton, 1994)
39. C.E. Jordan, A model of rapid-start swimming at intermediate Reynolds number: undulatory locomotion in the *chaetognath Sagitta elegans*. J. Exp. Biol. **169**, 119–137 (1992)
40. M. LaBarbera, Why the wheels won't go. Am. Nat. **121**(3), 395–408 (1983)
41. A. Biewener, Scaling body support in mammals: limb posture and muscle mechanics. Science **245**, 45–48 (1989)
42. A.A. Biewener, Biomechanical consequences of scaling. J. Exp. Biol. **208**(9), 1665–1676 (2005)
43. A. McNeil, *Elastic Mechanisms in Animal Movement* (Cambridge University, Cambridge, 1998)
44. T. Dawson, C. Taylor, Energetic cost of locomotion in kangaroos. Nature **246**(5431), 313–314 (1973)
45. D. Rus, M.T. Tolley, Design, fabrication and control of soft robots. Nature **521**(7553), 467–475 (2015)
46. Y. Tang, Q. Zhang, G. Lin, J. Yin, Switchable adhesion actuator for amphibious climbing soft robot. Soft Robot. **5**(5), 592–600 (2018)
47. A.A.M. Faudzi, M.R.M. Razif, G. Endo, H. Nabae, K. Suzumori, Soft-amphibious robot using thin and soft McKibben actuator, in *Proceedings of the 2017 IEEE International Conference on Advanced Intelligent Mechatronics* (2017), pp. 981–986
48. X. Liang, M. Xu, L. Xu, P. Liu, X. Ren, Z. Kong, J. Yang, S. Zhang, The AmphiHex: a novel amphibious robot with transformable leg-flipper composite propulsion mechanism, in *Proceedings of the 2012 IEEE/RSJ International Conference on Intelligent Robots and Systems* (2012), pp. 3667–3672
49. L. Shi, S. Guo, S. Mao, C. Yue, M. Li, K. Asaka, Development of an amphibious turtle-inspired spherical mother robot. J. Bionic Eng. **10**(4), 446–455 (2013)
50. A.J. Ijspeert, A. Crespi, Online trajectory generation in an amphibious snake robot using a lamprey-like central pattern generator model, in *Proceedings of the 2007 IEEE International Conference on Robotics and Automation* (2007), pp. 262–268
51. W. Wang, J. Yu, R. Ding, M. Tan, Bio-inspired design and realization of a novel multimode amphibious robot, in *Proceedings of the 2009 IEEE International Conference on Automation and Logistics* (2009), pp. 140–145
52. A. Crespi, K. Karakasiliotis, A. Guignard, A.J. Ijspeert, Salamandra robotica II: an amphibious robot to study salamander-like swimming and walking gaits. IEEE Trans. Robot. **29**(2), pp. 308–320 (2013)
53. J. Yu, Y. Tang, X. Zhang, C. Liu, Design of a wheel-propeller-leg integrated amphibious robot, in *Proceedings of the 2010 11th International Conference on Control Automation Robotics Vision* (2010), pp. 1815–1819
54. Y. Yi, Z. Geng, Z. Jianqing, C. Siyuan, F. Mengyin, Design, modeling and control of a novel amphibious robot with dual-swing-legs propulsion mechanism, in *Proceedings of the 2015 IEEE/RSJ International Conference on Intelligent Robots and Systems* (2015), pp. 559–566
55. F. Fish, Aerobic energetics of surface swimming in the muskrat *Ondatra Zibethicus*. Phys. Zool. **55**, 180–189 (1982)
56. T. Lode, Comparative measurements of terrestrial and aquatic locomotion in *Mustela lutreola* and M. putorius. Int. J. Mamm. Biol. **64**, 110–115 (1999)
57. T.M. Williams, Swimming by sea otters: adaptations for low energetic cost locomotion. J. Comp. Physiol. **164**(6), 815–824 (1989)
58. R. Shine, S, Shetty, Moving in two worlds: aquatic and terrestrial locomotion in sea snakes (*Laticauda Colubrina*, Laticaudidae): sea snake locomotion. J. Evol. Biol. **14**(2), 338–346 (2001)
59. T.M. Williams, W.A. Friedl, J.E. Haun, The physiology of bottlenose dolphins (Tursiops truncatus): heart rate, metabolic rate and plasma lactate concentration during exercise. J. Exp. Biol. **179**, 31–46 (1993)

60. *Amphibious 4WD WiFi Robotics SuperDroid*. https://www.superdroidrobots.com/shop/item.aspx/ig42-sb4-ea-amphibious-4wd-wifi-robot/2121/
61. J. Katz, *Race Car Aerodynamics* (Bentley Publishers, Cambridge, 1995)
62. S. Yamada, S. Hirose, G. Endo, K. Suzumori, H. Nabae, R-Crank: amphibious all terrain mobile robot, in *Proceedings of the 2016 IEEE/RSJ International Conference on Intelligent Robots and Systems* (2016), pp. 1067–1072
63. V. Kaznov, M. Seeman, Outdoor navigation with a spherical amphibious robot, in *Proceedings of the 2010 IEEE/RSJ International Conference on Intelligent Robots and Systems* (2010), pp. 5113–5118
64. K. Kawasaki, M. Zhao, K. Okada, M. Inaba, MUWA: multi-field universal wheel for air-land vehicle with quad variable-pitch propellers, in *Proceedings of the 2013 IEEE/RSJ International Conference on Intelligent Robots and Systems* (2013), pp. 1880–1885
65. M.G. Bekker, *Theory of Land Locomotion* (University of Michigan Press, Michigan, 1956)
66. J.Y. Wong, *Theory of Ground Vehicles* (Wiley, New York, 1978)
67. G. Meirion-Griffith, M. Spenko, An empirical study of the terramechanics of small unmanned ground vehicles, in *Proceedings of the IEEE Aerospace Conference* (2010), pp. 1–6
68. Y. Sun, S. Ma, ePaddle mechanism: towards the development of a versatile amphibious locomotion mechanism, in *International Conference on Intelligent Robots and Systems* (2011), pp. 5035–5040
69. Y. Sun, S. Ma, Y. Yang, H. Pu, Towards stable and efficient legged race-walking of an ePaddle-based robot. Mechatronics **23**(1), 108–120 (2013)
70. Y. Shen, Y. Sun, H. Pu, S. Ma, Experimental verification of the oscillating paddling gait for an ePaddle-EGM amphibious locomotion mechanism. IEEE Rob. Autom. Lett. **2**(4), 2322–2327 (2017)
71. N.B. Ignell, N. Rasmusson, J. Matsson, An overview of legged and wheeled robotic locomotion, in *Mini-Conference on Interesting Results in Computer Science and Engineering*, vol. 21 (2012)
72. C. Bernstein, M. Connolly, M. Gavrilash, D. Kucik, S. Threatt, Demonstration of surf-zone crawlers: results from AUV Fest 01, in *Surf Zone Crawler Group, Naval Surface Warfare Center, Panama City, FL* (2001)
73. M.A. Klein, A.S. Boxerbaum, R.D. Quinn, R. Harkins, R. Vaidyanathan, Seadog: a rugged mobile robot for surf-zone applications, in *IEEE RAS and EMBS International Conference on Biomedical Robotics and Biomechatronics* (2012), pp. 1335–1340
74. M.F. Silva, J.T. Machado, A literature review on the optimization of legged robots. J. Vib. Control **18**(12), 1753–1767 (2012)
75. J. Ayers, J. Witting, C. Olcott, N. McGruer, D. Massa, Lobster robots, in *Proceedings of the 2000 International Symposium on Aqua Biomechanisms* (2000)
76. S. Floyd, T. Keegan, J. Palmisano, M. Sitti, A novel water running robot inspired by basilisk lizards, in *Proceedings of the 2006 IEEE/RSJ International Conference on Intelligent Robots and Systems* (2006), pp. 5430–5436
77. H.S. Park, M. Sitti, Compliant footpad design analysis for a bio-inspired quadruped amphibious robot, in *Proceedings of the 2009 IEEE/RSJ International Conference on Intelligent Robots and Systems* (2009), pp. 645–651
78. H.S. Park, S. Floyd, M. Sitti, Roll and pitch motion analysis of a biologically inspired quadruped water runner robot. Int. J. Robot. Res. **29**(10), 1281–1297 (2010)
79. H. Kim, D. Lee, K. Jeong, T. Seo, Water and ground-running robotic platform by repeated motion of six spherical footpads. IEEE/ASME Trans. Mechatron. **21**(1), 175–183 (2015)
80. Y. Chen, N. Doshi, B. Goldberg, H. Wang, R.J. Wood, Controllable water surface to underwater transition through electrowetting in a hybrid terrestrial-aquatic microrobot. Nat. Commun. **9**(1), 2495 (2018)
81. B. Kwak, J. Bae, Locomotion of arthropods in aquatic environment and their applications in robotics. Bioinspir. Biomim. **13**(4), 041002 (2018)
82. G. Dudek, M. Jenkin, C. Prahacs, A. Hogue, J. Sattar, P. Giguere, A. German, H. Liu, S. Saunderson, A. Ripsman, S. Simhon, L. Torres, E. Milios, P. Zhang, I. Rekletis, A visually

guided swimming robot, in *Proceedings of the 2005 IEEE/RSJ International Conference on Intelligent Robots and Systems* (2005), pp. 3604–3609
83. G. Dudek, P. Giguere, C. Prahacs, S. Saunderson, J. Sattar, L.-a. Torres-Mendez, M. Jenkin, A. German, A. Hogue, A. Ripsman, J. Zacher, E. Milios, H. Liu, P. Zhang, M. Buehler, C. Georgiades, AQUA: an amphibious autonomous robot. Computer **40**(1), 46–53 (2007)
84. C. Prahacs, A. Saudners, M.K. Smith, D. McMordie, M. Buehler, Towards legged amphibious mobile robotics, in *Proceedings of the Canadian Engineering Education Association* (2011)
85. R. Quinn, J. Offi, D. Kingsley, R. Ritzmann, Improved mobility through abstracted biological principles, in *Proceedings of the IEEE/RSJ International Conference on Intelligent Robots and System*, vol. 3 (2002), pp. 2652–2657
86. A. Boxerbaum, P. Werk, R. Quinn, R. Vaidyanathan, Design of an autonomous amphibious robot for surf zone operation: part i mechanical design for multi-mode mobility, in *Proceedings of the 2005 IEEE/ASME International Conference on Advanced Intelligent Mechatronics* (2005), pp. 1459–1464
87. R. Harkins, J. Ward, R. Vaidyanathan, A. Boxerbaum, R. Quinn, Design of an autonomous amphibious robot for surf zone operations: part II–hardware, control implementation and simulation, in *Proceedings of the 2005 IEEE/ASME International Conference on Advanced Intelligent Mechatronics* (2005), pp. 1465–1470
88. A.S. Boxerbaum, M.A. Klein, J.E. Kline, S.C. Burgess, R.D. Quinn, R. Harkins, R. Vaidyanathan, Design, simulation, fabrication and testing of a bio-inspired amphibious robot with multiple modes of mobility. J. Robot. Mechatron. **24**(4), pp. 629–641 (2012)
89. B.B. Dey, S. Manjanna, G. Dudek, Ninja legs: Amphibious one degree of freedom robotic legs, in *2013 Proceedings of the IEEE/RSJ International Conference on Intelligent Robots and Systems* (2013), pp. 5622–5628
90. A.R. Vogel, K.N. Kaipa, G.M. Krummel, H.A. Bruck, S.K. Gupta, Design of a compliance assisted quadrupedal amphibious robot, in *Proceedings of the 2014 IEEE International Conference on Robotics and Automation* (2014), pp. 2378–2383
91. S. Zhang, X. Liang, L. Xu, M. Xu, Initial development of a novel amphibious robot with transformable fin-leg composite propulsion mechanisms. J. Bionic Eng. **10**(4), 434–445 (2013)
92. S. Zhang, Y. Zhou, M. Xu, X. Liang, J. Liu, J. Yang, AmphiHex-I: locomotory performance in amphibious environments with specially designed transformable flipper legs. IEEE/ASME Trans. Mechatron. **21**(3), 1720–1731 (2016)
93. S. Guo, S. Mao, L. Shi, M. Li, C. Yue, Development of a spherical amphibious mother robot, in *Proceedings of the 2013 ICME International Conference on Complex Medical Engineering* (2013), pp. 614–619
94. A. Crespi, A. Badertscher, A. Guignard, A.J. Ijspeert, Swimming and crawling with an amphibious snake robot, in *Proceedings of the International Conference on Robotics and Automation* (2005), pp. 3024–3028
95. T. Matsuo, T. Yokoyama, D. Ueno, K. Ishii, Biomimetic motion control system based on a CPG for an amphibious multi-link mobile robot. J. Bionic Eng. **5**, 91–97 (2008)
96. S. Yu, S. Ma, B. Li, Y. Wang, An amphibious snake-like robot: design and motion experiments on ground and in water, in *Proceedings of the 2009 International Conference on Information and Automation* (2009), pp. 500–505
97. S. Yu, S. Ma, B. Li, Y. Wang, An amphibious snake like robot with terrestrial and aquatic gaits, in *Proceedings of the 2011 IEEE International Conference on Robotics and Automation* (2011), pp. 2960–2961
98. J. Yu, R. Ding, Q. Yang, M. Tan, W. Wang, J. Zhang, On a bio-inspired amphibious robot capable of multimodal motion. IEEE/ASME Trans. Mechatron. **17**(5), 847–856 (2012)
99. J. Yu, R. Ding, Q. Yang, M. Tan, J. Zhang, Amphibious pattern design of a robotic fish with wheel-propeller fin mechanisms. J. Field Robot. **30**(5), 702–716 (2013)
100. *Robotics Pliant Energy Systems*. https://www.pliantenergy.com/
101. Robot traversing sea, sand and snow, in *Pliant Energy Systems*. https://www.youtube.com/watch?v=2pVsaWwAOh0

102. A.A. Transeth, K.Y. Pettersen, L. Påll, A survey on snake robot modeling and locomotion. Robotica **27**(7), 999–1015 (2009)
103. D. Trivedi, C.D. Rahn, W.M. Kier, I.D. Walker, Soft robotics: biological inspiration, state of the art, and future research. Appl. Bionics Biomech. **5**(3), 99–117 (2008)
104. F. Corucci, N. Cheney, F. Giorgio-Serchi, J. Bongard, C. Laschi, Evolving soft locomotion in aquatic and terrestrial environments: effects of material properties and environmental transitions. Soft Robot. **5**(4), 475–495 (2018)
105. A.D. Marchese, C.D. Onal, D. Rus, Autonomous soft robotic fish capable of escape maneuvers using fluidic elastomer actuators. Soft Robot. **1**(1) 75–87 (2014)
106. R.F. Shepherd, F. Ilievski, W. Choi, S.A. Morin, A.A. Stokes, A.D. Mazzeo, X. Chen, M. Wang, G.M. Whitesides, Multigait soft robot, in *Proceedings of the National Academy of Sciences*, vol. 108(51), pp. 20,400–20,403 (2011)
107. T. Paschal, M.A. Bell, J. Sperry, S. Sieniewicz, R.J. Wood, J.C. Weaver, Design, fabrication, and characterization of an untethered amphibious sea urchin-inspired robot. IEEE Robot. Autom. Lett. **4**(4), 3348–3354 (2019)
108. L. Hines, K. Petersen, G.Z. Lum, M. Sitti, Soft actuators for small-scale robotics. Adv. Mater. **29**(13), 1603483 (2017)
109. L.A. Hirano, L.S. Martins-Filho, R.O. Duarte, J.F. de Paiva, Development of an amphibious robotic propulsor based on electroactive polymers, in *Proceedings of the 2009 4th International Conference on Autonomous Robots and Agents* (2009), pp. 284–289
110. N.W. Bartlett, M.T. Tolley, J.T.B. Overvelde, J.C. Weaver, B. Mosadegh, K. Bertoldi, G.M. Whitesides, R.J. Wood, A 3d-printed, functionally graded soft robot powered by combustion. Science **349**(6244), 161–165 (2015)
111. J.L.C. Santiago, I.S. Godage, P. Gonthina, I.D. Walker, Soft robots and kangaroo tails: modulating compliance in continuum structures through mechanical layer jamming. Soft Robot. **3**(2), 54–63 (2016)
112. N.G. Cheng, M.B. Lobovsky, S.J. Keating, A.M. Setapen, K.I. Gero, A.E. Hosoi, K.D. Iagnemma, Design and analysis of a robust, low-cost, highly articulated manipulator enabled by jamming of granular media, in *Proceedings of the 2012 IEEE International Conference on Robotics and Automation* (2012), pp. 4328–4333
113. A.R. Deshpande, Z.T.H. Tse, H. Ren, Origami-inspired bi-directional soft pneumatic actuator with integrated variable stiffness mechanism, in *Proceedings of the 2017 18th International Conference on Advanced Robotics (ICAR)* (2017), pp. 417–421
114. I.D. Falco, M. Cianchetti, A. Menciassi, A soft multi-module manipulator with variable stiffness for minimally invasive surgery. Bioinspir. Biomim. **12**(5), 056008 (2017)
115. R.L. Baines, J.W. Booth, F.E. Fish, R. Kramer-Bottiglio, Toward a bio-inspired variable-stiffness morphing limb for amphibious robot locomotion, in *Proceedings of the 2019 2nd IEEE International Conference on Soft Robotics (RoboSoft)* (2019), pp. 704–710
116. F.E. Fish, Advantages of aquatic animals as models for bio-inspired drones over present AUV technology. Bioinspir. Biomim. **15**(2), 025001 (2020)
117. S. Seok, A. Wang, M.Y. Chuah, D. Otten, J. Lang, S. Kim, Design principles for highly efficient quadrupeds and implementation on the MIT cheetah robot, in *IEEE International Conference on Robotics and Automation* (2013), p. 3307–3312
118. A.J. Ijspeert, Amphibious and sprawling locomotion: From biology to robotics and back. Annu. Rev. Control Robot. Auton. Syst. **3**(1), 173–193 (2020)
119. C. Li, T. Zhang, D.I. Goldman, A terradynamics of legged locomotion on granular media. Science **339**(5452), 1408–1412 (2013)
120. H. Marvi, C. Gong, N. Gravish, H. Astley, M. Travers, R. Hatton, J. Mendelson, H. Choset, D. Hu, D. Goldman, Sidewinding with minimal slip: Snake and robot ascent of sandy slopes. Science **346**, 224–229 (2014)
121. B. Zhong, Y. Zhou, X. Li, M. Xu, S. Zhang, Locomotion performance of the amphibious robot on various terrains and underwater with flexible flipper legs. J. Bionic Eng. **13**(4), 525–536 (2016)

122. R. Baines, S. Freeman, F. Fish, R. Kramer-Bottiglio, Variable stiffness morphing limb for amphibious legged robots inspired by chelonian environmental adaptations. Bioinspir. Biomim. **15**(2), 025002 (2020)
123. J. Bertin, R.M. Cummings, *Aerodynamics for Engineers*, 5th edn. (Pearson Prentice-Hall, Upper Saddle River, 2009)

Chapter 4
Artificial Muscles for Underwater Soft Robotic System

Zijun Wang, Qiguang He, and Shengqiang Cai

4.1 Introduction

Traditional robots are mainly composed of rigid materials. Inspired by creatures in nature (especially marine animals), robots or machines constructed from soft materials have recently been explored intensively [1–4]. Representative examples include soft grippers [5], soft crawlers [3], and soft active lenses [6]. Particularly, soft robots that can operate underwater have also been constructed. For instance, a translucent soft swimming robot has been built from dielectric elastomer actuators [7]. Soft hydraulic-driven grippers have been fabricated for collecting fragile coral [8]. Various forms of light-powered soft robots have also been created using liquid crystal elastomers [9, 10] and optically responsive hydrogels [11].

The water environment has many uncertainties. Complex terrains, including wavy water, uneven sandy ground, and confined space raise challenges for the development of underwater robotic systems. Robots that have been constructed using compliant materials show many advantages including better adaptability,

Z. Wang
Materials Science and Engineering Program, University of California, La Jolla, CA, USA
e-mail: z2wang@eng.ucsd.edu

Q. He
Department of Mechanical and Aerospace Engineering, University of California, La Jolla, CA, USA
e-mail: qih042@eng.ucsd.edu

S. Cai (✉)
Materials Science and Engineering Program, University of California, La Jolla, CA, USA

Department of Mechanical and Aerospace Engineering, University of California, La Jolla, CA, USA
e-mail: shqcai@ucsd.edu

D. A. Paley, N. M. Wereley (eds.), *Bioinspired Sensing, Actuation, and Control in Underwater Soft Robotic Systems*, https://doi.org/10.1007/978-3-030-50476-2_4

impact resistance, and less disturbing operation compared to their rigid counterparts [1, 12].

To make a robot completely soft, its actuating components must be compliant. Possessing similar functions to muscles, soft actuators are often referred as artificial muscles. In the past several decades, diverse materials and structures have been broadly explored for constructing soft actuators. For instance, dielectric elastomer membrane has been shown to produce large deformation when subjected to electrical potential [13]; responsive hydrogel can change its volume by more than a hundred times when exposed to certain stimuli [14]; liquid crystal elastomer can generate significant thermally or optically driven contraction along the mesogen alignment direction [15]. Soft pneumatic actuators that can achieve diverse actuation modes have also been designed and fabricated [4].

This chapter briefly reviews a variety of soft actuators that have already been used for the design of soft robots. Herein, we mainly focus on their fabrication/synthesis, modelling, and demonstrations. The end of the chapter discusses special requirements for soft actuators in underwater robotic systems.

4.2 Soft Actuators

4.2.1 Performance Metrics of Soft Actuators

The performance of a soft actuator is mainly determined by the following parameters: actuating stress, actuating strain, response time/work bandwidth, energy efficiency, and work density. Table 4.1 provides a summary of the performance of several types of soft actuators.

4.2.2 Pneumatic/Hydraulic Artificial Muscle

Fluid-driven elastomeric actuators (FEAs) have been intensively studied during the past two decades [4, 22]. Most FEAs are essentially compliant balloon-like structures that can be inflated by pressurized gas or liquid as shown in Fig. 4.1a.

Early in the 1950s, a pneumatic contractive actuator mimicking animal muscle was developed using a rubber bladder and a fiber-woven sleeve, later known as a McKibben actuator or pneumatic artificial muscle (PAM) [23]. Since then, PAMs of various types have been developed such as a pleated PAM [24] and vacuum-based actuators [25–28]. Like real muscles, most of those PAMs only contract and extend, whereas diverse actuation modes can be achieved by the combination of several PAMs [29, 30]. More recently, PAMs with more complex geometries have been created. A single PAM can also exhibit various actuation modes such as bending and twisting [31–33].

Table 4.1 Comparison of performance metrics of soft actuators

	Stress (MPa)	Strain (%)	Bandwidth (Hz)	Typical frequency (Hz)	Work density (J/kg)	Efficiency (%)
Mammalian muscle	0.1–0.35	20–40	20	/	8–40	40
Fluidic elastomer actuator (FEA)	<5	<15	<10	2	0.5	<20
Dielectric elastomer actuator (DEA)	<7.2	<380 (area)	<1000	1–10	<3.5	<90
Ionic polymer-metal composite (IPMC)	<5	/	<100	5	<4	<3
Liquid crystal elastomer (LCE)	<2	<60	<5	0.1–0.5	<100	<1
Responsive hydrogel	<0.1	<500	<1	0.01–0.1	/	/
Shape memory polymer (SMP)	<20	<800	<1	0.5	/	<1
Twisted nylon Fiber	<100	<50	<10	1	2500	<1
Magneto-active elastomer (MAE)	<20	<5	<200	1–10	/	/

We briefly review the fabrication and modelling of PAMs. Several examples of recently developed PAM-based soft actuators are also discussed.

4.2.2.1 Fabrication

The fabrication process of PAMs varies among specific examples. Typically, fabricating a PAM involves the following steps: mold preparation, casting, and enclosing [34, 35]. First, a customized mold is designed and built. Second, uncured elastomeric material such as silicone rubber is poured into the mold and then cured inside. The final step is to seal the structure using carefully selected adhesive materials.

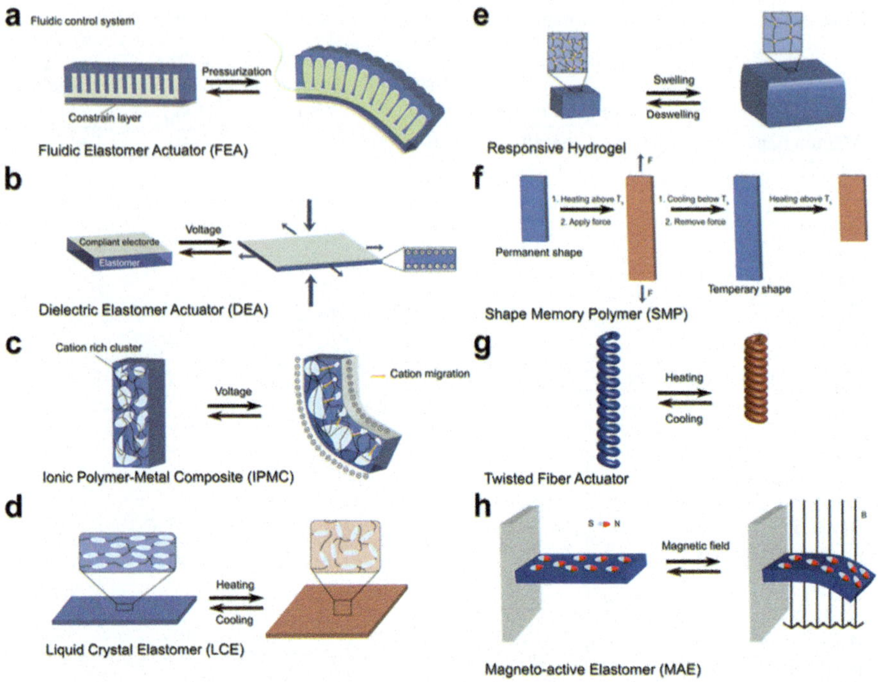

Fig. 4.1 Operation principle of soft actuators. (**a**) Fluidic elastomer actuator (FEA). (**b**) Dielectric elastomer actuator (DEA). (**c**) Ionic polymer-metal composite (IPMC). (**d**) Liquid crystal elastomer (LCE). (**e**) Responsive hydrogel. (**f**) Shape memory polymer (SMP). (**g**) Fiber-based artificial muscle. (**h**) Magneto-active elastomer (MAE)

To achieve versatile actuation, inextensible fibers have been integrated with inflatable elastomeric structures. Connolly et al. have further studied the relationship between fiber alignment and actuation response of PAMs [31, 32]. By carefully aligning the constraining fibers, multiple actuation modes such as bending, twisting, contracting, and extending can be realized [36].

Thanks to the rapid progress of 3D printing techniques, additive manufacturing has been successfully used to directly print PAMs within one step [37, 38]. Those 3D printed PAMs can have complex and customized shapes as well as multi-material combinations [39].

4.2.2.2 Modelling

The working mechanism of a PAM is straightforward: actuation of a PAM is caused by the change of its internal pressure, which can either increase or decrease (vacuum). As a PAM is composed of compliant and stretchable elastomeric membranes, large deformation can be achieved.

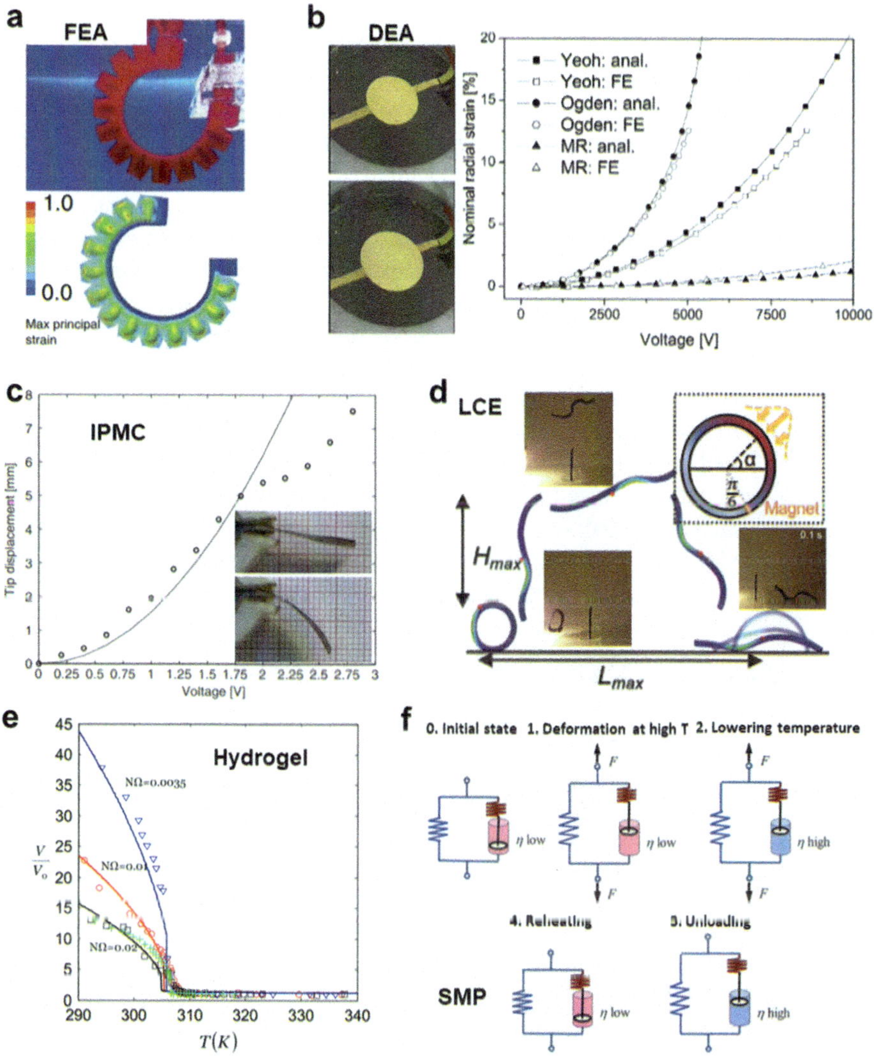

Fig. 4.2 Modelling of soft actuators. (**a**) Fluidic elastomer actuator (FEA) [16]. (**b**) Dielectric elastomer actuator (DEA) [17]. (**c**) Ionic polymer-metal composite (IPMC) [18]. (**d**) Liquid crystal elastomer (LCE) [19]. (**e**) Responsive hydrogel [20]. (**f**) Shape memory polymer (SMP) [21]

To quantitatively predict the response of a PAM to the change of its internal pressure or to design new PAMs with customized actuation modes, finite element simulations with appropriate hyperelastic material models have been often conducted (Fig. 4.2a) [16, 40]. More recently, a comprehensive simulation tool was presented by Paik et al. to guide the design of PAMs [40].

4.2.2.3 Demonstrations

Among all the soft actuators, PAMs are one of the most widely explored actuators for constructing soft robots and machines. For example, Ilievski et al. fabricated a starfish-like gripper with embedded pneumatic networks, which has shown adaptability when manipulating fragile objects with irregular shapes [35]. Vacuum-driven gripping has also been achieved through structural buckling [41]. To improve the adhesion between grippers and objects, soft grippers with a gecko-inspired surface were fabricated [42, 43]. Recently, Truby et al. reported a soft somatosensitive gripper, in which a sensory system was incorporated into the actuator through multi-material 3D printing [39]. More comprehensive reviews on pneumatic soft grippers can be found in [5].

In addition to soft grippers, a soft quadrupedal crawler has been designed by Shepherd et al., which can sustain large deformations and walk on varied terrains [44, 45]. Untethered and autonomous soft robots have also been fabricated, in which the internal pressurized gas was generated through fuel decomposition [46]. Recently, a snake-inspired soft robot was constructed by covering a pneumatic actuator with kirigami skin with highly tunable frictional properties [47].

4.2.3 Electroactive Polymer (EAP)

4.2.3.1 Dielectric Elastomers Actuator (DEA)

Dielectric elastomer actuator (DEA) has been explored for decades and is emerging as one of the most promising actuators for soft robotics because of its low cost, fast response, large actuating strain, high energy efficiency, and low level of noise [13, 48, 49]. A typical structure of DEA is a piece of dielectric elastomer (DE) film sandwiched by two compliant electrodes. With electrical potential along the thickness direction of the DE film, the elastomer expands in area due to the Maxwell stresses as shown in Fig. 4.1b. A DEA can generate large actuating stress (up to several MPa) and actuating strain (up to 380% in terms of area change) [50]. The actuating bandwidth of DE can be as high as 1000 Hz [51], which is probably the highest among all the currently available soft actuators. It should be noted that, due to the viscoelastic nature of most dielectric elastomers, the active strain generated by a DE can decrease significantly with the increase of the frequency of applied voltage. High voltage up to tens of kV and pre-stretch are often required for DEA to produce large actuation, which have limited their broader applications [48].

Recently, it has been reported that dielectric gels (e.g., PVC gel) can also deform in response to voltage [52]. The working mechanism of a dielectric gel has lots of similarities to DEAs, though the actuating voltage is one magnitude lower [52–54]. With the application of a bias voltage, the dielectric gel creeps to the positively charged electrode [53]. Due to its high transparency, the applications of PVC gel in optical devices have been explored [55–57].

Fabrication

To make a DEA, an elastomer film usually needs to be pre-stretched and then coated by compliant electrodes on both surfaces. VHB tape (3M) is the most commonly used elastomer film, and carbon grease is often used as compliant electrode [50]. Silicone rubber has been sometimes used for cases with high frequency actuation [48].

New materials have recently been explored as novel dielectric elastomer and compliant electrodes. For example, silicone rubber mixed with graphene or carbon nanotubes [58, 59] and hydrogels containing mobile ions have been used as transparent electrodes for DEAs [60]. Super soft and resilient bottlebrush elastomers [61] and interpenetrating polymer networks [62, 63] have been used as DE film. Interestingly, hydraulically amplified self-healable DEA has been recently constructed [64].

Modelling

Continuum mechanics models of a DE subjected to electromechanical loads have been developed by various researchers [65–69]. Based on the ideal DE assumption, Maxwell stress generated by the electric field inside the material plays an equivalent role as mechanical stress for deforming the elastomer [65]. For a thin film DEA, the deformation, the applied voltage, and the nominal stress can be related by the equations $s_1 = \frac{\partial W_s(\lambda_1,\lambda_2)}{\partial \lambda_1} - \frac{\epsilon E^2}{\lambda_1}$ and $s_2 = \frac{\partial W_s(\lambda_1,\lambda_2)}{\partial \lambda_2} - \frac{\epsilon E^2}{\lambda_2}$, where s_1 and s_2 are the nominal stresses in the two lateral directions; λ_1 and λ_2 are the stretches in the two principle directions, respectively; E is the true electric field; ϵ is electrical permittivity of the elastomer; and W_s is the free energy of the elastomer associated with stretching, the explicit form of which can be one of the commonly used hyperelastic models such as the Ogden model, Gent model, Neo-Hookean model, or Arruda-Boyce model (Fig. 4.2b) [17].

More general constitutive models of DE under electromechanical loads have been implemented as user material subroutines in commercially available finite element software such as ABAQUS [70, 71].

Demonstrations

Various soft robots actuated by DEAs have been demonstrated [72–74]. Spring roll and stacked layers are two common configurations for DEAs. Earlier work by Perine et al. demonstrated a six-leg walking robot with spring roll actuator [73, 74]. Interestingly, Shian et al. demonstrated a DE-based thin-film gripper, in which patterned fibers were used to program the deformation [75]. Recently, Kellaris et al. developed a self-healing DEA by embedding liquid dielectric and hydrogel electrodes in plastic pouches, which do not suffer from permanent damage due to electrical breakdown [64]. DEAs for other applications such as adaptive lenses [76–79], fluidic pumps [80, 81], and tactile devices [82] have also been reported.

4.2.3.2 Ionic Polymer-Metal Composite (IPMC)

Ionic polymer-metal composite (IPMC) is another type of electroactive polymer that has been extensively explored [83–86]. Bending motion of IPMC is induced by the ion migration in the composite, driven by an externally applied electric field (Fig. 4.1c) [87].

Fabrication

Typical fabrication for IPMC is sandwiching an ion-rich polymer film (such as Nafion or Flemion) between two compliant electrodes [83]. Recently, attempts have been made to explore new materials for either electrode or ionic polymer [88–92].

Modelling

When a voltage is applied to the electrodes, ions in the polymer film migrate from one side to the other, causing local swelling (ion-rich side) and contraction (ion-poor side) [87]. Both microscopic and macroscopic mechanic models have been developed to predict the deformation of IPMC [18, 93–95]. Early work by Nemat-Nasser studied the micromechanics of the actuation of IPMC [93]. A macroscopic continuum model was developed by Branco et al. [18]. A model given by Brunetto et al. considered the effects of a water environment [96].

Most IPMC actuators are in the form of beams. The key equation governing the voltage-induced bending of an IPMC beam can be empirically given by [96] $M = \frac{dhbE}{2}V$, where M is the bending moment generated by the applied voltage, V is the magnitude of the voltage, E is the Young's modulus of the material, d is the electromechanical coupling coefficient, and h and b are the height and width of the cross-section of the beam.

Demonstrations

IPMCs have been used as artificial fins in robotic systems due to bidirectional bending actuation. Aquatic robotics such as an untethered swimming fish robot and a snake-like robot have been constructed using IPMCs [97–99]. Grippers [100], micropumps, and valves [101] based on IPMC actuators have been widely demonstrated. Additionally, because of the deformation-induced electromotive voltage, IPMCs have also been explored as flexible motion sensors [102].

4.2.4 Liquid Crystal Elastomer (LCE)

Liquid crystal elastomer (LCE), which contains liquid crystal molecules (mesogens) and polymer networks, has attracted a lot of attention because of its superior actuation behavior and unique mechanical properties and optical properties. When LCE is exposed to thermal, optical, or chemical stimuli, its mesogens can change from a liquid crystal phase to amorphous phase, often accompanied by large and reversible deformation of the material as shown in Fig. 4.1d. Thermally driven LCEs show comparable actuating performance as mammalian muscle in terms of actuating stress (0.1–0.35 MPa), strain (50%), and work density.

4.2.4.1 Synthesis

As an actuating material, LCE with aligned mesogens is often required. The synthesis of LCE with certain mesogen alignment can be classified into two categories: one-pot polymerization [108] and two-step polymerization [109, 110]. For one-pot polymerization, mesogens, chain extenders, and cross-linkers are all mixed together, and the alignment of mesogens can be achieved by their contact with specifically treated surfaces or an externally applied electrical or magnetic field. Because a high electric or magnetic field is required to align mesogens and the treated surface can only affect the orientation of mesogens adjacent to it, the one-pot polymerization method has been often used to synthesize LCEs of small size. Two-step polymerizations, however, can been easily used to synthesize LCEs of arbitrary sizes. During the synthesis, a loosely cross-linked polymer network is formed first and then external mechanical stretch is applied to induce the alignment of mesogens, which can be fixed through the subsequent second-step cross-linking reaction.

By incorporating dynamic covalent bonds into polymer networks, multiple research groups have shown that the LCEs can be self-healable, reprogrammable, and reprocessable [111–113].

Additionally, recent work has shown that the direct ink writing (DIW) method can be used to print LCE structures with certain shapes as well as predesigned actuating behaviors [113–116]. It was found that the shear stress generated by the extrusion process could induce the alignment of mesogens in printed LCE filaments, endowing the actuating capabilities to printed LCE structures.

4.2.4.2 Modelling

The actuation of LCEs can be thermally driven via Joule heating. Embedding heating elements such as heating wire [103, 105, 117] and conductive ink [118] is a simple way to generate Joule heat (Fig. 4.3a), which makes LCE actuators compatible with most low-cost and well-developed electronic devices and batteries.

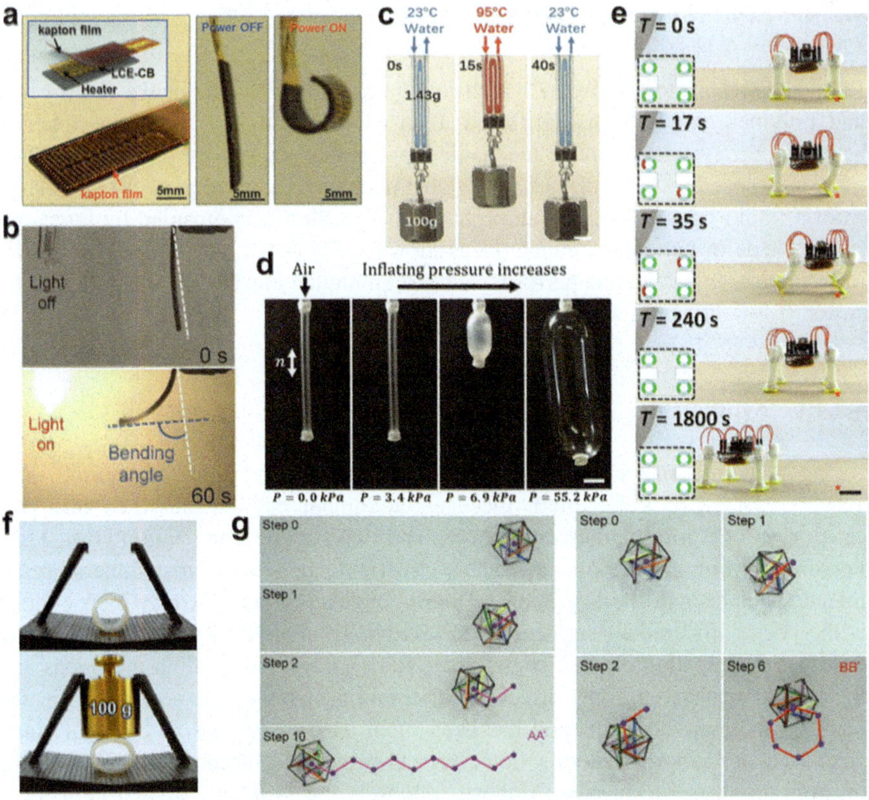

Fig. 4.3 Liquid crystal elastomer (LCE) as soft actuators can be driven by (**a**) electrical potential [103], (**b**) light [19], (**c**) high temperature fluid [104], and (**d**) pressurized gas [236]. Demonstrations of LCE actuators include (**e**) untethered robot [105], (**f**) biomimetic stent [106], and (**g**) tensegrity robot [107]

However, shape recovery of LCE usually takes a long time because of the slow and passive cooling process. Inspired by the vascular architectures in plants and animals, fluidic channels have been successfully introduced into LCE-based actuators [104]. The temperature change in LCE can be then controlled by alternatively injecting hot or cool fluid, which enables rapid contraction and recovery as shown in Fig. 4.3c.

Light has also been extensively used to trigger actuation of LCEs due to its remote and precise control [119, 120]. Light-induced actuation is often based on either photothermal or photochemical effects [19, 107, 121–123]. The former depends on the direct conversion of absorbed photon to heat (Fig. 4.3b), whereas the latter often relies on photoisomerization reaction of mesogens (trans and cis isomers transformation of specific mesogens such as azobenzene and its derivatives) [10].

In addition to heat and light, actuation of LCE can be also triggered by other stimuli. For instance, organic solvents such as toluene and THF have been reported to trigger nematic-isotropic phase transition of LCE [124, 125]; pressurized gas can cause inflation of long cylindrical LCE balloons accompanied by axial contraction as shown in Fig. 4.3d [236].

As shown in Fig. 4.2d, to simulate complex actuations of LCE structures, simplified thermomechanical coupling constitutive models have been often used [19].

4.2.4.3 Demonstrations

With spatially patterned mesogens in LCE, diverse actuation modes have been demonstrated (see Fig. 4.3d–f). Specifically, LCE-based soft robots have been built to exhibit many functionalities such as locomotion [123, 126], gripping [105, 127], jumping [19], rolling [9, 128, 129], and swimming [122, 130, 131] under various stimuli. In addition, LCE actuators have also been used to build novel biomimetic devices such as tendrils [132, 133], an iris [117, 121], and a flytrap [134].

4.2.5 Responsive Hydrogel

A polymer network swollen with water can result in a hydrogel. With the integration of different functional groups into polymer chains, a hydrogel can swell or shrink greatly when it is exposed to certain external stimuli such as temperature change [135], light [136], magnetic field [137], pH change [138], or ionic strength (Fig. 4.1e) [139]. Diverse responsive hydrogels have been widely used as soft actuators in various structures and devices. Hydrogel actuators can be made with excellent biocompatibility, although their applications can be limited by their poor mechanical properties, slow response, and the requirement of a water environment. To enhance toughness, responsive hydrogels with interpenetrating networks or multifunctional cross-linkers have been successfully synthesized [140, 141]. Responsive hydrogels with porous structures have been fabricated to increase their response speed [142, 143].

4.2.5.1 Synthesis

Some of the responsive hydrogels can be synthesized from commercially available monomers and cross-linkers within one or two steps such as PNIPAM, P(AAm-co-AAc), and PDMAAm [144–147]. The synthesis of some other responsive hydrogels may involve multiple steps of chemical reactions in a specific environment [148].

4.2.5.2 Modelling

Both continuum mechanics modelling and molecular-level modelling for understanding and predicting the active deformation of responsive hydrogels have been intensively developed [149, 150]. The continuum mechanics models for responsive hydrogels are often based on the modifications of Flory-Huggins theory [151]. Examples include the models for pH-sensitive hydrogels [152], temperature-sensitive hydrogels (Fig. 4.2e) [20], and light-responsive hydrogels [153]. The key equations for describing the response of those hydrogels can be given by $\sigma_{ij} = \frac{\partial W_s(\boldsymbol{F})}{\partial F_{iK}} \frac{F_{jK}}{\det(\boldsymbol{F})} - \Pi\left(c_s, c_i\right) \delta_{ij}$, where σ_{ij} is the Cauchy stress applied on the hydrogel, $\boldsymbol{F}$ is the deformation gradient with components F_{iK}, and Π is the osmosis pressure, which can be dependent on temperature and may include contributions from water and mobile ions, with c_s, c_i as the concentrations of water molecules and mobile ions in the hydrogel, respectively. For most of these continuum-level models, corresponding numerical programs based on finite element simulation or the phase field method have also been developed, enabling the computation of complex deformations of these responsive hydrogels when subjected to different thermal, chemical, optical, and mechanical loadings [154].

In addition to continuum mechanics modelling, molecular-level modelling has also been created for understanding different phenomena of responsive hydrogels [150]. Molecular dynamic simulation has been adopted to provide additional microscopic insights into specific observed phenomena [155].

4.2.5.3 Demonstrations

Myriad novel hydrogel-based structures have been designed and fabricated. Herein, we only briefly discuss the following three representative examples. The first example is a hydrogel robot powered by the oscillatory Belousov–Zhabotinsky (BZ) reaction. The periodic oxidation and reduction cycles during the BZ reaction can induce periodic swelling and shrinking of a hydrogel without any variation of external stimuli, which has been harnessed to design a self-walking robot and biomimetic self-sustained peristaltic motion [156]. The second example is a soft robotic walker composed of polyelectrolyte gels [157]. A polyelectrolyte gel can deform in an ionic solution when it is subjected to an electric field. An electric field can induce the redistribution of ions, leading to a change of osmotic pressure in the gel. A polyelectrolyte gel robot has exhibited various modes of motion with the application of an electric field [158]. The third example is a soft composite containing hydrogel and cofacially oriented electrolyte nanosheets [159]. When temperature changes, the hydrogel composite can change its shape while conserving its volume. Since the shape change of the gel does not involve water absorption and desorption, the hydrogel can respond rapidly to the temperature change.

4.2.6 *Shape Memory Polymer (SMP)*

Most glassy or semicrystalline polymers can memorize their shapes after plastic deformation to a certain extent [160]. Those polymers are often referred as shape memory polymers (SMPs). In applications, SMPs are usually first plastically deformed to certain shapes by mechanical force at high temperature and then cooled down to fix temporary shapes. When the deformed SMPs are heated above a certain temperature (such as glass transition temperature), they recover back to their initial shapes before plastic deformation (Fig. 4.1f). The thermally induced shape recovery process has been often harnessed for actuation. Recently, SMPs that can memorize more than two different configurations (multi-SMP) and achieve reversible actuation have been synthesized [161]. Those reversible SMPs are also known as two-way SMPs; they are particularly useful for the applications where cyclic actuation is needed [162–165].

The most commonly used stimulus to trigger the actuation of SMPs is heat, which can be applied through direct heating, photothermal effect [166, 167], Joule heating effect [168–171], and inducive heating effect [172]. Additionally, the shape memory effects in polymers can also be induced by the presence of solvent [173, 174].

4.2.6.1 Synthesis

Essentially, all the polymers in glassy state can have certain shape memory effects. Currently, the most commonly used SMPs include epoxy-based SMPs [175, 176], polyurethane-based SMPs [177, 178], and other polymer networks with modified functional segments [163, 179, 180]. Traditional polymer processing techniques such as molding can be used to process SMPs into different structures. Recently, additive manufacturing technique has been explored to achieve complex shapes of SMPs [181–183].

4.2.6.2 Modelling

A spring dashpot model (Fig. 4.2f) has been often used to quantitatively predict complex thermomechanical behaviors of shape memory polymers [184–188]. Constitutive models of SMPs involving various arrangements and combinations of springs and dashpots have been developed [21]. Rheological properties of springs and dashpots are usually temperature-dependent and nonlinear. The volume fractions of glassy phase, crystalline phase, or rubbery phase in SMP are typically regarded as internal variables in the model, which can evolve with the changes of temperature as well as mechanical load. Several different constitutive models of SMPs have also been successfully implemented as user material subroutines in finite element simulation software such as ABAQUS [189, 190].

4.2.6.3 Demonstrations

Many attempts have been made to fabricate SMP-based active structures. Inspired by origami, deployable devices have been constructed with SMPs [183, 191]. Jin et al. have developed a soft robot using crystalline SMPs, which can achieve either permanent shape change when thermally programmed or reversible actuation when programmed by light [192]. Takashima et al. fabricated a McKibben actuator using SMPs, which can fix its shape after depressurization [193]. Zarek et al. developed a flexible electronic device based on a thermally responsive SMP and demonstrated its application as a shape memory switch [181].

4.2.7 Twisted Fiber Actuator

With highly aligned polymer chains, nylon fiber shows anisotropic thermal expansion behavior [194, 195]. When a nylon fiber is heated up, it expands in diameter and contracts in length, and thus can be fabricated into linear/torsional actuators. However, the thermally induced contractive strain of a straight nylon fiber is too small for most practical applications. In 2014, Haines et al. developed a type of novel artificial muscle by simply twisting a nylon fiber to a highly coiled shape, which can contract as much as 49% upon heating through geometrical amplification [194]. The most salient features of the twisted fiber actuators include easy fabrication, low cost and large actuation stress and strain.

4.2.7.1 Fabrication

Typically, the fabrication of twisted fiber actuators consists of two main steps: coiling and annealing [194]. Twisting a nylon fiber with appropriate tensile loading can lead to a regularly coiled structure. Subsequent annealing process is required to fix the shape.

Twisted fiber actuators can be actuated through Joule heating or photothermal effects. Silver coating layer [196], multiwall-carbon-nanotubes (MWCNT) yarn [197, 198], and nichrome coating layer [199, 200] have been used as electrically resistive materials for generating Joule heating. Photothermal conversion can be achieved through incorporating material with high light absorption [201]. Chen et al. reported helical fiber actuators driven by solvent or its vapor, achieving 10% contractive strain [202].

In addition to nylon, twisted fiber actuators based on other materials have also been developed [203–205]. Hiraoka et al. reported coiled fiber actuators made of linear low-density polyethylene (LLDPE), which have shown higher power efficiency as compared to nylon-based counterparts [203]. Yang et al. fabricated a coiled actuator with spandex fibers, which achieved larger contractive strain (45%) under the same working temperature [204]. Recently, Yuan et al. developed a micro torsional engine using stiffened PVA fibers and achieved fast and high-torque rotations [205]. Fibers with special structures such as core-sheath have also been developed to further increase the energy density of the actuator [206].

4.2.7.2 Modelling

Mechanics models have been developed to study the relationship between the geometry, the thermomechanical properties of the fiber materials, and the actuation performance of the twisted fiber actuators [201, 207–209]. Those models can guide the design of twisted fiber actuators and further enhance performance. Other actuating modes of twisted fiber actuators have also been explored with the aid of numerical modelling tools [201].

4.2.7.3 Demonstrations

Robotic arms and fingers have been fabricated using twisted-fiber actuators, which can perform various grasping maneuvers [210, 211]. Kim et al. demonstrated that twisted nylon fiber can convert heat to electricity [212]. Wang et al. developed a refrigerator with high efficiency and low cost, which is realized by stretching coiled natural rubber fibers [213]. Recently, Kanik et al. constructed a thermally and optically controlled twisted-fiber actuator combined with a piezoresistive strain sensor [214].

4.2.8 Magneto-active Elastomer (MAE)

Soft actuators fabricated from magneto-active elastomers (MAEs) have drawn more attention recently as they can respond quickly, enable remote control, and can be easily made into small sizes (<1 mm), thus are promising for operations in confined spaces [2, 215, 216]. Magneto-active elastomer is usually composed of elastomeric matrix and magnetically responsive particles (typically ferromagnetic particles). When a magnetic field is applied, embedded particles tend to align with the external field, which results in macroscopic deformation, as well as field-dependent mechanical properties [2, 216, 217].

4.2.8.1 Fabrication

A commonly adopted method to fabricate MAE is simply mixing magnetizable particles such as carbonyl iron particles [218, 219], Barium hexaferrite [220], and Nd-Fe-B [221–223] with uncured elastomers such as silicone rubber and polyurethanes (PU). Subsequently, an external magnetic field is applied to orient magnetic particles before the curing of the elastomer. To obtain MAE with certain geometries, mold casting is typically used. This method usually produces MAEs with uniformly orientated particles.

Recently, 3D printing method has been explored to fabricate MAEs with complex geometry and alignment of magnetic particles, which can generate tailorable actuation mode [222, 224]. For instance, Kim et al. printed programmed ferromagnetic domains in silicone rubber and demonstrated various transformation modes [222].

4.2.8.2 Modelling

Continuum mechanics modelling has also been developed for MAEs [225, 226]. The constitutive model of MAE has a similar form as that of dielectric elastomer discussed previously. For a hard magneto-active elastomer, the applied magnetic field and the true stress can be related by [227] $\sigma_{ij} = \frac{1}{\det(\boldsymbol{F})}\frac{\partial W_s(\mathbf{F})}{\partial F_{iK}}F_{jK} - \frac{1}{\mu_0}B_i^{app}B_j^r$, where σ_{ij} is the Cauchy stress applied on the elastomer, F_{iK} is the deformation gradient, W_s is the free energy of the elastomer associated with stretching, μ_0 is magnetic permeability, B_i^{app} is applied magnetic flux density, and B_j^r is residual magnetic flux density in the elastomer.

4.2.8.3 Demonstrations

MAEs are especially suitable for applications that require small-size actuators, fast responsiveness, and remote control. Garstecki et al. fabricated a propulsion swimmer driven by a rotating magnetic field [228]. Soft-bodied robots that can crawl, swim, and jump have been demonstrated [221]. 3D printed, fast-transforming structures have been developed [222]. Recently, Ze et al. developed a composite of MAE and shape memory polymer, which can be actuated in a magnetic field and maintain the actuated shape through the shape-memory effect after removing the magnetic field [229].

MAE-based soft actuators can be promising in developing novel biomedical devices [230]. For example, Kim et al. demonstrated a ferromagnetic robot that can navigate through narrow spaces such as a tortuous vascular phantom mimicking a blood vessel in humans [231].

4.3 Underwater Soft Robotic Systems

4.3.1 Specific Requirements for the Actuators

Compared to soft robotic systems functioning in air, the design and fabrication of underwater soft robotic systems can be more challenging. For instance, for thermally driven soft actuators, including liquid crystal elastomers, shape memory polymers, and twisted fiber-based actuators, either large heating power or certain thermal insulations should be included in the design. To construct underwater soft robots using electrically powered actuators such as DEAs, dielectric gel-based actuators, and IPMC-based actuators, electrical insulation must be carefully considered.

In addition, the motion of soft robots in water often needs to overcome larger resistant forces, compared to their motion in air. Therefore, those soft actuators with large actuating stress or power may be better choices for underwater soft robotic systems.

4.3.2 *Examples of Underwater Soft Robotic Systems*

As shown in Fig. 4.4, most aforementioned actuators have been explored to construct underwater soft robotic systems. Soft grippers with pneumatic and hydraulic actuators can gently handle sea animals such as corals and fish [16]. It has also been demonstrated that PAMs can be integrated with micropumps, valves, and controllers to form an untethered and autonomous underwater soft robot [12, 232]. With careful designs, DEA has also been used to build underwater soft robots. For instance, in the design of an eel-like soft robot [7], salty water is used as electrodes for the DEA, which simplifies the fabrication of the structure and keeps the entire body transparent. For thermally driven LCE actuators working in water, though high heating power is needed to generate actuation, its recovery process by cooling can be greatly accelerated. Many forms of LCE-based underwater soft robots have been successfully fabricated in the past several years [122, 130]. IPMC and hydrogel-based actuators have been mostly explored in water environments to prevent material dehydration. Myriad examples of soft robots based on IPMCs and hydrogels can be found in the literature, such as a hydrogel underwater walker and swimmer and IPMC fish [11, 98, 235].

4.4 Outlook

Generally, it is more challenging to design and fabricate underwater soft robotic systems than air and ground ones. The criteria for selecting proper soft actuators for underwater soft robots need to be carefully evaluated. Specific modifications or improvements of existing soft actuators may also be needed to make them fit to underwater applications. Because of the buoyancy of water, the weight of an underwater soft robot provides much less of a constraint on their design, which is often one of the limiting factors in fabricating soft robots for use above water. Additionally, the applications of underwater and ground soft robotic systems can differ substantially. Consequently, it is reasonable to expect the emergence of new

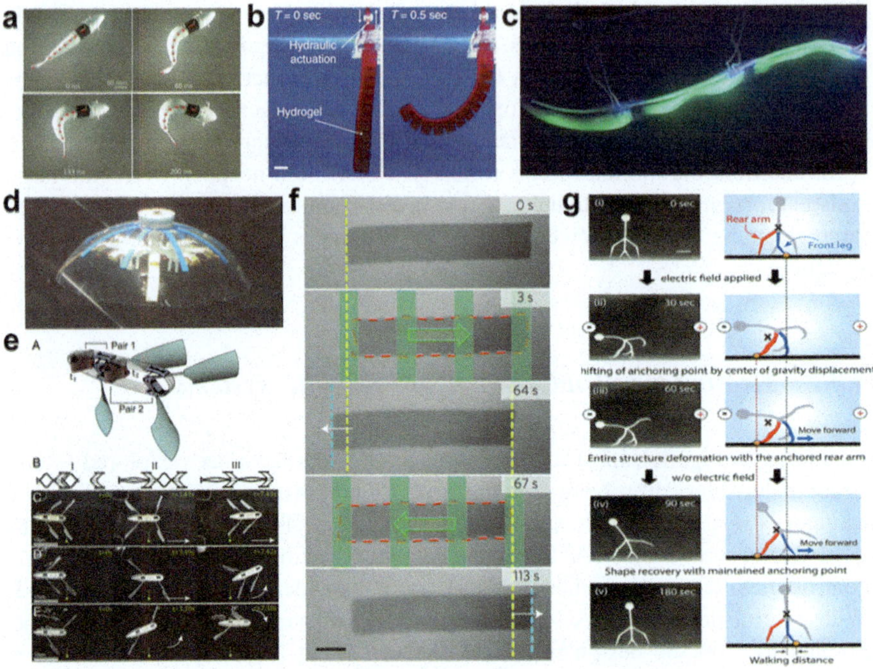

Fig. 4.4 Examples of underwater soft robots. (**a**) Autonomous fish driven by pneumatic actuator [232]. (**b**) Soft gripper driven by hydraulic pressure [16]. (**c**) Eel-like robot driven by DEA [7]. (**d**) Jellyfish-inspired underwater vehicle driven by IPMC [233]. (**e**) Untethered swimmer driven by SMP [234]. (**f**) LCE microrobot driven by light [122]. (**g**) Hydrogel walker [235]

forms of underwater soft robotic systems that are completely different from the ground ones in terms of shape, size, and functionality.

Acknowledgment The authors acknowledge support from ONR Grant No. N00014-17-1-2062.

References

1. W.-S. Chu et al., Review of biomimetic underwater robots using smart actuators. Int. J. Precis. Eng. Manuf. **13**(7), 1281–1292 (2012)
2. L. Hines et al., Soft actuators for small-scale robotics. Adv. Mater. **29**(13), 1603483 (2017)
3. D. Rus, M.T. Tolley, Design, fabrication and control of soft robots. Nature **521**(7553), 467–475 (2015)
4. P. Polygerinos et al., Soft robotics: Review of fluid-driven intrinsically soft devices; manufacturing, sensing, control, and applications in human-robot interaction. Adv. Eng. Mater. **19**(12), 1700016 (2017)
5. J. Hughes et al., Soft manipulators and grippers: A review. Front. Robot. AI **3**, 69 (2016)
6. H. Ren, S.-T. Wu, Adaptive lenses based on soft electroactive materials. Appl. Sci. **8**(7), 1085 (2018)

7. C. Christianson et al., Translucent soft robots driven by frameless fluid electrode dielectric elastomer actuators. Sci Robot **3**(17), eaat1893 (2018)
8. K.C. Galloway et al., Soft robotic grippers for biological sampling on deep reefs. Soft Robot **3**(1), 23–33 (2016)
9. C. Ahn, K. Li, S. Cai, Light or thermally powered autonomous rolling of an elastomer rod. ACS Appl. Mater. Interfaces **10**(30), 25689–25696 (2018)
10. A.H. Gelebart et al., Making waves in a photoactive polymer film. Nature **546**(7660), 632 (2017)
11. Y. Zhao et al., Soft phototactic swimmer based on self-sustained hydrogel oscillator. Sci Robot **4**(33), eaax7112 (2019)
12. R.K. Katzschmann et al., Exploration of underwater life with an acoustically controlled soft robotic fish. Sci Robot **3**(16), eaar3449 (2018)
13. R.E. Pelrine, R.D. Kornbluh, J.P. Joseph, Electrostriction of polymer dielectrics with compliant electrodes as a means of actuation. Sensors Actuators A Phys. **64**(1), 77–85 (1998)
14. T. Miyata, N. Asami, T. Uragami, A reversibly antigen-responsive hydrogel. Nature **399**(6738), 766 (1999)
15. M.-H. Li, P. Keller, Artificial muscles based on liquid crystal elastomers. Philos. Trans. R. Soc. A Math. Phys. Eng. Sci. **364**(1847), 2763–2777 (2006)
16. H. Yuk et al., Hydraulic hydrogel actuators and robots optically and sonically camouflaged in water. Nat. Commun. **8**, 14230 (2017)
17. M. Wissler, E. Mazza, Modeling of a pre-strained circular actuator made of dielectric elastomers. Sensors Actuators A Phys. **120**(1), 184–192 (2005)
18. P.C. Branco, J. Dente, Derivation of a continuum model and its electric equivalent-circuit representation for ionic polymer–metal composite (IPMC) electromechanics. Smart Mater. Struct. **15**(2), 378 (2006)
19. C. Ahn, X. Liang, S. Cai, Bioinspired design of light-powered crawling, squeezing, and jumping untethered soft robot. Adv. Mater. Technol. **4**(7), 1900185 (2019)
20. S. Cai, Z. Suo, Mechanics and chemical thermodynamics of phase transition in temperature-sensitive hydrogels. J. Mech. Phys. Solids **59**(11), 2259–2278 (2011)
21. K. Yu et al., Mechanisms of multi-shape memory effects and associated energy release in shape memory polymers. Soft Matter **8**(20), 5687–5695 (2012)
22. B. Gorissen et al., Elastic inflatable actuators for soft robotic applications. Adv. Mater. **29**(43), 1604977 (2017)
23. C.-P. Chou, B. Hannaford, Measurement and modeling of McKibben pneumatic artificial muscles. IEEE Trans. Robot. Autom. **12**(1), 90–102 (1996)
24. F. Daerden, D. Lefeber, The concept and design of pleated pneumatic artificial muscles. International Journal of Fluid Power **2**(3), 41–50 (2001)
25. D. Yang et al., Buckling pneumatic linear actuators inspired by muscle. Advanced Materials Technologies **1**(3), 1600055 (2016)
26. E. Brown et al., Universal robotic gripper based on the jamming of granular material. Proc. Natl. Acad. Sci. **107**(44), 18809–18814 (2010)
27. M.A. Robertson, J. Paik, New soft robots really suck: Vacuum-powered systems empower diverse capabilities. Sci Robot **2**, eaan6357 (2017)
28. S. Li et al., Fluid-driven origami-inspired artificial muscles. Proc. Natl. Acad. Sci. **114**(50), 13132–13137 (2017)
29. F. Daerden, D. Lefeber, Pneumatic artificial muscles: Actuators for robotics and automation. Eur. J. Mech. Environ. Eng. **47**(1), 11–21 (2002)
30. B. Tondu et al., A seven-degrees-of-freedom robot-arm driven by pneumatic artificial muscles for humanoid robots. Int. J. Robot. Res. **24**(4), 257–274 (2005)
31. F. Connolly et al., Mechanical programming of soft actuators by varying fiber angle. Soft Robotics **2**(1), 26–32 (2015)
32. F. Connolly, C.J. Walsh, K. Bertoldi, Automatic design of fiber-reinforced soft actuators for trajectory matching. Proc. Natl. Acad. Sci. **114**(1), 51–56 (2017)

33. S.Y. Kim et al., Reconfigurable soft body trajectories using unidirectionally stretchable composite laminae. Nat. Commun. **10**(1), 1–8 (2019)
34. Y. Xia, G.M. Whitesides, Soft lithography. Annu. Rev. Mater. Sci. **28**(1), 153–184 (1998)
35. F. Ilievski et al., Soft robotics for chemists. Angew. Chem. Int. Ed. **50**(8), 1890–1895 (2011)
36. S.Y. Kim et al., Reconfigurable soft body trajectories using unidirectionally stretchable composite laminae. Nat. Commun. **10**(1), 3464 (2019)
37. D. Drotman et al., 3D printed soft actuators for a legged robot capable of navigating unstructured terrain, in *2017 IEEE International Conference on Robotics and Automation (ICRA)*, (IEEE, 2017)
38. O.D. Yirmibesoglu et al., Direct 3D printing of silicone elastomer soft robots and their performance comparison with molded counterparts, in *2018 IEEE International Conference on Soft Robotics (RoboSoft)*, (IEEE, 2018)
39. R.L. Truby et al., Soft somatosensitive actuators via embedded 3D printing. Adv. Mater. **30**(15), 1706383 (2018)
40. P. Moseley et al., Modeling, design, and development of soft pneumatic actuators with finite element method. Adv. Eng. Mater. **18**(6), 978–988 (2016)
41. D. Yang et al., Buckling of elastomeric beams enables actuation of soft machines. Adv. Mater. **27**(41), 6323–6327 (2015)
42. P. Glick et al., A soft robotic gripper with gecko-inspired adhesive. IEEE Robot. Autom. Lett. **3**(2), 903–910 (2018)
43. S. Song, C. Majidi, M. Sitti, Geckogripper: A soft, inflatable robotic gripper using gecko-inspired elastomer micro-fiber adhesives, in *2014 IEEE/RSJ International Conference on Intelligent Robots and Systems*, (IEEE, 2014)
44. R.F. Shepherd et al., Multigait soft robot. Proc. Natl. Acad. Sci. **108**(51), 20400–20403 (2011)
45. S.A. Morin et al., Camouflage and display for soft machines. Science **337**(6096), 828–832 (2012)
46. M. Wehner et al., An integrated design and fabrication strategy for entirely soft, autonomous robots. Nature **536**(7617), 451 (2016)
47. A. Rafsanjani et al., *Kirigami skins make a simple soft actuator crawl*. Sci Robot **3**(15), eaar7555 (2018)
48. P. Brochu, Q. Pei, Advances in dielectric elastomers for actuators and artificial muscles. Macromol. Rapid Commun. **31**(1), 10–36 (2010)
49. U. Gupta et al., Soft robots based on dielectric elastomer actuators: A review. Smart Mater. Struct. **28**(10), 103002 (2019)
50. R. Pelrine et al., High-speed electrically actuated elastomers with strain greater than 100%. Science **287**(5454), 836–839 (2000)
51. X. Ji et al., An autonomous untethered fast soft robotic insect driven by low-voltage dielectric elastomer actuators. Sci Robot **4**(37) (2019)
52. M.Z. Uddin et al., Electrically induced creeping and bending deformation of plasticized poly (vinyl chloride). Chem. Lett. **30**(4), 360–361 (2001)
53. H. Xia, M. Takasaki, T. Hirai, Actuation mechanism of plasticized PVC by electric field. Sensors Actuators A Phys. **157**(2), 307–312 (2010)
54. Y. Li, M. Hashimoto, PVC gel based artificial muscles: Characterizations and actuation modular constructions. Sensors Actuators A Phys. **233**, 246–258 (2015)
55. J.W. Bae et al., High-performance PVC gel for adaptive micro-lenses with variable focal length. Sci. Rep. **7**(1), 2068 (2017)
56. T. Hirai et al., Electrically active artificial pupil showing amoeba-like pseudopodial deformation. Adv. Mater. **21**(28), 2886–2888 (2009)
57. P. Huang, Z. Zhou, H. Ren, Fresnel zone plate fabricated using a polyvinyl chloride gel. Opt. Eng. **57**(11), 117101 (2018)
58. N. Ning et al., Tailoring dielectric and actuated properties of elastomer composites by bioinspired poly (dopamine) encapsulated graphene oxide. ACS Appl. Mater. Interfaces **7**(20), 10755–10762 (2015)

59. T.A. Kim et al., Single-walled carbon nanotube/silicone rubber composites for compliant electrodes. Carbon **50**(2), 444–449 (2012)
60. C. Keplinger et al., Stretchable, transparent, ionic conductors. Science **341**(6149), 984–987 (2013)
61. M. Vatankhah-Varnoosfaderani et al., Bottlebrush elastomers: A new platform for freestanding electroactuation. Adv. Mater. **29**(2), 1604209 (2017)
62. S.M. Ha et al., Interpenetrating networks of elastomers exhibiting 300% electrically-induced area strain. Smart Mater. Struct. **16**(2), S280 (2007)
63. X. Niu et al., Bistable large-strain actuation of interpenetrating polymer networks. Adv. Mater. **24**(48), 6513–6519 (2012)
64. N. Kellaris et al., Peano-HASEL actuators: Muscle-mimetic, electrohydraulic transducers that linearly contract on activation. Sci Robot **3**(14), eaar3276 (2018)
65. Z. Suo, Theory of dielectric elastomers. Acta Mechanica Solida Sinica **23**(6), 549–578 (2010)
66. S.J.A. Koh et al., Mechanisms of large actuation strain in dielectric elastomers. J Polym Sci B **49**(7), 504–515 (2011)
67. S.M. Jiménez, R.M. McMeeking, Deformation dependent dielectric permittivity and its effect on actuator performance and stability. Int. J. Non Lin. Mech. **57**, 183–191 (2013)
68. A. Dorfmann, R. Ogden, Nonlinear electroelasticity. Acta Mech. **174**(3–4), 167–183 (2005)
69. A. Dorfmann, R. Ogden, Nonlinear electroelastic deformations. J. Elast. **82**(2), 99–127 (2006)
70. X. Zhao, Z. Suo, Method to analyze electromechanical stability of dielectric elastomers. Appl. Phys. Lett. **91**(6), 061921 (2007)
71. X. Zhao, Z. Suo, Theory of dielectric elastomers capable of giant deformation of actuation. Phys. Rev. Lett. **104**(17), 178302 (2010)
72. T. Li et al., Fast-moving soft electronic fish. Sci. Adv. **3**(4), e1602045 (2017)
73. R. Pelrine et al., Dielectric elastomer artificial muscle actuators: toward biomimetic motion, in *Smart Structures and Materials 2002: Electroactive polymer actuators and devices (EAPAD)*, (International Society for Optics and Photonics, 2002)
74. Q. Pei et al., Recent progress on electroelastomer artificial muscles and their application for biomimetic robots, in *Smart Structures and Materials 2004: Electroactive Polymer Actuators and Devices (EAPAD)*, (International Society for Optics and Photonics, 2004)
75. S. Shian, K. Bertoldi, D.R. Clarke, Dielectric elastomer based "grippers" for soft robotics. Adv. Mater. **27**(43), 6814–6819 (2015)
76. S. Shian, R.M. Diebold, D.R. Clarke, Tunable lenses using transparent dielectric elastomer actuators. Opt. Express **21**(7), 8669–8676 (2013)
77. F. Carpi et al., Bioinspired tunable lens with muscle-like electroactive elastomers. Adv. Funct. Mater. **21**(21), 4152–4158 (2011)
78. T. Lu et al., Computational model of deformable lenses actuated by dielectric elastomers. J. Appl. Phys. **114**(10), 104104 (2013)
79. J. Li et al., A biomimetic soft lens controlled by electrooculographic signal. Adv. Funct. Mater., 1903762 (2019)
80. P. Lotz, M. Matysek, H.F. Schlaak, Peristaltic pump made of dielectric elastomer actuators, in *Electroactive Polymer Actuators and Devices (EAPAD), 2009*, (International Society for Optics and Photonics, 2009)
81. D. McCoul, Q. Pei, Tubular dielectric elastomer actuator for active fluidic control. Smart Mater. Struct. **24**(10), 105016 (2015)
82. I.M. Koo et al., Development of soft-actuator-based wearable tactile display. IEEE Trans. Robot. **24**(3), 549–558 (2008)
83. M. Shahinpoor, Ionic polymer–conductor composites as biomimetic sensors, robotic actuators and artificial muscles—A review. Electrochim. Acta **48**(14–16), 2343–2353 (2003)
84. B. Bhandari, G.-Y. Lee, S.-H. Ahn, A review on IPMC material as actuators and sensors: Fabrications, characteristics and applications. Int. J. Precis. Eng. Manuf. **13**(1), 141–163 (2012)

85. R. Tiwari, E. Garcia, The state of understanding of ionic polymer metal composite architecture: A review. Smart Mater. Struct. **20**(8), 083001 (2011)
86. C. Jo et al., Recent advances in ionic polymer–metal composite actuators and their modeling and applications. Prog. Polym. Sci. **38**(7), 1037–1066 (2013)
87. M. Shahinpoor, Mechanoelectrical phenomena in ionic polymers. Mathematics and Mechanics of Solids **8**(3), 281–288 (2003)
88. Y. Lian et al., Enhanced electromechanical performance of graphite oxide-nafion nanocomposite actuator. J. Phys. Chem. C **114**(21), 9659–9663 (2010)
89. L. Kong, W. Chen, Carbon nanotube and Graphene-based bioinspired electrochemical actuators. Adv. Mater. **26**(7), 1025–1043 (2014)
90. C.-K. Chung et al., A novel fabrication of ionic polymer-metal composites (IPMC) actuator with silver nano-powders. Sens. Actuators B **117**(2), 367–375 (2006)
91. Y. Tang et al., Novel sulfonated polysulfone ion exchange membranes for ionic polymer–metal composite actuators. Sens. Actuators B **202**, 1164–1174 (2014)
92. J.-W. Lee, Y.-T. Yoo, Preparation and performance of IPMC actuators with electrospun Nafion®–MWNT composite electrodes. Sens. Actuators B **159**(1), 103–111 (2011)
93. S. Nemat-Nasser, J.Y. Li, Electromechanical response of ionic polymer-metal composites. J. Appl. Phys. **87**(7), 3321–3331 (2000)
94. Z. Chen et al., A dynamic model for ionic polymer–metal composite sensors. Smart Mater. Struct. **16**(4), 1477 (2007)
95. P. Nardinocchi, M. Pezzulla, L. Placidi, Thermodynamically based multiphysic modeling of ionic polymer metal composites. J. Intell. Mater. Syst. Struct. **22**(16), 1887–1897 (2011)
96. P. Brunetto et al., A model of ionic polymer–metal composite actuators in underwater operations. Smart Mater. Struct. **17**(2), 025029 (2008)
97. N. Kamamichi et al., A snake-like swimming robot using IPMC actuator/sensor, in *Proceedings 2006 IEEE International Conference on Robotics and Automation, 2006. ICRA 2006*, (IEEE, 2006)
98. X. Ye, Y. Su, S. Guo, A centimeter-scale autonomous robotic fish actuated by IPMC actuator, in *2007 IEEE International Conference on Robotics and Biomimetics (ROBIO). 2007*, (IEEE)
99. M. Aureli, V. Kopman, M. Porfiri, Free-locomotion of underwater vehicles actuated by ionic polymer metal composites. IEEE/ASME Trans. Mechatron. **15**(4), 603–614 (2009)
100. R.K. Jain, S. Majumder, A. Dutta, SCARA based peg-in-hole assembly using compliant IPMC micro gripper. Robot. Auton. Syst. **61**(3), 297–311 (2013)
101. J.J. Pak et al., Fabrication of ionic-polymer-metal-composite (IPMC) micropump using a commercial Nafion, in *Smart Structures and Materials 2004: Electroactive Polymer Actuators and Devices (EAPAD)*, (International Society for Optics and Photonics, 2004)
102. Y. Bahramzadeh, M. Shahinpoor, Dynamic curvature sensing employing ionic-polymer–metal composite sensors. Smart Mater. Struct. **20**(9), 094011 (2011)
103. C. Wang et al., Soft ultrathin electronics innervated adaptive fully soft robots. Adv. Mater. **30**(13), 1706695 (2018)
104. Q. He et al., Bioinspired design of vascular artificial muscle. Advanced Materials Technologies **4**(1), 1800244 (2019)
105. Q. He et al., Electrically controlled liquid crystal elastomer–based soft tubular actuator with multimodal actuation. Sci. Adv. **5**(10), eaax5746 (2019)
106. M.O. Saed et al., High strain actuation liquid crystal elastomers via modulation of mesophase structure. Soft Matter **13**(41), 7537–7547 (2017)
107. Z. Wang et al., A light-powered Ultralight Tensegrity robot with high deformability and load capacity. Adv. Mater. **31**(7), 1806849 (2019)
108. T.H. Ware et al., Voxelated liquid crystal elastomers. Science **347**(6225), 982–984 (2015)
109. C. Yakacki et al., Tailorable and programmable liquid-crystalline elastomers using a two-stage thiol–acrylate reaction. RSC Adv. **5**(25), 18997–19001 (2015)
110. J. Küpfer, H. Finkelmann, Nematic liquid single crystal elastomers. Makromol. Chem. Rapid **12**(12), 717–726 (1991)

111. Z. Wang et al., Reprogrammable, reprocessible, and self-healable liquid crystal elastomer with exchangeable disulfide bonds. ACS Appl. Mater. Interfaces **9**(38), 33119–33128 (2017)
112. Z. Wang et al., Programmable actuation of liquid crystal elastomers via "living" exchange reaction. Soft Matter **15**(13), 2811–2816 (2019)
113. E.C. Davidson et al., 3D printable and reconfigurable liquid crystal elastomers with light-induced shape memory via dynamic bond exchange. Adv. Mater. **32**(1), e1905682 (2019)
114. A. Kotikian et al., 3D printing of liquid crystal elastomeric actuators with spatially programed nematic order. Adv. Mater. **30**(10), 1706164 (2018)
115. M. López-Valdeolivas et al., 4D printed actuators with soft-robotic functions. Macromol. Rapid Commun. **39**(5), 1700710 (2018)
116. C.P. Ambulo et al., Four-dimensional printing of liquid crystal elastomers. ACS Appl. Mater. Interfaces **9**(42), 37332–37339 (2017)
117. S. Schuhladen et al., Iris-like Tunable aperture employing liquid-crystal elastomers. Adv. Mater. **26**(42), 7247–7251 (2014)
118. C. Yuan et al., 3D printed reversible shape changing soft actuators assisted by liquid crystal elastomers. Soft Matter **13**(33), 5558–5568 (2017)
119. Z.C. Jiang, Y.Y. Xiao, Y. Zhao, Shining light on liquid crystal polymer networks: Preparing, reconfiguring, and driving soft actuators. Advanced Optical Materials **7**(16), 1900262 (2019)
120. T. Ube, T. Ikeda, Photomobile polymer materials with complex 3D deformation, continuous motions, self-regulation, and enhanced processability. Adv. Opt. Mater. **7**(16), 1900380 (2019)
121. H. Zeng et al., Self-regulating iris based on light-actuated liquid crystal elastomer. Adv. Mater. **29**(30), 1701814 (2017)
122. S. Palagi et al., Structured light enables biomimetic swimming and versatile locomotion of photoresponsive soft microrobots. Nat. Mater. **15**(6), 647 (2016)
123. M. Rogóż et al., Light-driven soft robot mimics caterpillar locomotion in natural scale. Advanced Optical Materials **4**(11), 1689–1694 (2016)
124. J.M. Boothby, H. Kim, T.H. Ware, Shape changes in chemoresponsive liquid crystal elastomers. Sens. Actuators B **240**, 511–518 (2017)
125. C. Ahn, X. Liang, S. Cai, Inhomogeneous stretch induced patterning of molecular orientation in liquid crystal elastomers. Extr. Mech. nics Lett. **5**, 30–36 (2015)
126. X. Qian et al., Untethered recyclable tubular actuators with versatile locomotion for soft continuum robots. Adv. Mater. **30**(29), 1801103 (2018)
127. X. Lu et al., Liquid-crystalline dynamic networks doped with gold nanorods showing enhanced photocontrol of actuation. Adv. Mater. **30**(14), 1706597 (2018)
128. A. Kotikian et al., Untethered soft robotic matter with passive control of shape morphing and propulsion. Sci Robot **4**(33), eaax7044 (2019)
129. X. Lu et al., Tunable photocontrolled motions using stored strain energy in malleable azobenzene liquid crystalline polymer actuators. Adv. Mater. **29**(28), 1606467 (2017)
130. H. Tian et al., Polydopamine-coated main-chain liquid crystal elastomer as optically driven artificial muscle. ACS Appl. Mater. Interfaces **10**(9), 8307–8316 (2018)
131. M. Camacho-Lopez et al., Fast liquid-crystal elastomer swims into the dark. Nat. Mater. **3**(5), 307 (2004)
132. S. Iamsaard et al., Conversion of light into macroscopic helical motion. Nat. Chem. **6**(3), 229 (2014)
133. M. Wang, B.-P. Lin, H. Yang, A plant tendril mimic soft actuator with phototunable bending and chiral twisting motion modes. Nat. Commun. **7**, 13981 (2016)
134. O.M. Wani, H. Zeng, A. Priimagi, A light-driven artificial flytrap. Nat. Commun. **8**, 15546 (2017)
135. A. Richter et al., Micropumps operated by swelling and shrinking of temperature-sensitive hydrogels. Lab Chip **9**(4), 613–618 (2009)
136. E. Lee et al., Photothermally driven fast responding photo-actuators fabricated with comb-type hydrogels and magnetite nanoparticles. Sci. Rep. **5**, 15124 (2015)

137. R. Fuhrer et al., Crosslinking metal nanoparticles into the polymer backbone of hydrogels enables preparation of soft, magnetic field-driven actuators with muscle-like flexibility. Small **5**(3), 383–388 (2009)
138. S.-C. Chen et al., A novel pH-sensitive hydrogel composed of N, O-carboxymethyl chitosan and alginate cross-linked by genipin for protein drug delivery. J. Control. Release **96**(2), 285–300 (2004)
139. R. Zhang et al., A novel pH-and ionic-strength-sensitive carboxy methyl dextran hydrogel. Biomaterials **26**(22), 4677–4683 (2005)
140. J.W. Lee et al., Synthesis and characteristics of interpenetrating polymer network hydrogel composed of chitosan and poly (acrylic acid). J. Appl. Polym. Sci. **73**(1), 113–120 (1999)
141. E. Karadağ, D. SARAYDIN, Swelling of superabsorbent acrylamide/sodium acrylate hydrogels prepared using multifunctional crosslinkers. Turk. J. Chem. **26**(6), 863–876 (2002)
142. J.H. Kim et al., Rapid temperature/pH response of porous alginate-g-poly (N-isopropylacrylamide) hydrogels. Polymer **43**(26), 7549–7558 (2002)
143. B. Sun et al., Porous double network gels with high toughness, high stretchability and fast solvent-absorption. Soft Matter **13**(38), 6852–6857 (2017)
144. Q. Tang et al., Two-step synthesis of polyacrylamide/polyacrylate interpenetrating network hydrogels and its swelling/deswelling properties. J. Mater. Sci. **43**(17), 5884–5890 (2008)
145. H. Tang, B. Zhang, P. Wu, On the two-step phase transition behavior of the poly (N-isopropylacrylamide)(PNIPAM) brush: Different zones with different orders. Soft Matter **10**(37), 7278–7284 (2014)
146. S. Nesrinne, A. Djamel, Synthesis, characterization and rheological behavior of pH sensitive poly (acrylamide-co-acrylic acid) hydrogels. Arab. J. Chem. **10**(4), 539–547 (2017)
147. L. Toman et al., Amphiphilic conetworks. II. Novel two-step synthesis of poly [2-(dimethylamino) ethyl methacrylate]–polyisobutylene, poly (N-isopropylacrylamide)–polyisobutylene, and poly (N, N-dimethylacrylamide)–polyisobutylene hydrogels. J. Polym. Sci. A Polym. Chem. **44**(21), 6378–6384 (2006)
148. E.M. Ahmed, Hydrogel: Preparation, characterization, and applications: A review. J. Adv. Res. **6**(2), 105–121 (2015)
149. S.K. De et al., Equilibrium swelling and kinetics of pH-responsive hydrogels: Models, experiments, and simulations. J. Microelectromech. Syst. **11**(5), 544–555 (2002)
150. S. Deshmukh et al., Molecular modeling of thermo-responsive hydrogels: Observation of lower critical solution temperature. Soft Matter **5**(7), 1514–1521 (2009)
151. P.J. Flory, J. Rehner Jr., Statistical mechanics of cross-linked polymer networks I. rubberlike elasticity. J. Chem. Phys. **11**(11), 512–520 (1943)
152. R. Marcombe et al., A theory of constrained swelling of a pH-sensitive hydrogel. Soft Matter **6**(4), 784–793 (2010)
153. M. Dehghany et al., A thermodynamically-consistent large deformation theory coupling photochemical reaction and electrochemistry for light-responsive gels. J. Mech. Phys. Solids **116**, 239–266 (2018)
154. W. Hong, Z. Liu, Z. Suo, Inhomogeneous swelling of a gel in equilibrium with a solvent and mechanical load. Int. J. Solids Struct. **46**(17), 3282–3289 (2009)
155. G. Longhi et al., Molecular dynamics simulation of a model oligomer for poly (N-isopropylamide) in water. Chem. Phys. Lett. **386**(1–3), 123–127 (2004)
156. S. Maeda et al., Self-walking gel. Adv. Mater. **19**(21), 3480–3484 (2007)
157. D. Morales et al., Electro-actuated hydrogel walkers with dual responsive legs. Soft Matter **10**(9), 1337–1348 (2014)
158. E. Palleau et al., Reversible patterning and actuation of hydrogels by electrically assisted ionoprinting. Nat. Commun. **4**, 2257 (2013)
159. M. Liu et al., An anisotropic hydrogel with electrostatic repulsion between cofacially aligned nanosheets. Nature **517**(7532), 68 (2015)
160. T. Xie, Tunable polymer multi-shape memory effect. Nature **464**(7286), 267 (2010)
161. Q. Zhao, H.J. Qi, T. Xie, Recent progress in shape memory polymer: New behavior, enabling materials, and mechanistic understanding. Prog. Polym. Sci. **49**, 79–120 (2015)

162. T. Chung, A. Romo-Uribe, P.T. Mather, Two-way reversible shape memory in a semicrystalline network. Macromolecules **41**(1), 184–192 (2008)
163. M. Behl et al., Reversible bidirectional shape-memory polymers. Adv. Mater. **25**(32), 4466–4469 (2013)
164. J. Zotzmann et al., Reversible triple-shape effect of polymer networks containing polypentadecalactone-and poly (ε-caprolactone)-segments. Adv. Mater. **22**(31), 3424–3429 (2010)
165. J. Zhou et al., Shapeshifting: Reversible shape memory in semicrystalline elastomers. Macromolecules **47**(5), 1768–1776 (2014)
166. Z. Li et al., Polydopamine coated shape memory polymer: Enabling light triggered shape recovery, light controlled shape reprogramming and surface functionalization. Chem. Sci. **7**(7), 4741–4747 (2016)
167. J.R. Kumpfer, S.J. Rowan, Thermo-, photo-, and chemo-responsive shape-memory properties from photo-cross-linked metallo-supramolecular polymers. J. Am. Chem. Soc. **133**(32), 12866–12874 (2011)
168. J. Leng et al., Electroactivate shape-memory polymer filled with nanocarbon particles and short carbon fibers. Appl. Phys. Lett. **91**(14), 144105 (2007)
169. J. Leng et al., Significantly reducing electrical resistivity by forming conductive Ni chains in a polyurethane shape-memory polymer/carbon-black composite. Appl. Phys. Lett. **92**(20), 204101 (2008)
170. X. Luo, P.T. Mather, Conductive shape memory nanocomposites for high speed electrical actuation. Soft Matter **6**(10), 2146–2149 (2010)
171. Y. Liu et al., Review of electro-active shape-memory polymer composite. Compos. Sci. Technol. **69**(13), 2064–2068 (2009)
172. R. Mohr et al., Initiation of shape-memory effect by inductive heating of magnetic nanoparticles in thermoplastic polymers. Proc. Natl. Acad. Sci. **103**(10), 3540–3545 (2006)
173. H. Du, J. Zhang, Solvent induced shape recovery of shape memory polymer based on chemically cross-linked poly (vinyl alcohol). Soft Matter **6**(14), 3370–3376 (2010)
174. D. Quitmann et al., Solvent-sensitive reversible stress-response of shape memory natural rubber. ACS Appl. Mater. Interfaces **5**(9), 3504–3507 (2013)
175. T. Xie, I.A. Rousseau, Facile tailoring of thermal transition temperatures of epoxy shape memory polymers. Polymer **50**(8), 1852–1856 (2009)
176. N. Zheng et al., High strain epoxy shape memory polymer. Polym. Chem. **6**(16), 3046–3053 (2015)
177. W. Huang et al., Water-driven programmable polyurethane shape memory polymer: Demonstration and mechanism. Appl. Phys. Lett. **86**(11), 114105 (2005)
178. W. Huang et al., Thermo-moisture responsive polyurethane shape-memory polymer and composites: A review. J. Mater. Chem. **20**(17), 3367–3381 (2010)
179. A. Lendlein et al., Shape-memory polymer networks from oligo (ϵ-caprolactone) dimethacrylates. J. Polym. Sci. A Polym. Chem. **43**(7), 1369–1381 (2005)
180. L. Xiao et al., Novel triple-shape PCU/PPDO interpenetrating polymer networks constructed by self-complementary quadruple hydrogen bonding and covalent bonding. Polym. Chem. **5**(7), 2231–2241 (2014)
181. M. Zarek et al., 3D printing of shape memory polymers for flexible electronic devices. Adv. Mater. **28**(22), 4449–4454 (2016)
182. Q. Ge et al., Multimaterial 4D printing with tailorable shape memory polymers. Sci. Rep. **6**, 31110 (2016)
183. Y. Mao et al., Sequential self-folding structures by 3D printed digital shape memory polymers. Sci. Rep. **5**, 13616 (2015)
184. H.J. Qi, M.L. Dunn, *Thermomechanical behavior and modeling approaches, in shape-memory polymers and multifunctional composites* (CRC Press, 2010), pp. 75–100
185. T.D. Nguyen et al., A thermoviscoelastic model for amorphous shape memory polymers: Incorporating structural and stress relaxation. J. Mech. Phys. Solids **56**(9), 2792–2814 (2008)

186. T.D. Nguyen, Modeling shape-memory behavior of polymers. Polym. Rev. **53**(1), 130–152 (2013)
187. H.J. Qi et al., Finite deformation thermo-mechanical behavior of thermally induced shape memory polymers. J. Mech. Phys. Solids **56**(5), 1730–1751 (2008)
188. Y. Liu et al., Thermomechanics of shape memory polymers: Uniaxial experiments and constitutive modeling. Int. J. Plast. **22**(2), 279–313 (2006)
189. J. Diani et al., Predicting thermal shape memory of crosslinked polymer networks from linear viscoelasticity. Int. J. Solids Struct. **49**(5), 793–799 (2012)
190. O. Balogun, C. Mo, Shape memory polymers: Three-dimensional isotropic modeling. Smart Mater. Struct. **23**(4), 045008 (2014)
191. X. Lan et al., Fiber reinforced shape-memory polymer composite and its application in a deployable hinge. Smart Mater. Struct. **18**(2), 024002 (2009)
192. B. Jin et al., Programming a crystalline shape memory polymer network with thermo-and photo-reversible bonds toward a single-component soft robot. Sci. Adv. **4**(1), eaao3865 (2018)
193. K. Takashima, J. Rossiter, T. Mukai, McKibben artificial muscle using shape-memory polymer. Sens. Actuators A Phys. **164**(1–2), 116–124 (2010)
194. C.S. Haines et al., Artificial muscles from fishing line and sewing thread. Science **343**(6173), 868–872 (2014)
195. S.M. Mirvakili, I.W. Hunter, Artificial muscles: Mechanisms, applications, and challenges. Adv. Mater. **30**(6), 1704407 (2018)
196. S.M. Mirvakili et al., Simple and strong: twisted silver painted nylon artificial muscle actuated by Joule heating, in *Electroactive Polymer Actuators and Devices (EAPAD), 2014*, (International Society for Optics and Photonics, 2014)
197. J. Foroughi et al., Torsional carbon nanotube artificial muscles. Science **334**(6055), 494–497 (2011)
198. M.D. Lima et al., Electrically, chemically, and photonically powered torsional and tensile actuation of hybrid carbon nanotube yarn muscles. Science **338**(6109), 928–932 (2012)
199. S.M. Mirvakili, I.W. Hunter, *Bending artificial muscle from nylon filaments. in Electroactive Polymer Actuators and Devices (EAPAD), 2016* (International Society for Optics and Photonics, 2016)
200. S.M. Mirvakili et al., Niobium nanowire yarns and their application as artificial muscles. Adv. Funct. Mater. **23**(35), 4311–4316 (2013)
201. S.M. Mirvakili, I.W. Hunter, Multidirectional artificial muscles from nylon. Adv. Mater. **29**(4), 1604734 (2017)
202. P. Chen et al., Hierarchically arranged helical fibre actuators driven by solvents and vapours. Nat. Nanotechnol. **10**(12), 1077 (2015)
203. M. Hiraoka et al., Power-efficient low-temperature woven coiled fibre actuator for wearable applications. Sci. Rep. **6**, 36358 (2016)
204. S.Y. Yang et al., High performance twisted and coiled soft actuator with spandex fiber for artificial muscles. Smart Mater. Struct. **26**(10), 105025 (2017)
205. J. Yuan et al., Shape memory nanocomposite fibers for untethered high-energy microengines. Science **365**(6449), 155–158 (2019)
206. J. Mu et al., Sheath-run artificial muscles. Science **365**(6449), 150–155 (2019)
207. S. Sharafi, G. Li, A multiscale approach for modeling actuation response of polymeric artificial muscles. Soft Matter **11**(19), 3833–3843 (2015)
208. Q. Yang, G. Li, A top-down multi-scale modeling for actuation response of polymeric artificial muscles. J. Mech. Phys. Solids **92**, 237–259 (2016)
209. A. Abbas, J. Zhao, A physics based model for twisted and coiled actuator, in *2017 IEEE International Conference on Robotics and Automation (ICRA)*, (IEEE, 2017)
210. M.C. Yip, G. Niemeyer, High-performance robotic muscles from conductive nylon sewing thread, in *2015 IEEE International Conference on Robotics and Automation (ICRA)*, (IEEE, 2015)

211. K.H. Cho et al., A robotic finger driven by twisted and coiled polymer actuator, in *Electroactive Polymer Actuators and Devices (EAPAD) 2016*, (International Society for Optics and Photonics, 2016)
212. S.H. Kim et al., Harvesting temperature fluctuations as electrical energy using torsional and tensile polymer muscles. Energ. Environ. Sci. **8**(11), 3336–3344 (2015)
213. R. Wang et al., Torsional refrigeration by twisted, coiled, and supercoiled fibers. Science **366**(6462), 216–221 (2019)
214. M. Kanik et al., Strain-programmable fiber-based artificial muscle. Science **365**(6449), 145–150 (2019)
215. X. Zhao, Y. Kim, Soft microbots programmed by nanomagnets. Nature **575**(7781), 58–59 (2019)
216. R.M. Erb et al., Actuating soft matter with magnetic torque. Adv. Funct. Mater. **26**(22), 3859–3880 (2016)
217. J.M. Ginder et al., Magnetorheological elastomers: properties and applications, in *Smart Structures and Materials 1999: Smart Materials Technologies*, (International Society for Optics and Photonics, 1999)
218. J.-P. Pelteret, P. Steinmann, *Magneto-Active Polymers: Fabrication, characterisation, modelling and simulation at the micro-and macro-scale* (Walter de Gruyter GmbH & Co KG, 2019)
219. G. Stepanov et al., Motion of ferroparticles inside the polymeric matrix in magnetoactive elastomers. J. Phys. Condens. Matter **20**(20), 204121 (2008)
220. S. Ahmed et al., Multi-field responsive origami structures: Preliminary modeling and experiments, in *ASME 2013 international design engineering technical conferences and computers and information in engineering conference*, (American Society of Mechanical Engineers Digital Collection, 2013)
221. W. Hu et al., Small-scale soft-bodied robot with multimodal locomotion. Nature **554**(7690), 81–85 (2018)
222. Y. Kim et al., Printing ferromagnetic domains for untethered fast-transforming soft materials. Nature **558**(7709), 274–279 (2018)
223. Z. Ren et al., Multi-functional soft-bodied jellyfish-like swimming. Nat. Commun. **10**(1), 1–12 (2019)
224. D. Kokkinis, M. Schaffner, A.R. Studart, Multimaterial magnetically assisted 3D printing of composite materials. Nat. Commun. **6**(1), 1–10 (2015)
225. W.F. Brown, *Magnetoelastic interactions*, vol 9 (Springer, 1966)
226. L. Liu, An energy formulation of continuum magneto-electro elasticity with applications. J. Mech. Phys. Solids **63**, 451–480 (2014)
227. R. Zhao et al., Mechanics of hard-magnetic soft materials. J. Mech. Phys. Solids **124**, 244–263 (2019)
228. P. Garstecki et al., Propulsion of flexible polymer structures in a rotating magnetic field. J. Phys. Condens. Matter **21**(20), 204110 (2009)
229. Q. Ze et al., Magnetic shape memory polymers with integrated multifunctional shape manipulation. Adv. Mater. **32**(4), 1906657 (2020)
230. S. Martel, Beyond imaging: Macro-and microscale medical robots actuated by clinical MRI scanners. Sci Robot **2**(3), eaam8119 (2017)
231. Y. Kim et al., Ferromagnetic soft continuum robots. Sci Robot **4**(33), eaax7329 (2019)
232. A.D. Marchese, C.D. Onal, D. Rus, Autonomous soft robotic fish capable of escape maneuvers using fluidic elastomer actuators. Soft Robotics **1**(1), 75–87 (2014)
233. J. Najem et al., Biomimetic jellyfish-inspired underwater vehicle actuated by ionic polymer metal composite actuators. Smart Mater. Struct. **21**(9), 094026 (2012)
234. T. Chen et al., Harnessing bistability for directional propulsion of soft, untethered robots. Proc. Natl. Acad. Sci. **115**(22), 5698–5702 (2018)
235. D. Han et al., Soft robotic manipulation and locomotion with a 3d printed electroactive hydrogel. ACS Appl. Mater. Interfaces **10**(21), 17512–17518 (2018)
236. Q. He, Y. Zheng, Z. Wang, et al. Anomalous Inflation of a Nematic Balloon[J]. J. Mech. Phys. Solids, 104013 (2020)

Chapter 5
Bioinspired Sensors and Actuators Based on Stimuli-Responsive Hydrogels for Underwater Soft Robotics

Chiao-Yueh Lo, Yusen Zhao, Yanfei Ma, Shuwang Wu, Yousif Alsaid, Matthew M. Peet, Rebecca E. Fisher, Hamidreza Marvi, Daniel M. Aukes, Spring Berman, and Ximin He

5.1 Introduction

The capability of automatically detecting environmental and bodily conditions, then rapidly adapting appearance and motion to respond, is a characteristic feature of living organisms. Realizing these abilities with less complex systems has been a central goal for robotics and would be exceptionally advantageous for underwater soft robots. Traditionally, mimicking these capabilities with man-made devices requires integration of semiconductor-based sensors and actuators coupled with control algorithms, oftentimes also requiring additional cameras for monitoring motion. Delamination at the interface can arise from a modulus mismatch between the sensor and actuator. Sensors must be flexible and soft to avoid interference with

C.-Y. Lo · Y. Zhao · Y. Ma · S. Wu · Y. Alsaid · X. He (✉)
Department of Material Science and Engineering, University of California, Los Angeles, Los Angeles, CA, USA
e-mail: ximinhe@ucla.edu

M. M. Peet
Department of Mechanical and Aerospace Engineering, Arizona State University, Tempe, AZ, USA

R. E. Fisher
School of Life Science, Arizona State University, Tempe, AZ, USA

Department of Basic Medical Sciences, University of Arizona College of Medicine—Phoenix, Phoenix, AZ, USA

H. Marvi · S. Berman
School for Engineering of Matter, Transport and Energy, Arizona State University, Tempe, AZ, USA

D. M. Aukes
The Polytechnic School, Ira A. Fulton Schools of Engineering, Arizona State University, Mesa, AZ, USA

D. A. Paley, N. M. Wereley (eds.), *Bioinspired Sensing, Actuation, and Control in Underwater Soft Robotic Systems*, https://doi.org/10.1007/978-3-030-50476-2_5

the material properties for precise strain detection, whereas actuators are required to be stiff to exert sufficient force [1]. It also presents engineering challenges when considering scale-matching of the actuator and the sensors, the number of sensors required, and the limited power supply to the sensors [2].

One of the solutions to the aforementioned problems is designing a soft material with high actuation force that can also simultaneously self-monitor its local environmental condition and self-motion. Locomotive organs, such as animal muscles and entirely soft structures (e.g., octopus arms and elephant trunks), perform complex tasks through serial movements via local sensing and interactive actuation from neuromuscular systems. Inspired by this, self-monitoring soft actuators that couple actuation and sensing in one bulk material have been developed [3, 4].

For actuation in soft robotics, stimuli-responsive polymers such as liquid crystalline elastomers (LCEs) [5], dielectric elastomer actuators (DEAs) [6], shape memory polymers (SMPs) [7], and stimuli-responsive hydrogels [8] are capable of realizing local actuation in response to environmental cues. This local and intrinsic response capability not only features non-tethered application potential but also resembles the muscle-reflex mechanism in animals that functions as a defense against danger by providing rapid actuation in response to simple and intense stimuli.

Typically, conductive polymers exhibiting strain sensing based on piezoresistance is a more accurate mechanism, but they lack the essential elasticity for soft robotics due to their rigid chemical structure. One strategy for creating conductive polymers with high elasticity and flexibility is to embed interconnected conductive polymers in soft, flexible matrices or substrates. For instance, the integration of conductive polymers, e.g., polyaniline (PANI) and polypyrrole (PPy), with responsive hydrogel matrices, such as polyacrylic acid (PAA) [9] and polyacrylamide (PAAm) [10], has made piezo-electrical strain sensors ultra-stretchable with wide detection ranges. Brandy et al. reported a PPy-coated polyurethane foam with naphthalene di-sulfonic acid (NDSA) as the dopant capable of pressure sensing [11]. PANI strengthened with polydimethylsiloxane (PDMS) [12] and PPy-deposited PDMS device coupled with Au microarray [13] have also been developed to realize piezoresistivity sensors.

Traditionally, robotics locomotion patterns are based on uniform material expansion or shrinkage, which are predefined and lack re-programmability. Soft robots with multiple degrees of freedom (DoF) usually consist of multivalve systems to control individual actuation units. Omnidirectional actuators with high DoF can be achieved by utilizing smart polymers, such as stimuli-responsive hydrogels. Smart hydrogels are a type of active polymer that can undergo significant volumetric or other property change in response to a variety of environmental stimuli, such as temperature [14, 15], pH [16, 17], electric field [18], and specific chemicals [19]. This multifunctionality originates from the hydrogel's ability to absorb and desorb water up to ten times its own mass in response to environmental changes. Their flexibility, biocompatibility, broad range of stimulus, high volume-change ratio, and mechanical tunability have promoted their applications in multidisciplinary fields ranging from tissue engineering [20] to soft electronics [21–23]. Smart hydrogels' unique and attractive properties as actuators also led to multiple applications in soft robotics at various length scales, ranging from micro-robots for drug delivery [24, 25] to macro actuators for object manipulation [26, 27]. It has been reported that muscle-like, ion-induced multilayered actuators, composed of cross-linked PAA and PAAm hydrogels, have been demonstrated for use in robot arms for their abrupt swelling behavior triggered by ionic strength or pH fluctuations [28].

Besides ion-induced osmotic actuation, electric fields have also been used to actuate robots made of PAA/PAAm hydrogel blended with PPy and carbon-black [29]. A thermally responsive hydrogel, poly(*N*-isopropylacrylamide) (PNIPAAm), has been broadly utilized as a smart hydrogel. PNIPAAm is known for its characteristic swelling-deswelling thermal behavior in aqueous media with a threshold temperature of around 32 °C. The actuation mechanism is based on the temperature-induced polymer structure change, which induces a hydrophilic-to-hydrophobic, coil-to-globule phase transition, resulting in drastic water desorption and abrupt volume collapse at elevated temperatures [30]. Thus, when temperature increases above PNIPAAm's lower critical solution temperature (LCST), a temperature-driven actuation can be induced. Because of its temperature-driven volumetric change, PNIPAAm can be given controllable actuation capabilities through local heating with embedded flexible Joule heaters [31] or remotely with light illumination via the photothermal effect [32, 33]. Utilizing this photothermal effect, remotely controlled and untethered robots have been made by incorporating photo-absorbers into thermo-responsive materials, achieving photoactuation. Remote photoactuation is advantageous for enabling unhindered robotic movement and higher DoF.

In practice, hydrogel actuators demonstrate ultrahigh tensile strains and good compatibility with underwater environments; however they usually exhibit slow response (minutes to hours timescale) and low elastic modulus fracture strength [34, 35], determined by the intrinsic polymer structure.

By utilizing all the strengths of hydrogels while overcoming the limitations, it is possible to develop more versatile smart materials possessing living matters' local adaptability and mechanical robustness, resulting in unprecedented and previously unachievable functions for next-generation underwater soft robotics.

5.2 Responsive Hydrogel as Camouflage and Visual Sensors in Aqueous Environment

Living organisms ubiquitously display colors that adapt to environmental changes, a process that relies on soft layers of cells or proteins. For example, iridophores produce iridescent colors generated from light reflecting off embedded platelet stacks; the color can be tuned by the thickness change of these platelets [36, 37].

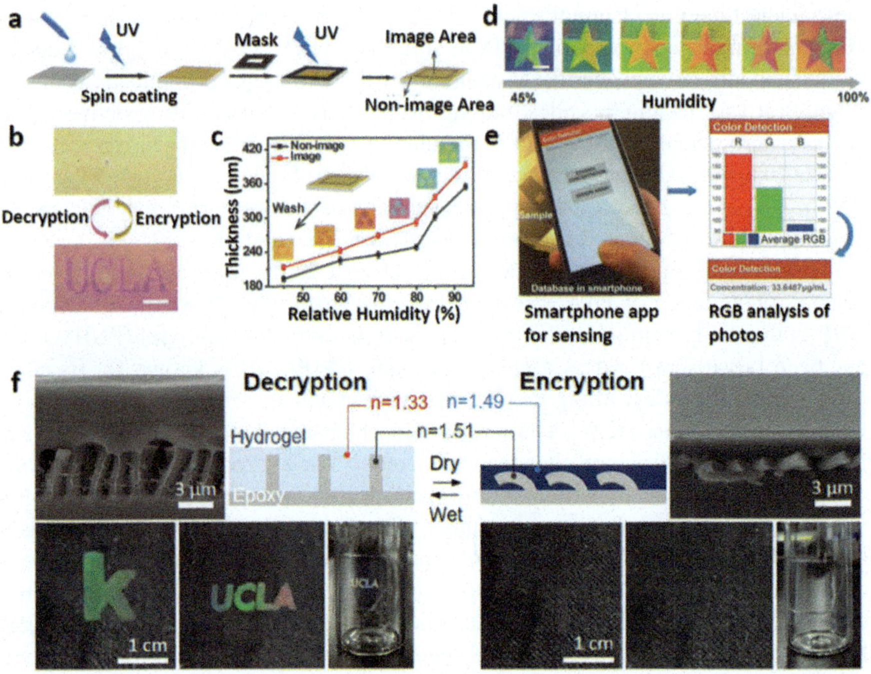

Fig. 5.1 (**a**) Schematic of the patterning process of the hydrogel interferometer. (**b**) An information encryption device: the decryption is triggered instantly by moisture, whereas the information is hidden immediately after moisture removal. (**c**) Thickness and color dependency on relative humidity. (**d**) A humidity indicator based on a multicolor image pattern. (**e**) Smartphone-based sensing system for both quantitative estimation and multi-analyte identification. (**f**) Quantitative sensing, mechanism, and SEM images of ethanol vapor using hydrocipher. The colored patterns at hydrated state disappeared after drying (Adapted (**a**–**e**) from [38], copyright by Advanced Materials, and (**f**) from [40], copyright by Advanced Optical Materials)

The hydrogel interferometer [38–40], inspired by the soft layers in color-changing animals, mimics those structures via rapid hydrogel thickness change in response to external stimuli, resulting in instant interference color switching (Fig. 5.1a–d). This active color switching mechanism provides a visual and quantifiable means of observing a variety of environmental metrics. Furthermore, the resulting structural color is usually glittering and iridescent, never fading as long as the structures are maintained. The versatility of such an adaptive coloration material is significant, and it has been demonstrated with a broad range of applications: as a volatile-vapor sensor with highly accurate quantitative detection, a colorimetric sensor array for multi-analyte recognition, a breath-controlled information encryptor, and a colorimetric humidity indicator. The hydrogel interferometer can also be fabricated into easy-to-use and portable sensing devices, such as a smartphone-based colorimetric analyzer (Fig. 5.1e) [38]. A universal sensing platform based on the hydrogel interferometer and a local concentrating effect in the hydrogel has also been reported with ultrahigh, femtomolar-level sensitivity in detecting (bio)chemical molecules [39].

Structural colors can also be switched on and off actively and simply by adjusting the periodicity or refractive index contrast between the active layer and the surroundings. Specifically, a hydrocipher has been reported [41] based on a water-responsive photonic grating consisting of a micropillar array and a hydrogel overcoating, of which diffraction colors can be reversibly switched on by hydration and off by dehydration. When the hydrogel overcoating is swollen by water, the difference in refractive index between hydrogel and micropillar is significant, so the photonic structure shows a strong iridescence from angle-dependent diffraction (Fig. 5.1f). Upon hydrogel dehydration, the refractive index of the hydrogel overcoating changes and becomes similar to the micropillar, which dramatically reduces the diffraction intensity and renders the surface highly transparent. The dehydrated-to-hydrated state transition occurs within 1 s, showing the capability for fast decryption. This creative combination of stimulus-responsive hydrogel with 2D grating structure offers a new and simple strategy for realizing reversible, durable, and fast-response cryptography with potentially broad impact on the anti-counterfeiting technology market.

5.3 Muscle-Inspired Strong Contractile Hydrogel with High Energy Density for Underwater Actuators

Hydrogel-based actuators are advantageous when used in biomimetic robotics, especially in underwater and in vivo applications. Hydrogels have been utilized to mimic living organism parts and perform minimally invasive surgeries, due to their soft, wet, and biocompatible nature as well as their chemical functionality, shape-morphing ability, and adaptability [42]. However, existing hydrogel actuators, mostly osmotic-driven, are intrinsically low-speed and/or low-force. Therefore, researchers have attempted to circumvent the osmotic-actuation mechanism for more effective mechanical properties enhancement. For example, it was shown that implanting hydraulic actuations facilitated water diffusion into/out of the hydrogel body and thus provided soft actuators and robots with high-speed, high-force exertion for underwater applications [43].

Intrinsically, hydrogels are not capable of performing both rapid deformation and generating large force due to the limitation from their osmotic-pressure-driven framework [44–46]. Rapid response speed requires highly porous microstructure for improved solvent diffusion, which usually leads to decreased elastic modulus and force exertion. As a result, response speed and exerting force are negatively coupled rather than acting synergistically. Since mechanical power output is defined as the product of exerted force multiplied by the deformation speed, the mechanical power output of traditional hydrogels is limited by this negative correlation. In contrast, natural actuators such as muscle fibers from frog legs deliver over 100 kPa of force at extremely high actuation speed of 200%/s (Table 5.1) [47, 48]. Muscles utilize chemo-mechanical energy conversion, which stores and releases mechanical energy in and from certain compound molecules via chemical reactions (ATP $\rightarrow$ ADP + Pi) [49] (Fig. 5.2a). The energy conversion mechanism of natural muscles does not rely on solvent diffusion and thus endows them with high stretchability and toughness at high water content, high energy density, and high output power. However, in practice, it is difficult to replicate such a complicated, molecular-level mechanism in soft robotics. Therefore, even after many investigations in biomimetics and tough hydrogel materials, no hydrogel has been realized that is comparable in performance to natural muscles in these three properties.

Muscle-powered jumping for many animals, such as frogs, kangaroos, and rabbits, presents incredibly high speed and force. The catapult mechanism of muscles first converts chemical energy to mechanical energy and prestores potential energy in elastic structures. Then, the energy is released to produce kinetic energy for joint movements (Fig. 5.2a). Inspired by this energy conversion mechanism in the elastic actuation of muscle, a new universal design concept of material was proposed based on elastic energy storing and releasing (EESR) in an elastic hydrogel polymer network [50]. The detailed mechanism of this elastic-driven hydrogel (EDG) is shown in Fig. 5.2b. (1) Elastic potential energy is stored in the hydrogel by mechanical stress, followed by locking the hydrogel in the stretched state through forming new chemical bonds under external stimulus (e.g., pH change or illumination). (2) To release the pre-stored energy, these newly formed chemical bonds are broken via a different external stimulus, resulting in rapid and powerful contraction of the material on demand. The resulting material achieved both high contraction force (40 kPa) and high deformation speed (25%/min) simultaneously, and exhibited high energy density (15.3 kJ/m^3). Furthermore, the material outperformed biological muscles in four mechanical aspects (elastic modulus, 3.7 MPa; strain, 600%; toughness, 12.7 MJ/m^3; energy density, 15.3 kJ/m^3), which were 10, 15, 16, and 2 times higher than that of mammalian skeletal muscle, respectively (~0.35 MPa, ~40%, 0.8 MJ/m^3, and 8 kJ/m^3), while containing high water content (~80%) and energy conversion efficiency (~49%) comparable to muscles [47].

Compared with the osmotic-pressure-driven hydrogels in which force and deformation speed are negatively correlated, the EDGs based on EESR delivered both stronger and faster contraction with better mechanical properties. The simultaneous increase of both force and speed led to a remarkably high energy density up to 15.3 kJ/m^3, which outperformed all commonly studied osmotic-driven hydrogels and even biological muscles (Table 5.1). The EDG material could also be fabricated into strips to mimic not only the shape but also the function of mammalian skeletal muscle (Fig. 5.2c). It was shown that an EDG-based hydrogel actuator is capable of generating a strong actuation force to lift a weight. (Fig. 5.2d).

In summary, the responsive performance and energy density of hydrogels can be greatly improved by the innovative and universally applicable EESR mechanism at a molecular level. The simultaneous improvement in three mechanical properties, namely contraction force, deformation speed, and elastic modulus, has broken the

Table 5.1 Actuation performance comparison of elastic-driven hydrogels (EDG), osmotic-driven hydrogels, and mammalian skeletal muscle

		Osmotic-driven hydrogels (PAAc)		
Actuation performances	Elastic-driven hydrogels (EDG)	High modulus	Low modulus	Mammalian skeletal muscle
Stress (kPa)	3700	40	28	100–350
Strain (%)	600	107	2300	20–40
Elastic modulus (MPa)	0.7	0.071	0.014	10–60
Toughness (MJ/m^3)	12.7	0.4	0.0275	0.1–0.8
Contraction force (kPa)	~40	1.36	0.4	>100
Deformation speed (%/min)	25	1	3	200%/s
Energy density (kJ/m^3)	15.3	0.013	0.0053	8
Average power (W/m^3)	51 (25.5 mW/kg)	0.0072	0.0029	10^5 (50–100 W/kg)
Efficiency (%)	49	—	—	40
Water content	72–80%	85%	90%	75%–78%

Data retrieved from [47, 48, 51–53]

conflicting relationship between them for conventional osmotic-driven hydrogels. This innovative actuating material design offered the outstanding combination of softness, high water content, and programmability, while also demonstrating high contraction force with no loss in deformation speed. This has effectively solved the long-term force-speed dilemma in hydrogel actuators, further enabling the use of hydrogels as smart actuators for underwater soft robotics.

5.4 All-in-One Soft Robots with Self-Sensing Actuation and Self-Regulating Capability

5.4.1 Multifunctional Hydrogels as Self-Sensing Robots

To avoid the shortcomings of heterogeneous soft-robot designs, a class of all-in-one soft robotic materials have been reported, which possess both life-like tactile sensations and actuating capabilities. These soft materials couple independent actuation with strain-/temperature-sensing capabilities from two separate mechanisms in one soft material, specifically photothermally stimulated actuation and electrical-resistance-based sensing [54]. A unique light-responsive actuator was designed to simultaneously work as a motion sensor via its own resistance change. Another

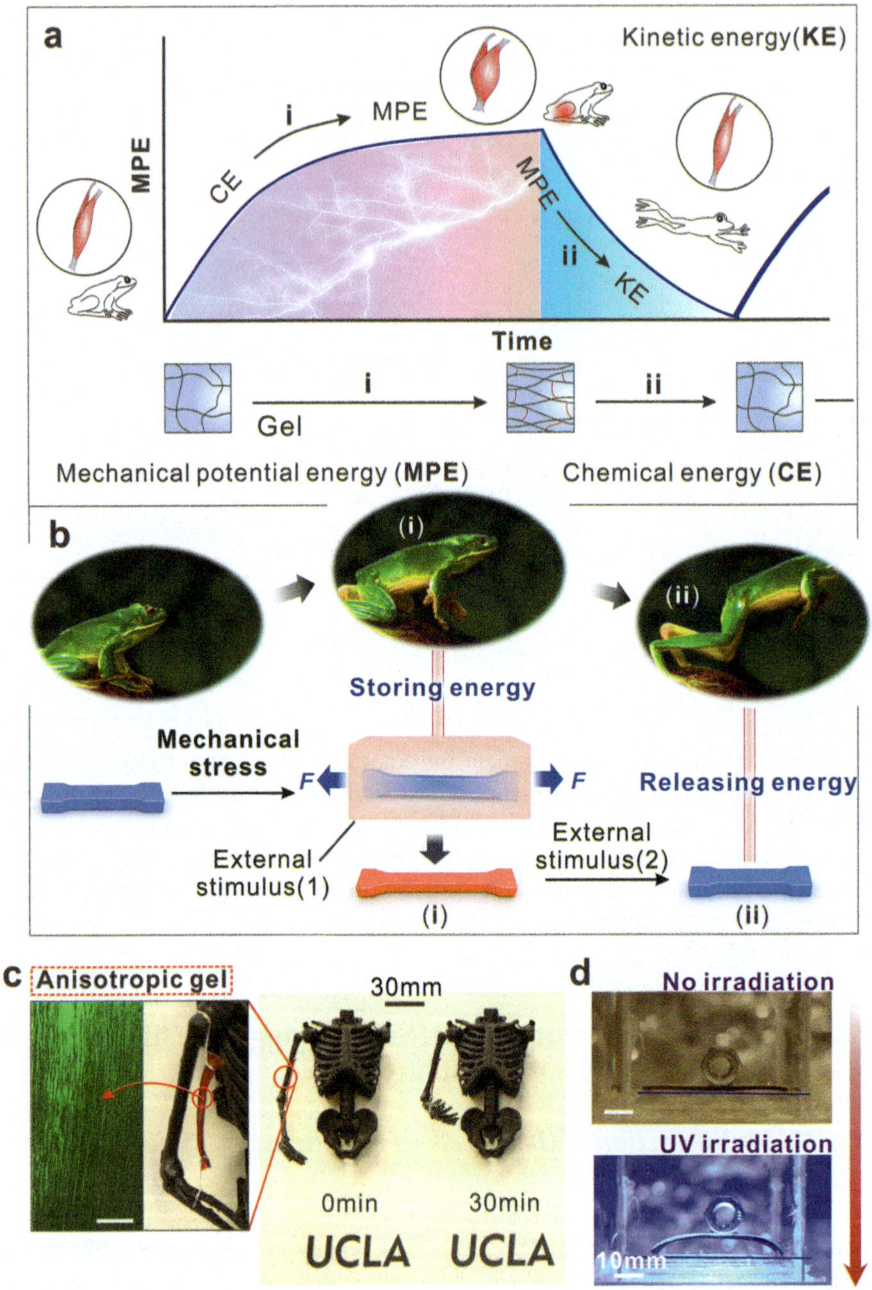

Fig. 5.2 (**a**) Leaping of a frog through elastic potential energy storage and subsequent releasing from chemical energy. The muscle is initially held at a stretched state, where the chemical energy is converted and stored as elastic potential energy (red area). Then, the muscle is instantaneously released (blue area), and the potential energy stored is converted to kinetic energy. (**b**) Schematics of the design strategy of on-demand super-contractive hydrogel. (**c**) The EDG strip was attached on a 3D-printed arm to realize lifting. (**d**) An EDG-based bimorph actuator (right) could lift a weight (4 g) through bending showing high actuation force (Figure adapted from unpublished work [50])

characteristic feature of this self-sensing actuator is the ability to actuate omnidirectionally with local deformation due to its homogeneous, isotropic nature, in contrast to conventional actuators with restricted directionality, anisotropic structure, and uniform deformation. It has been reported previously that these isotropic, responsive materials can deform in dynamic equilibrium and can exhibit controllable and tunable motion based on negative feedback loops [55, 56]. Additionally, to avoid potential delamination and filler aggregation, this all-in-one self-sensing actuator is fabricated via constructing a homogeneous double-network structure to replace the conventional design: elastomers embedded with immiscible fillers. In detail, this homogeneous double-network is composed of a thermal-responsive, soft, and elastic hydrogel PNIPAAm as the actuator component and a black conductive polymer PPy as the mechanical strengthener, light absorber, and piezo-resistivity-based sensing component.

The sensing mechanism is based on the piezo-resistive effect; the resistivity of the hydrogel is significantly reduced when the PPy network is shrunken. When the gel is heated up gradually to its LCST, the resistance drops abruptly as the cross-sectional area shrinks, in contrast to the expected geometric effect where an increase in resistance is expected. This unusual behavior can be attributed to the conductive PPy network density increase upon shrinkage (Fig. 5.3a). According to percolation conduction theory, above some threshold density, the percolation conduction is turned on by the formation of a fully continuous PPy network [57]. The conductive PPy transitions from a low density, disconnected network to high density, connected network that facilitates the overlap of delocalized π orbitals; this results in a higher probability for electrons to migrate via intra-chain migration and inter-chain hopping, leading to a significant increase in conductivity. This percolation conductivity subsequently dominates over the geometric effect, resulting in a decreased apparent resistance at larger shrinkage levels. When the hydrogel is uniaxially stretched, the resistance also decreases because the elongation along the tensile direction causes Poisson contraction of the cross-sectional area, thus increasing the PPy density along the lateral direction.

In addition, the PPy network serves not only as a conductive filler but also as a mechanical strengthener; the fracture stress and strain of the composite hydrogel after introducing the PPy network improved from 13 kPa to 30 kPa and from 115% to 210%, respectively. Since pure PPy itself is rigid and lacks elasticity, this enhanced elastic behavior can be attributed to the interpenetration of PPy and PNIPAAm networks.

The PNIPAAm gel was designed to be nanostructured by cross-linking the PNIPAAm network via nanogels instead of direct chemical cross-linking [58]. The resulting double-network hydrogel exhibits both larger and more rapid volumetric change comparing to conventional material. The equilibrium volumetric change when heated up from 25 to 50 °C increased from $\Delta V/V = 57\%$ to $\Delta V/V = 70\%$. Furthermore, more abrupt phase transition when heated to near LCST (30–35 °C) was observed at a remarkable equilibrium volumetric change of 13%/°C, compared to a chemically cross-linked sample at only 6%/°C. Dynamically, it was also observed that initial volumetric shrinkage rate within the first 30 s at 50 °C are

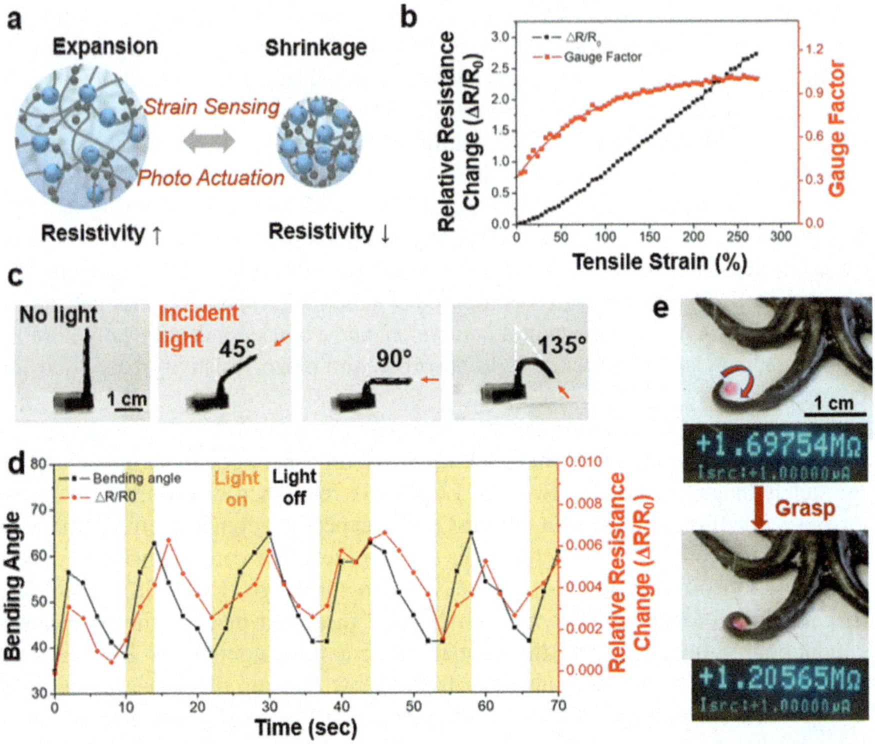

Fig. 5.3 (**a**) Schematic of the self-sensing actuator mechanism: Resistivity change arose from the density change of the conducting polymer network when exposed to external stress or thermal stimulated volumetric change. (**b**) The self-sensing hydrogel performed ultrahigh strain sensing range (up to 280%), with decent sensitivity at high strain (G/F = 1). (**c**) Side view of original pillar standing upright on the bottom, light-induced pillar bending at 45°, 90 °, and 135 °. (**d**) The bending angle pattern of an underwater pillar robot coincides with the resistance change pattern. (**e**) Self-motion-sensing grasping by NIR manipulation (Figure adapted from unpublished work [54])

significantly faster (23.6%) than conventional chemically cross-linked hydrogels (4.0%).

All-in-one self-sensing capability has been demonstrated by remotely controlling the material with near-infrared (NIR) laser illumination and achieved self-sensing by measuring resistance change in real-time, under various locomotion including tracking, bending, weight-lifting, grasping, and object moving. The ultra-stretchability sensor capability was realized, with a gauge factor (GF, $\Delta R/R_0/\varepsilon$) of 1.0 at >200% tensile strain (Fig. 5.3b), which is of typical range among high-strain soft sensors [59].

Here, rapid and omnidirectional light-tracking performance has been demonstrated by a cylindrical pillar manipulated by NIR light source. This fast and omnidirectional light-tracking behavior provides the versatility of advanced design for a photo-responsive actuator. The pillar also showed precise control and a wide operation window of incident angles from 45° to 135° (Fig. 5.3c) while the bending speed reached as high as 34.6°/s at 90° incident angle. The diffusion time scale was measured to be in the range of seconds, which was an order of magnitude faster than conventional PNIPAAm gel [60]. The corresponding temperature gradients on the pillar were also simulated and presented. The symmetric cylindrical hydrogel pillar design allows omnidirectional light-tracking capability covering the entire sphere.

It has been further demonstrated that the self-sensing actuation is stable and completely reversible if provided with a water reservoir (Fig. 5.3d). The moving pattern recorded in terms of bending angle coincided with the resistance change pattern, demonstrating a feedback delay below 1 s, comparable to typical hydrogel strain sensors. The self-sensing hydrogel can be fabricated into octopus arm-like shapes, which respond to NIR light and can curl, grasp, and move an object with real-time motion sensing via tracking the resistance change between two attached electrodes (Fig. 5.3e). The capability of self-sensing and actuation provides the potential for closed-loop control, which actively predicts and manipulates the robot's motion based on sensing the signal.

5.4.2 Self-Regulated Movements for Autonomous Soft Robots

As a further step towards mimicking self-contained biological organisms, stimuli-responsive hydrogels are also capable of realizing autonomous, self-regulated actuation with a built-in feedback loop, such as light-tracking shown in Fig. 5.4 [61]. A sunflower-like biomimetic omnidirectional tracker (SunBOT) has been reported with biological sensorimotor-like behavior, controlled by a built-in feedback loop in the intricately coupled photochemo-mechanical interactions. An array of SunBOTs was used in a solar vapor generation device, achieving up to a 400% solar energy-harvesting enhancement over non-tropistic materials. Practically, the SunBOT, as a novel smart material, provides an untethered, controller-free design to achieve light-tracking function, and can be a long-sought solution to enhanced energy harvesting via the autonomous maximization of the input power density.

5.4.3 Environment-Powered and -Dictated Hydrogel Oscillators as Swimming Robots

In soft robotics, a repeating locomotion is typically generated by an oscillating stimulus or intermittent power input, such as by modulating a piezoelectric transducer with an AC signal. In other words, most man-made oscillations require nonconstant energy input and cannot perform environment-dictated movement. Thus, a tethered or battery-connected power source and complicated actuation

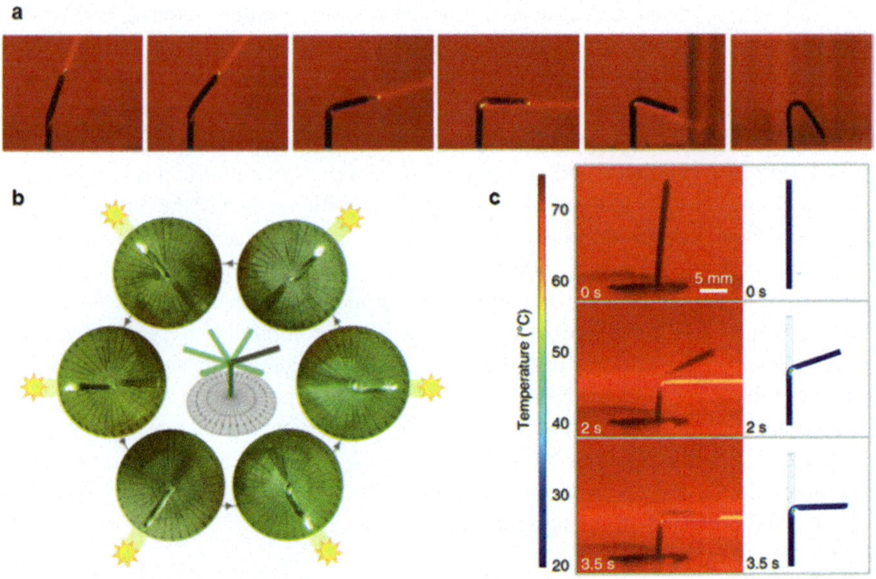

Fig. 5.4 (**a**) Photos of an AuNP–PNIPAAm SunBOT in water aiming at a laser that illuminates from various zenith angles in the range 20–150°. (**b**) Top-view snapshots of a SunBOT continuously tracking light that revolves azimuthally around the SunBOT. (**c**) The kinetics of a TDD tracking process, evaluated by experiment (left) and finite element analysis (FEA) simulation (right), which synced perfectly (Adapted from [61], copyright by Nature Nanotechnology)

mechanisms are usually required in these systems. In contrast, living organisms are capable of harvesting energy from a static ambient environment to produce autonomous oscillations such as heartbeats and neuron impulses [62]. Therefore, it is extremely advantageous for robotics, especially untethered autonomous robots, if human-made oscillators could be actuated and powered in a static ambient environment.

Visible-light triggered, self-sustaining oscillation with a photo-responsive hydrogel has been developed, capable of swimming in high-damping-coefficient fluids (Fig. 5.5a) [56]. The long-term and self-sustained oscillation from the material enabled the robot to maintain periodic actuation under constant external input energy (i.e., without intermittent input). The hydrogel oscillator exhibited broad available size range (0.36–1.10 mm in thickness) and broad light intensity window (120–1125 mW light). The symmetric cylindrical hydrogel pillar design allows for omnidirectional oscillation covering the entire episphere: a wide range of deflection angles (the angle between the deflected direction and the initial direction) from ±90° to nearly 0° (zenith angles) and 0–360° azimuthal angles (Fig. 5.5b). The oscillation properties were found to be dependent on sample geometry, such as arm length and diameter, which enabled the tunability to fulfill various requirements (Fig. 5.5c).

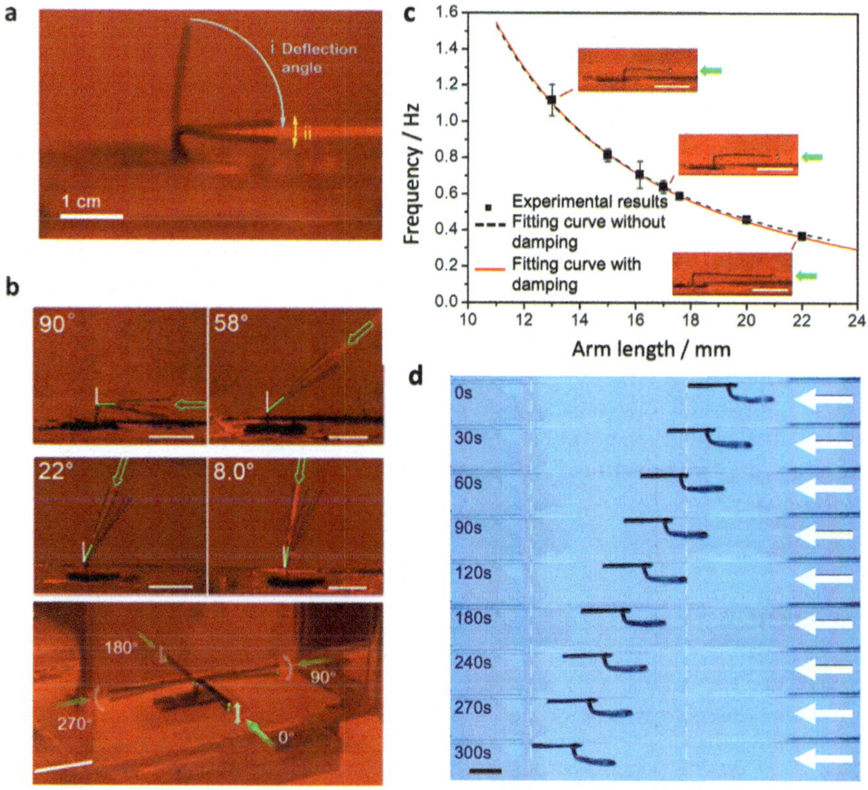

Fig. 5.5 (**a**) Superimposed image showing light tracking followed by oscillation of a light-responsive hydrogel pillar. (**b**) Omnidirectional tracking and oscillation. (**c**) The characteristic frequency of the oscillator is dependent on geometry such as arm length. (**d**) Sequential snapshots of a swimmer under static light source illumination (Adapted from [56], copyright by Science Robotics)

An example of underwater application for this hydrogel oscillator is a swimming robot powered by propulsion-enabled locomotion. As a soft swimming robot, namely OsciBot, its efficient self-propulsion is powerful to generate a force large enough to overcome water resistance for directional movement, moving away from a light stimulus. Notably, this hydrogel oscillator generates remarkable force, which is large enough to overcome the damping from water in an aqueous environment (Fig. 5.5d). The OsciBot performed a highly controllable oscillating motion with outstanding output work and hence a fast and controllable swimming speed (>1.15 body lengths/min), which outperforms the state-of-the-art soft swimmers made of liquid crystal elastomers and other hydrogels [44, 63–65]. This superior performance indicated that hydrogel oscillators demonstrate great potential for automated systems, soft robotics, miniaturized transport, solar sails, drug delivery, and surgical applications.

5.5 Conclusions

This chapter summarizes recent research progress on stimulus-responsive hydrogels as sensors and actuators, including new optical properties, mechanical properties, response mechanisms, and underwater applications. Through chemical functionalization, environmental/strain sensing, power enhancement, self-regulation, and simultaneous self-sensing, both steady stimulus-tracking and autonomous oscillation can be achieved. With a homogeneous structure nature, delicate and non-predefined motions can be addressed post-manufacture. The presented hydrogels are able to convert local strain, curvature, bending angle, and heat into more quantitative electric signals, meeting the demand for a homogeneous hydrogel-based active tactile material platform for soft robotics. These strategies can be widely applied for unlimited soft materials and composites, which will provide broad choices of building blocks for novel soft robots. The encouraging progress in smart soft materials may lead to critical contributions in the rapid advancement of underwater soft robotics.

References

1. J.C. Yeo, H.K. Yap, W. Xi, Z.P. Wang, C.H. Yeow, C.T. Lim, Flexible and stretchable strain sensing actuator for wearable soft robotic applications. Advanced Materials Technologies **1**(3), 1600018 (2016)
2. M.A. Mcevoy, N. Correll, Materials that couple sensing, actuation, computation, and communication. Science **347**(6228), 1261689 (2015)
3. S. Kim, C. Laschi, B. Trimmer, Soft robotics: A bioinspired evolution in robotics. Trends Biotechnol. **31**(5), 23–30 (2013)
4. D. Trivedi, C.D. Rahn, W.M. Kier, I.D. Walker, Soft robotics: Biological inspiration, state of the art, and future research. Appl. Bionics Biomech. **5**(3) (2008)
5. H. Zeng, P. Wasylczyk, D.S. Wiersma, A. Priimagi, Light robots: Bridging the gap between microrobotics and photomechanics in soft materials. Adv. Mater. **30**(24) (2018)
6. A. O'Halloran, F. O'Malley, P. Mchugh, A review on dielectric elastomer actuators, technology, applications, and challenges. J. Appl. Phys. **104**(7) (2008)
7. C. Liu, H. Qin, P.T. Mather, Review of progress in shape-memory polymers. J. Mater. Chem. **17**(16) (2007)
8. S.-J. Jeon, A.W. Hauser, R.C. Hayward, Shape-morphing materials from stimuli-responsive hydrogel hybrids. Acc. Chem. Res. **50**(2), 161–169 (2017)
9. T. Wang, Y. Zhang, Q.C. Liu, W. Cheng, X.R. Wang, L.J. Pan, B.X. Xu, H.X. Xu, A self-healable, highly stretchable, and solution processable conductive polymer composite for ultrasensitive strain and pressure sensing. Adv. Funct. Mater. **28**(7), 1705551 (2018)
10. J.J. Duan, X.C. Liang, J.H. Guo, K.K. Zhu, L.N. Zhang, Ultra-stretchable and force-sensitive hydrogels reinforced with chitosan microspheres embedded in polymer networks. Adv. Mater. **28**(36), 8037–8044 (2016)
11. S. Brady, D. Diamond, K.T. Lau, Inherently conducting polymer modified polyurethane smart foam for pressure sensing. Sens. Actuators A Phys **119**(2), 398–404 (2005)
12. T. Del Castillo-Castro, M.M. Castillo-Ortega, J.C. Encinas, P.J.H. Franco, H.J. Carrillo-Escalante, Piezo-resistance effect in composite based on cross-linked Polydimethylsiloxane and Polyaniline: Potential pressure sensor application. J. Mater. Sci. **47**(4), 1794–1802 (2012)

13. Q. Shao, Z.Q. Niu, M. Hirtz, L. Jiang, Y.J. Liu, Z.H. Wang, X.D. Chen, High-performance and Tailorable pressure sensor based on ultrathin conductive polymer film. Small **10**(8), 1466–1472 (2014)
14. L. Chen, M.J. Liu, L. Lin, T. Zhang, J. Ma, Y.L. Song, L. Jiang, Thermal-responsive hydrogel surface: Tunable wettability and adhesion to oil at the water/solid Interface. Soft Matter **6**(12), 2708–2712 (2010)
15. X.B. Zhang, C.L. Pint, M.H. Lee, B.E. Schubert, A. Jamshidi, K. Takei, H. Ko, A. Gillies, R. Bardhan, J.J. Urban, M. Wu, R. Fearing, A. Javey, Optically- and thermally-responsive programmable materials based on carbon nanotube-hydrogel polymer composites. Nano Lett. **11**(8), 3239–3244 (2011)
16. D. Schmaljohann, Thermo- and pH-responsive polymers in drug delivery. Adv. Drug Deliv. Rev. **58**(15), 1655–1670 (2006)
17. S.K. De, N.R. Aluru, B. Johnson, W.C. Crone, D.J. Beebe, J. Moore, Equilibrium swelling and kinetics of pH-responsive hydrogels: Models, experiments, and simulations. J. Microelectromech. Syst. **11**(5), 544–555 (2002)
18. D. Han, C. Farino, C. Yang, T. Scott, D. Browe, W. Choi, J.W. Freeman, H. Lee, Soft robotic manipulation and locomotion with a 3D printed electroactive hydrogel. ACS Appl. Mater. Interfaces **10**(21), 17512–17518 (2018)
19. A. Matsumoto, R. Yoshida, K. Kataoka, Glucose-responsive polymer gel bearing Phenylborate derivative as a glucose-sensing moiety operating at the physiological pH. Biomacromolecules **5**(3), 1038–1045 (2004)
20. K.Y. Lee, D.J. Mooney, Hydrogels for tissue engineering. Chem. Rev. **101**(7), 1869–1879 (2001)
21. C. Larson, B. Peele, S. Li, S. Robinson, M. Totaro, L. Beccai, B. Mazzolai, R. Shepherd, Highly stretchable electroluminescent skin for optical Signaling and tactile sensing. Science **351**(6277), 1071–1074 (2016)
22. S. Bauer, S. Bauer-Gogonea, I. Graz, M. Kaltenbrunner, C. Keplinger, R. Schwodiauer, 25th anniversary article: A soft future: From robots and sensor skin to energy harvesters. Adv. Mater. **26**(1), 149–162 (2014)
23. C. Keplinger, J.Y. Sun, C.C. Foo, P. Rothemund, G.M. Whitesides, Z.G. Suo, Stretchable, transparent, ionic conductors. Science **341**(6149), 984–987 (2013)
24. Y. Qiu, K. Park, Environment-sensitive hydrogels for drug delivery. Adv. Drug Deliv. Rev. **53**(3), 321–339 (2001)
25. K.L. Ang, S. Venkatraman, R.V. Ramanujan, Magnetic PNIPA hydrogels for hyperthermia applications in cancer therapy. Mater. Sci. Eng. C Biomim. Supramol. Syst **27**(3), 347–351 (2007)
26. Y. Mori, H. Tokura, M. Yoshikawa, Properties of hydrogels synthesized by freezing and thawing aqueous polyvinyl alcohol solutions and their applications. J. Mater. Sci. **32**(2), 491–496 (1997)
27. L. Zhao, J.H. Huang, Y.C. Zhang, T. Wang, W.X. Sun, Z. Tong, Programmable and bidirectional bending of soft actuators based on Janus structure with sticky tough PAA-clay hydrogel. ACS Appl. Mater. Interfaces **9**(13), 11866–11873 (2017)
28. Z.S. Liu, P. Calvert, Multilayer hydrogels as muscle-like actuators. Adv. Mater. **12**(4), 288–291 (2000)
29. E.A. Moschou, S.F. Peteu, L.G. Bachas, M.J. Madou, S. Daunert, Artificial muscle material with fast electroactuation under neutral pH conditions. Chem. Mater. **16**(12), 2499–2502 (2004)
30. Z.Y. Lei, P.Y. Wu, A highly transparent and ultra-stretchable conductor with stable conductivity during large deformation. Nat. Commun. **10**, 3429 (2019)
31. C.J. Yu, Z. Duan, P.X. Yuan, Y.H. Li, Y.W. Su, X. Zhang, Y.P. Pan, L.L. Dai, R.G. Nuzzo, Y.G. Huang, H.Q. Jiang, J.A. Rogers, Electronically programmable, reversible shape change in two- and three-dimensional hydrogel structures. Adv. Mater. **25**(11), 1541–1546 (2013)
32. R.C. Luo, J. Wu, N.D. Dinh, C.H. Chen, Gradient porous elastic hydrogels with shape-memory property and anisotropic responses for programmable locomotion. Adv. Funct. Mater. **25**(47), 7272–7279 (2015)

33. E. Wang, M.S. Desai, S.W. Lee, Light-controlled Graphene-elastin composite hydrogel actuators. Nano Lett. **13**(6), 2826–2830 (2013)
34. M.C. Koetting, J.T. Peters, S.D. Steichen, N.A. Peppas, Stimulus-responsive hydrogels: Theory, modern advances, and applications. Mater. Sci. Eng. R Rep **93**, 1–49 (2015)
35. W. Francis, A. Dunne, C. Delaney, L. Florea, D. Diamond, Spiropyran based hydrogels actuators—Walking in the light. Sens. Actuators B **250**, 608–616 (2017)
36. R.M. Kramer, W.J. Crookes-Goodson, R.R. Naik, The self-organizing properties of squid reflectin protein. Nat. Mater. **6**(7), 533–538 (2007)
37. L. Phan, W.G. Walkup, D.D. Ordinario, E. Karshalev, J.M. Jocson, A.M. Burke, A.A. Gorodetsky, Reconfigurable infrared camouflage coatings from a cephalopod protein. Adv. Mater. **25**(39), 5621 (2013)
38. M. Qin, M. Sun, R.B. Bai, Y.Q. Mao, X.S. Qian, D. Sikka, Y. Zhao, H.J. Qi, Z.G. Suo, X.M. He, Bioinspired hydrogel interferometer for adaptive coloration and chemical sensing. Adv. Mater. **30**(21), 1800468 (2018)
39. M. Sun, R.B. Bai, X.Y. Yang, J.Q. Song, M. Qin, Z.G. Suo, X.M. He, Hydrogel interferometry for ultrasensitive and highly selective chemical detection. Adv. Mater. **30**(46), 1804916 (2018)
40. M. Qin, M. Sun, M.T. Hua, X.M. He, Bioinspired structural color sensors based on responsive soft materials. Curr. Opin. Solid State Mater. Sci. **23**(1), 13–27 (2019)
41. J. Choi, M. Hua, S.Y. Lee, W. Jo, C.Y. Lo, S.H. Kim, H.T. Kim, X.M. He, Hydrocipher: Bioinspired dynamic structural color-based cryptographic surface. Adv. Opt. Mater. **8**(1), 1901259 (2020)
42. H. Banerjee, M. Suhail, H.L. Ren, Hydrogel actuators and sensors for biomedical soft robots: Brief overview with impending challenges. Biomimetics **3**(3), 15 (2018)
43. H. Yuk, S.T. Lin, C. Ma, M. Takaffoli, N.X. Fang, X.H. Zhao, Hydraulic hydrogel actuators and robots optically and sonically camouflaged in water. Nat. Commun. **8**, 14230 (2017)
44. Y.S. Kim, M.J. Liu, Y. Ishida, Y. Ebina, M. Osada, T. Sasaki, T. Hikima, M. Takata, T. Aida, Thermoresponsive actuation enabled by permittivity switching in an electrostatically anisotropic hydrogel. Nat. Mater. **14**(10), 1002–1007 (2015)
45. Y. Takashima, S. Hatanaka, M. Otsubo, M. Nakahata, T. Kakuta, A. Hashidzume, H. Yamaguchi, A. Harada, Expansion-contraction of photoresponsive artificial muscle regulated by host-guest interactions. Nat. Commun. **3**, 1270 (2012)
46. S.M. Chin, C.V. Synatschke, S.P. Liu, R.J. Nap, N.A. Sather, Q.F. Wang, Z. Alvarez, A.N. Edelbrock, T. Fyrner, L.C. Palmer, I. Szleifer, M.O. De La Cruz, S.I. Stupp, Covalent-supramolecular hybrid polymers as muscle-inspired anisotropic actuators. Nat. Commun. **9**, 2395 (2018)
47. T. Mirfakhrai, J.D.W. Madden, R.H. Baughman, Polymer artificial muscles. Mater. Today **10**(4), 30–38 (2007)
48. R.E. Godt, B.D. Lindley, Influence of temperature upon contractile activation and isometric force production in mechanically skinned muscle-fibers of the frog. J. Gen. Physiol. **80**(2), 279–297 (1982)
49. L.A. Cole, *Biology of Life: Biochemistry, Physiology and Philosophy* (Academic Press, London, 2016), pp. 1–184
50. Y.F. Ma, M.T. Hua, S.W. Wu, Y.J. Du, X.W. Pei, F. Zhou, X.M. He, Bioinspired high-power-density strong contractile hydrogel by programmable elastic recoil, Sci. Adv. (2020). (accepted)
51. G.J. Lutz, L.C. Rome, Built for jumping—The design of the frog muscular system. Science **263**(5145), 370–372 (1994)
52. T.J. Roberts, R.L. Marsh, Probing the limits to muscle-powered accelerations: Lessons from jumping bullfrogs. J. Exp. Biol. **206**(15), 2567–2580 (2003)
53. D. Taylor, N. O'Mara, E. Ryan, M. Takaza, C. Simms, The fracture toughness of soft tissues. J. Mech. Behav. Biomed. Mater. **6**, 139–147 (2012)
54. Y.S. Zhao, C.Y. Lo, L.C. Ruan, C.H. Pi, C.G. Kim, Y. Alsaid, I. Frenkel, R. Rico, T.C. Tsao, X.M. He, Somatosensory Actuator Based on Stretchable Conductive Photothermally-Responsive Hydrogel. (Submitted)

55. X.M. He, M. Aizenberg, O. Kuksenok, L.D. Zarzar, A. Shastri, A.C. Balazs, J. Aizenberg, Synthetic homeostatic materials with chemo-mechano-chemical self-regulation. Nature **487**(7406), 214–218 (2012)
56. Y.S. Zhao, C. Xuan, X.S. Qian, Y. Alsaid, M.T. Hua, L.H. Jin, X.M. He, Soft phototactic swimmer based on self-sustained hydrogel oscillator. Sci. Robot. **4**(33), Eaax7112 (2019)
57. H. Liu, Q.M. Li, S.D. Zhang, R. Yin, X.H. Liu, Y.X. He, K. Dai, C.X. Shan, J. Guo, C.T. Liu, C.Y. Shen, X.J. Wang, N. Wang, Z.C. Wang, R.B. Wei, Z.H. Guo, Electrically conductive polymer composites for smart flexible strain sensors: A critical review. J. Mater. Chem. C **6**(45), 12121–12141 (2018)
58. L.W. Xia, R. Xie, X.J. Ju, W. Wang, Q.M. Chen, L.Y. Chu, Nano-structured smart hydrogels with rapid response and high elasticity. Nat. Commun. **4**, 2226 (2013)
59. W. Jayathilaka, K. Qi, Y.L. Qin, A. Chinnappan, W. Serrano-Garcia, C. Baskar, H.B. Wang, J.X. He, S.Z. Cui, S.W. Thomas, S. Ramakrishna, Significance of nanomaterials in wearables: A review on wearable actuators and sensors. Adv. Mater. **31**(7), 1805921 (2019)
60. E.S. Gil, S.H. Park, L.W. Tien, B. Trimmer, S.M. Hudson, D.L. Kaplan, Mechanically robust, rapidly actuating, and biologically functionalized macroporous poly(N-Isopropylacrylamide)/silk hybrid hydrogels. Langmuir **26**(19), 15614–15624 (2010)
61. X.S. Qian, Y.S. Zhao, Y. Alsaid, X. Wang, M.T. Hua, T. Galy, H. Gopalakrishna, Y.Y. Yang, J.S. Cui, N. Liu, M. Marszewski, L. Pilon, H.Q. Jiang, X.M. He, Artificial phototropism for omnidirectional tracking and harvesting of light. Nat. Nanotechnol. **14**(11), 1048 (2019)
62. K. Kruse, F. Jülicher, Oscillations in cell biology. Curr. Opin. Cell Biol. **17**(1), 20–26 (2005)
63. C. Christianson, N.N. Goldberg, D.D. Deheyn, S. Cai, M.T. Tolley, Translucent soft robots driven by frameless fluid electrode dielectric elastomer actuators. Sci. Robot. **3**(17), Eaat1893 (2018)
64. S. Palagi, A.G. Mark, S.Y. Reigh, K. Melde, T. Qiu, H. Zeng, C. Parmeggiani, D. Martella, A. Sanchez-Castillo, N. Kapernaum, Structured light enables biomimetic swimming and versatile locomotion of photoresponsive soft microrobots. Nat. Mater. **15**(6), 647–653 (2016)
65. B.J. Williams, S.V. Anand, J. Rajagopalan, M.T.A. Saif, A self-propelled biohybrid swimmer at low Reynolds number. Nat. Commun. **5**(1), 1–8 (2014)

Chapter 6
Ionic Polymer-Metal Composite (IPMC) Artificial Muscles in Underwater Environments: Review of Actuation, Sensing, Controls, and Applications to Soft Robotics

Nazanin Minaian, Zakai J. Olsen, and Kwang J. Kim

6.1 Introduction

Electroactive polymers (EAPs) are multifunctional smart polymers that have unique electromechanical (electrical to mechanical) and mechanoelectrical (mechanical to electrical) transduction properties that result in reproducible actuation and sensing capabilities. Their responsive shape-changing nature when provided an external electrical stimulus has been comparable to that of biological muscles, hence acquiring the descriptor artificial muscles [1, 2]. As an actuator, an EAP's inherent compliant nature, large actuation strain, low noise, and scalability allow them to serve in unique applications where current rigid-body devices fail to meet design criteria.

EAPs are typically divided into two categories, electronic and ionic, where the fundamental distinction between the two is their primary mechanism of activation. The driving force for electronic EAPs is mainly due to an induced electric field or Coulomb forces, whereas ionic EAPs are actuated by ion mobility or diffusion [3]. This chapter discusses the background and various applications of the ionic EAP known as an ionic polymer-metal composite (IPMC). An IPMC comprises an ion exchange membrane, typically Nafion, with two discrete surface electrodes, commonly made of a noble metal. The ionomer is synthesized to allow the diffusion of mobile cations that migrate under an applied electric field, transporting with them solvent molecules due to hydration effects. This accumulation of solvent molecules and osmotic pressure at the cathode of the IPMC is the main mechanism of actuation. These materials also exhibit thermally activated shape-memory properties that allow for more complex motions to be produced. On the other hand, mechanical

N. Minaian · Z. J. Olsen · K. J. Kim (✉)
Active Materials and Smart Living Laboratory, Mechanical Engineering Department, University of Nevada, Las Vegas, NV, USA
e-mail: kwang.kim@unlv.edu

D. A. Paley, N. M. Wereley (eds.), *Bioinspired Sensing, Actuation, and Control in Underwater Soft Robotic Systems*, https://doi.org/10.1007/978-3-030-50476-2_6

deformations of an IPMC induce solvent flow and ionic redistribution within the membrane. This in turn gives rise to a detectable voltage across the electrodes and is the mechanism behind the material's sensing capability.

Where IPMCs have the advantage over other EAPs are their low-driving voltage, relatively fast response speed, and a preference for operation in underwater environments [4–6]. These qualities have brought IPMCs to the forefront of the research and development of underwater soft robotics.

6.2 Fundamentals

IPMCs have been shown to be a highly functional material, having been used as both actuators and sensors in a wide range of soft robotic and biomimetic applications. Here a brief overview of IPMCs is provided, highlighting their history, fabrication and materials, and underlying physical mechanisms. Having first been discussed as a potential electroactive polymer-based actuator in a 1993 patent by Oguro and his colleagues [7], IPMCs have a long and rich history of research and development. The seminal four-part series on IPMC fundamentals by Shahinpoor and Kim has served as a key reference in the decades since it was published [8–11]. Throughout these years, many reviews on IPMC have been published, keeping up with the rapidly changing landscape of EAP materials and soft robotics. In this chapter, we will cover a broad range of fundamental IPMC concepts and provide ample demonstrations of their capabilities as actuators and sensors in the field of underwater robotic systems.

6.2.1 Structure and Manufacturing of IPMCs

The IPMC is a highly complex material, the study of which is a rich, interdisciplinary endeavor that draws on aspects from the fields of electrochemistry, porous materials, composite materials, polymer science, and many others. As such, their material makeup and fabrication are important aspects for understanding their actuation, sensing, and controllability. These all come together to form the broader picture of how IPMC materials may be introduced into the field of underwater soft robotics, and here we present an overview of the materials and techniques used to manufacture IPMC for implementation as actuators and sensors.

6.2.1.1 Ionomer Materials and Manufacturing

IPMCs are unique electroactive polymers in that they exhibit both electromechanical and mechanoelectrical transduction, like that of piezoelectric materials, and are constructed from flexible polymers that can sustain large deformations and strains

[8, 12]. The core structure of an IPMC is an ionomeric-polymer (ionomer), which consists of strong ionic groups attached to a charge-neutral polymer backbone [13]. These ionic groups aggregate to form a network of fully connected ionic clusters that gives the material its water permeability and selective ion transport properties [14–17]. The structure of these clusters is understood to be a function of the electrostatic interactions between the polymer chains and their ionic end groups.

In standard IPMC fabrication, purchased or manufactured thin sheets of ionomeric material are used for the membrane component. Nafion and Flemion are the most common throughout the literature. Recent studies have examined other materials, such as Aquivion, for use in IPMCs and demonstrated increases in performance for both actuation and sensing due to relatively higher ionic conductivity and ion exchange capacity [18, 19]. Ionomers are typically available in either a precursor form or activated form, where the former is preferred for more nontraditional fabrication techniques such as 3D printing or injection molding. When working with an ionomer in precursor form, prior to plating, it is imperative to activate the ionomer through a hydrolysis process, which will exchange the initial end group for a sulfonated end group. Activating the ionomer will allow ion conduction, but will no longer retain the ability to melt, disallowing any further thermal-based fabrication processes [20].

In recent years, researchers have produced new and innovative techniques for the fabrication of membrane materials into complex and 3D shapes. Breaking away from the traditional rectangular cantilever IPMC membranes, Trabia et al. [21] developed a spray fabrication method in which a solution of Nafion in a solvent is deposited into stenciled shapes across a substrate using an airbrush. The solvent is then allowed to evaporate. Membrane thickness may be controlled by the stencil thickness and the combination of stencil and substrate shape dictates the final geometry of the ionomeric membrane. These membranes can then be chemically plated and turned into IPMCs of novel shapes for characterization and application in both actuation and sensing.

With the advancement in additive manufacturing, a great deal of effort has been made to leverage this technology in the fabrication and development of soft robotic systems, particularly IPMCs. Carrico et al. first demonstrated a method for manufacturing precursor Nafion filament that could be used in a custom-designed 3D printer to fabricate ionomer membranes for use in IPMCs. Their approach paved the wave for rapid advancement in developing new IPMC actuator geometries and IPMC-based soft robots [22–24].

Shortly after the first 3D-printed Nafion membranes were produced, a search began for new materials better suited to the thermal processing that is core to additive manufacturing. Trabia et al. showed that Aquivion, an ionomeric material similar in structure to Nafion, has thermal properties in its precursor form that allow for easier melt processing into filament as well as during the actual 3D printing [18, 19]. This allowed for their research group to produce new planar geometries as well as traditional cantilever rectangle IPMC actuators. These Aquivion-based IPMC exhibit better performance when compared to equivalent Nafion counterparts.

6.2.1.2 Electrode Fabrication Techniques

With this ionomer structure, the IPMC is formed by compositing an electrode material, typically a noble metal such as platinum (also gold, palladium, and silver) [8, 12], onto the surface of a thin ionomer membrane. This electrode formation consists of an initial compositing process followed by a chemical deposition process [8, 12] and results in a composite material wherein the metallic electrodes are chemically embedded into the ionomer structure. This process forms the final material, a combined structure of an ionomer and metal electrodes, hence forming an ionic polymer-metal composite. The foundational method for fabricating IPMCs was detailed in [9] and has since been expanded and altered by researchers in attempts to enhance material properties, IPMC performance, and ease of fabrication.

One of the most common modifications to the traditional and straightforward electroding process is the use of masking and selective electroding. The goal of this process is to pattern the electrodes across the surface of the ionomer in such a way as to create a multifunctional device, as opposed to the rather simple cantilever bending type IPMC. The intent may be to provide a multi-degree-of-freedom actuator that utilizes different electrode sections in conjunction to create complex bending curvatures within a single actuator or to provide two distinct electrode regions wherein one electrode pair may carry an actuation signal, while the other carries a sensing response of the IPMC for self-sensing. Later it will also be discussed how by using a segmented electrode across a single IPMC ionomer the phenomena of back-relaxation may be mitigated. Examples of such masking and selective electroding process can be found in [25–27].

Aside from the traditional electroless plating process aforementioned, alternative methods in electrode formation have also been explored by researchers. The pursuit of these alternative methods has been in response to the limitations of electroless deposition such as the lack of control over physical morphology, challenges with reproducibility, and the extensive time dedication due to the requirement of multiple plating sessions. These alternative methods have in turn been developed with the intent to provide performance optimization and to advance the speed and quantity of manufacturing efforts. One of these methods detailed by Akle et al. is known as the Direct Assembly Process (DAP), which was motivated by the procedure of fuel cell electrode fabrication and was applied to IPMCs [28, 29]. The method involves painting large surface electrodes composed of a mixed electrically conductive powder and ionomer solution either directly on the membrane or to a decal surface that is later hot-pressed on the membrane.

The benefits of this method are in its simplicity and its direct control of the electrode dimensions and material morphology and composition. However, post-process, additional treatment may be required to increase the surface conductivity for actuation performance. This treatment can either involve the electroless plating process or through a conductive layer deposition such as by applying a conductive foil or sputter coating a conductive layer [28, 30].

The DAP electrode fabrication method also permits for a variety of electrode materials excluded by the impregnation-reduction process. Single-walled carbon nanotubes (SWNT), RuO_2, carbide-derived carbon (CDC), and other conductive powders have been substituted as a component in the painted electrode mixture [31, 32]. The inclusion of other conductive electrode materials through the DAP method can result in increased optimization potential of IPMC performance as well as decreased manufacturing costs.

Another alternative electrode fabrication method by Chung et al. was developed with the intent to improve cost, processing time, and reproducibility for large-scale production [33]. Like the DAP method, silver nano-powders are dissolved in a diluted Nafion solution to form a conductive mixture. The solution is then casted and embossed to form a cohesive structure. This step is followed by a single electroless plating procedure of silver and micro-electroforming of nickel [33]. The use of silver has its disadvantages as the oxidation of silver during actuation can degrade the bending performance of the IPMC. Aside from oxidation, metal-based electrodes also are prone to cracks due to fatigue, which highly degrades long-term performance and durability.

Researchers have delved into methods for increasing electrode durability through several novel methods such as with the implementation of graphene electrodes by Kim et al. [34]. In this process, functionalized reduced graphene oxide paper (rGOP) is adhered to the ionomer surface after roughening the inner surface of the rGOP by laser-scribing increased adhesion.

Additionally, the manufacturing of IPMCs via the polymerization in situ of intrinsically conducting polymers (ICPs), typically polypyrrole (PPy) or PEDOT:PSS, has also been reported in various literature [35–37]. ICPs unique alternating double-single bond structure allow them to function as an organic conducting polymer that can then be composited with filler material for improved performance. In summary, during the manufacturing of IPMCs, the choice of backbone ionomer, electrode material, mobile ion species, solvent, and hydration level all have a measurable impact on the behavior and performance of IPMC materials [14, 38–42], demonstrating their complex and variable nature.

6.2.2 *Underlying Physics*

Despite the complexity of these materials and their underlying chemical makeup, the scientific community has constructed a firm understanding of the key fundamentals involved in their actuation and sensing capabilities. Under an applied electric field, mobile ionic species migrate, transporting with them some of the solvent species due to hydration [43]. The redistribution of ionic and solvent species, in conjunction with electrostatic forces, causes a differential swelling of the material as a whole [44–46]. Conversely, when a mechanical deformation is applied, the volumetric strain of the polymer backbone induces a pressure gradient, causing solvent and, hence, mobile ions, to flow within the porous cluster network. This redistribution of mobile ionic species, along with the polymer-fixed ionic groups rearranging according to the strain of the backbone, generates a charge imbalance across the medium that may be detected across the surface electrodes [44, 47, 48]. While the underlying mechanisms for these transduction phenomena are understood, describing this behavior in such a way as to predict the deformation response or sensing signal of an IPMC is a difficult task, in part due to the variability in performance between samples of IPMC materials. Challenges also arise when considering the variable nature of individual IPMCs. Performance will often differ between samples due to minor inconsistencies during the fabrication process. These nuances can thus produce an impact on the exact behavior and performance of an individual IPMC, which is challenging to predict in a single all-encompassing model.

6.2.2.1 Actuation

Some early efforts in describing the physical underpinnings of IPMCs were provided by de Gennes et al. and Nemat-Nasser and Li [46, 49], where the two took very different approaches in their models. Postulating the primary deformation mechanism to arise from migration of entrained water carried by the ionic species, de Gennes provided a linear irreversible thermodynamics model that formed Onsager relations coupling the current density and solvent transport within the IPMC to the external electric field and pressure gradient. These equations can be used to explain both the electromechanical actuation and the mechanoelectrical sensing behavior of a static IPMC.

As the transduction principles of IPMCs are, at their core, based on electrochemical behavior of ions in solution, it is no surprise that many researchers turn to the classical Poisson-Nernst-Planck (PNP) equations for modeling the electric field formation and ionic species electromigration [43]. It is quite common to find physics-based models that use the PNP equations to describe the electrochemical aspects of an IPMC and then construct relations between the ionic concentration and internal stress of the polymer matrix. With the ion transport being determined by the PNP equations, the induced stress is often related to the osmotic pressure and at times electrostatic forces generated due to ion redistribution within the pores of the cluster network. At steady state, the stress in the polymer backbone, related to the strain of the material, balances the combined osmotic and electrostatic pressure [50, 51].

An alternative electromechanical coupling approach is to relate the charge, charge density, or powers of the charge density to an aspect of mechanical loading that may be applied to a structural model. Utilizing the charge across the IPMC as a basis for the coupling is typically done when the electrochemistry is modeled with an equivalent circuit [52, 53], whereas the charge density is more commonly used when using some kind of transport equation such as the PNP system. The mechanical loading is then postulated to be a function of charge density, or some power series expansion of the charge density, and typically related to a mechanical loading force [54], or the internal stress of the polymer [55]. Relating the charge density to surface traction loading has been shown to cause erroneous effects for unconstrained actuators, as discussed in [56].

Instead of relying on existing transport equations like the PNP system, some researchers opt to begin their analysis from a thermodynamics perspective [57–61]. This approach is particularly useful for a couple of reasons. The thermodynamic approach allows for researchers to cast their model in a broader context, relying on fewer assumptions from the outset, such as a dilute solution for the Nernst-Planck equation, and then imposing restrictions after the fact in a systematic way. Furthermore, the thermodynamic framework may be written in both spatial and material coordinates, and the deformation of the IPMC material is not restricted to small displacements or strains. This approach introduces more complex electromechanical and mechanoelectrical coupling, and allows for the stress, electrochemical potential, and electric field to be derived from thermodynamic potentials such as the Helmholtz free energy.

Further complicating factors for modeling IPMCs arise in relation to larger applied voltages and the electrochemical behavior of the equations near the electrodes. Dimensional analysis shows that the typical PNP system has a small dimensionless parameter associated with the highest spatial derivatives, indicating that the equations are what is called singularly perturbed [62]. These equations tend to display distinct layers near domain boundaries where the solution undergoes a rapid transition to accommodate the boundary conditions and, in the rest of the domain, the solution shows an entirely different behavior. Due to the existence of these boundary layers, solving the equations analytically or numerically requires some additional finesse. Finite element solvers may have difficulty in capturing the steep gradients present and hence require dense meshes near the boundaries, leading to increased computational effort.

Analytical solutions can benefit from the thinness of the boundary layer, and utilizing mathematical approaches such as the method of matched asymptotic expansion can provide a semi-analytical solution [51]. Complications in the model are only compounded with increasing voltage, where the classical PNP equations predict a continual increase in ionic concentration, inconsistent with the assumption of a dilute solution, and steric effects and other electrostatic correlations become important [63–67]. It has also been shown that electrode surface roughness and penetration of the electrode particles into the ionomer play major roles in IPMC behavior [63, 64]. During actuation, IPMCs may also display a low-frequency deformation, typically in the opposite direction as the primary actuation, that is termed back-relaxation. This phenomenon is still being understood and its occurrence varies with cation and ionomer type as well as voltage [38, 40, 41, 63]. The combination of osmotic pressure and back-relaxation phenomena govern the IPMC deformation from its initial transient behavior to its steady-state response [63, 68].

6.2.2.2 Sensing

The mechanisms of sensing in IPMC materials can be related to the inverse operations for that of actuation. Induced deformation to the material causes a redistribution of mobile cations in the solvent and fixed anions according to the material strain, which form a detectable voltage or current signal, depending on the state of the circuit at the electrodes where the signal is detected. We first highlight the model produced by Nemat-Nasser and Li. They derived their model on the basis of balancing the stress of the polymer backbone with an eigen-stress generated due to osmotic and electrostatic effects [46]. Conducting an in-depth analysis of the cluster-network morphology and the electrostatic interactions that take place, they describe the deformation-induced dipoles generated within the pores as the basis for the IPMCs sensing response model. This approach showed good agreement with experimental data for a wide range of tip displacements for a cantilever IPMC sample. Many other groups have since worked on understanding and modeling the transduction phenomena of sensing. Under an open-circuit condition, the current through the electrodes at the point of connection is zero and the redistribution of ions creates an electric potential difference across the surface electrodes [48, 61, 69–71]. In a short-circuit condition, both electrodes take on the same electric potential and the through-the-thickness current is detected as the sensing signal [48, 71–74].

Cha et al. [71] demonstrated that under open-circuit conditions for perfectly conducting electrodes, the ionic redistribution remains electroneutral throughout the deformation and hence no boundary-layer formation occurs, as in actuation modeling. Further, these authors note that in short-circuit conditions, the effects of composite layers, near the polymer-electrode interfaces where material and electrical properties vary from the bulk, play a major role in the short-circuit current. In these regions, boundary layers form as cation depletion and accumulation occur at the electrodes under compression and tension, respectively, and electroneutrality is no longer established. This build-up of charge and its rate of change is related to the short-circuit current signal that is produced under these conditions.

6.2.2.3 Reduced-Order Modeling

A common theme across sensing literature is the development of a complex multi-physics model followed by the reduction to a simplified form, whether that be beam theory for the mechanics or an equivalent circuit for the electrochemistry. Akle and Habchi used the PNP system to describe the electrochemical behavior of an IPMC, but related the charge density to the mechanical stress generated by an applied bending moment according to classical beam theory [75]. Similarly, one may start from the framework of continuum mechanics or the PNP system of equations and derive relations between the stress and charge density. From there, the electrochemical equations are translated into equivalent circuit elements and the mechanical equations into effective stresses or stress resultants, resulting in a model with a significantly reduced complexity and more tenable to real-time sensing applications [47, 48, 53, 72, 73, 76, 77].

6.2.3 Aspects of IPMC Control Strategies

Due to the highly nonlinear and stochastic nature of IPMC material and its transduction capabilities, predicting and controlling the behavior of an IPMC-based device can be considered a challenging task [78]. This difficulty is compounded further due to effects such as back-relaxation and nonlinear responses in the material due to viscoelastic effects and mechanical vibrations [79].

A great deal of effort has been made in developing a fundamental understanding of IPMC actuation and sensing in order to design methods for controlling them in real-world applications. Researchers have shown both predictive (feedforward) and corrective (feedback) schemes are capable to control IPMCs to a varying degree of success. This is shown in examples where a simple PID feedback control is implemented for a propulsive device [80], as well as for open-loop and closed-loop position and force control of an IPMC strip [81].

A problem arises when using a feedback control scheme on its own once the back-relaxation phenomena begins to occur. The urge for continual reverse deformation due to the back-relaxation drives the actuation voltage higher and, as the relaxation is linked to higher applied potentials, this phenomenon can create a cyclic response and reach a point where the applied voltage exceeds the electrolysis potential for the IPMC material. At this point, the performance degrades and energy is wasted in the electrolysis process [79].

One approach to combat this behavior uses a sectored IPMC, with the sections being used to generate an antagonistic-type deformation, the net result of which is controlled using a feedforward scheme and reduces the tracking error by a large margin, 85%. Using feedback control, this margin is improved even further, up to 97% [82]. This method is successful because since each section of IPMC is actuated in an antagonistic way, their back-relaxation phenomena are also occurring antagonistically. Hence, the controller is able to mitigate positional drift by simply shifting the relative input voltage to each section as opposed to fighting the back-relaxation through continuously increasing the actuation signal.

Other examples of both feedback and feedforward control of IPMCs are numerous, utilizing more substantial schemes such as LQR, robust control, fuzzy control, hysteresis modeling, and equivalent circuit-based large bandwidth models [77, 83–87]. The full breadth of possible control methods is beyond the scope of this discussion, but the referenced works provide a brief overview of the many approaches and advancements made in the control of these dynamic and complex EAP materials. As discussed previously, a major component to constructing an effective control scheme is to have a fast and accurate model of the IPMC actuator. For controls applications, a robust physics-based model obtained from analytical or numerical approximations of continuum mechanics can be simplified to reduced-ordered models such as with beam theory or a circuit-based approach.

6.3 Applications

As discussed at length, IPMC materials exhibit both electromechanical and mechanoelectrical transduction and as such they have applications and roles as both actuators and sensors. Here, we discuss some of the major areas of research regarding their applications, specifically as it relates to aquatic environments and underwater robotic systems.

6.3.1 IPMCs as Actuators and Artificial Muscles

Researchers are continuously exploring the many possible applications of using IPMCs in soft robotics. Because of their ability to operate in water, IPMCs naturally find use in aquatic systems and, specifically, biomimetic robotics. The field of biomimetic robotics is focused on the design, development, and applications of robotic systems that are inspired by and seek to replicate or mimic structures and functionality found in the natural world. When it comes to IPMCs, the influences from nature tend to come from aquatic animals. IPMC materials are not the only choice for biomimetic underwater robots, but they are a particularly well-suited one.

Rectangular cantilever IPMC actuators may be easily utilized as the basis for biomimetic fins. A prime example is found in [25], where a patterned electroding process was used to fabricate an IPMC with multiple active portions along the length of a monolithic membrane, giving the actuator multiple degrees of freedom (DOF). Each active portion of the actuator may be driven by a different input signal, causing a localized migration of ions and solvent and producing a richer range of deformation.

Using masking and etching techniques, Chen and Tan [27] fabricated an IPMC-based fin with active and passive portions of varying thickness to achieve complex deflection of the free end. This type of fin construction allows for not only bending of the actuator as obtained with a single IPMC but also actuating the active portions out of phase gives rise to twisting and traveling-wave deformation. Similar approaches for fin design have been used by other researchers. Embedding IPMCs into a passive sleeve is a common technique for achieving intricate actuator kinematics. Using four rectangular IPMCs inside a cast PDMS sleeve, [88] developed a pectoral fin actuator for aquatic robots inspired by the manta ray and demonstrated its free-swimming capabilities, with a speed of ~0.05 body lengths per second and consuming less than 1W of power.

A soft boot made of a castable silicon material, RTV 500, was used to embed three IPMC actuators and create a fin capable of complex twisting and bending

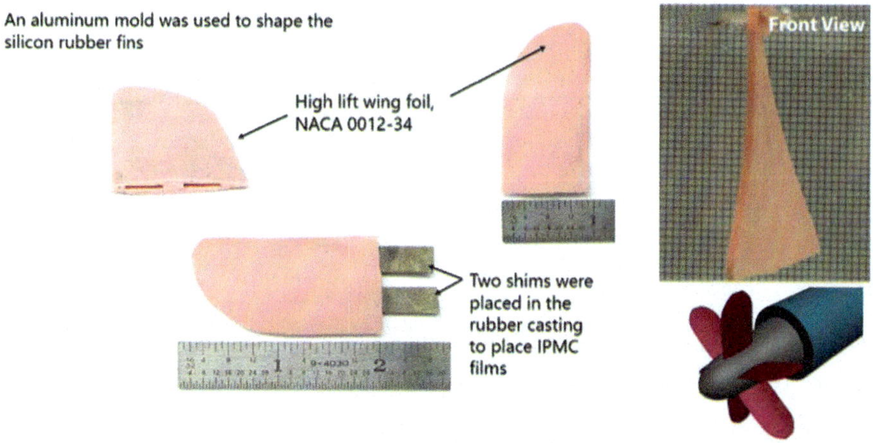

Fig. 6.1 Photographs of a soft boot made of a castable silicon material, RTV 500, used to embed three IPMC actuators to create a fin capable of complex twisting and bending deformation that is useful for underwater propulsion and control

deformation [89] (see Fig. 6.1).[1] Actuators such as these have been shown as promising propulsive and control devices in soft robotics over the last two decades. These examples demonstrate just one aspect of the field of biomimetic robots. The fin actuators discussed show promise for both steering and propulsion of small fish-like soft robots, examples of which are numerous throughout the literature.

IPMC-based pectoral and caudal fins are commonly utilized for propulsion and steering of robotic fish. The design of these fish-like robots is typically accompanied by a related hydrodynamic model to predict the swimming performance and behavior of the system and compare with experimental tests. The robots are often constructed with 3D-printed bodies that are waterproofed and house all of the control and power electronics, making for an untethered and self-contained robotic system. A few examples of such robots are provided in [90–92].

Exploitation of the shape-memory properties of the ionomer has provided an additional facet in IPMC-based actuators. Xie examined the multi-shape memory effect of Nafion in [93]; he was able to demonstrate annealed dry Nafion holding up to four distinct temporary shapes in a single shape memory cycle. This effect is unique to polymers possessing a broad single glass transition range such as perfluorosulfonic acid ionomers like Nafion. The broad glass transition range of Nafion (~55 °C to ~130 °C) can be considered a large collective contribution of numerous infinitely sharp transitions distributed within the aforementioned range and can theoretically allow for a vast amount of uniquely memorized shapes [93].

This unique material characteristic was later applied to IPMCs when Rossiter et al. were able to demonstrate shape-memory control under typical hydrated IPMC

[1]KJK thanks Promode R. Bandyopadhyay for suggesting this experiment.

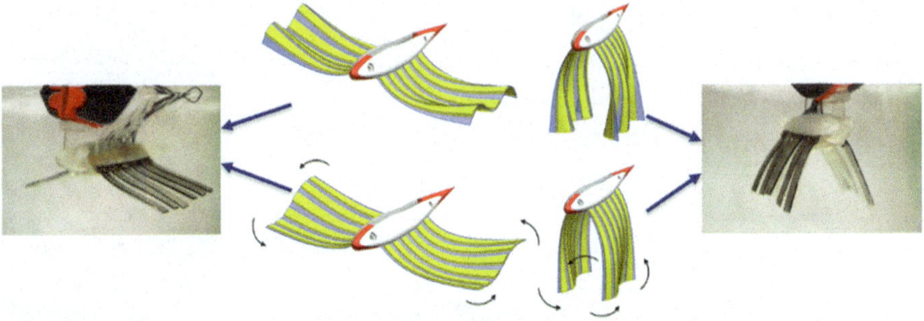

Fig. 6.2 A soft robot made with shape-memory IPMCs capable of switching between mobuliform (left) and gymnotiform (right) swimming modes

conditions [94]. The electromechanical control in conjunction with the thermomechanical actuation effect of the ionomer's multi-shape memory properties broadens the possibilities of IPMC actuation applications. Shen et al. demonstrated that through this combination of electromechanical and thermomechanical transduction, complex actuation schemes in which an IPMC is able to perform simultaneous twisting and bending configurations are possible [95]. These combined phenomena allow for simultaneous high-frequency bending motions, accompanied with stable complex motions over a larger time scale. Shen was able to later examine several variables that contribute to the number of programmable shapes such as the membrane and ion type, where it was determined that ion type contributed significantly to the glass transition range of the IPMC [96]. Membranes with mobile lithium cations were found to possess the largest glass transition range. The highly complex transformable nature of this actuator lends itself well into the design of bioinspired robotics. This benefit is apparent in the design of a robotic fish with the ability to transform its locomotion behavior from gymnotiform to mobuliform via memorized shape memory positions shown in Fig. 6.2 [97, 98].

Small fish and manta rays are not the only biological source of inspiration for IPMC-powered robotics. Due to the soft, highly deformable nature of the jellyfish body, these animals are often used as a target for the development of new biomimetic robots, particularly those that are enabled with flexible actuators such as IPMCs. Yeom and Oh created a biomimetic jellyfish robot that uses IPMC actuators thermally treated to retain a resting curvature when no electrical stimulus was applied [99]. Their design used eight of these curved actuators as the legs of the jellyfish with thin cellophane paper bridging the gap between actuators and adhesive tape along the length of the IPMCs to obtain mechanical connection between the tape and actuators. They then proceeded to characterize the robot for different input signals, one input being bioinspired and the other with an equivalent RMS value.

An alternative perspective to biomimetic design is presented in [100], where as opposed to directly replicating the structure and form of a biological jellyfish, the mechanism behind its locomotion was used as the main inspiration. In this work [100], the jet propulsion mechanism was explored and a new robotic design built around a remedy for one of its major drawbacks, namely that during the relaxation phase of the swimming cycle, a negative thrusting force is experienced that slows down the robot or jellyfish. The work demonstrated a new design for a jellyfish-inspired robot and constructed geometric and mechanics-based models to study the design's performance and swimming energetics.

This highlights an important aspect of biomimetic and bioinspired robotics: although inspiration is taken from the biological world, in the engineering world designers are free to modify and adapt structures or mechanism to better satisfy a given objective or constraint. As the field progresses, we should expect to see many other exciting examples of this as other creative interpretations of biomimetic robots emerge.

6.3.2 *IPMCs as Energy Harvesters*

The bidirectional transduction behavior of IPMC materials also lend itself to applications in energy harvesting. Many EAP materials have demonstrated capabilities for such purposes, but here we highlight some of the many works that utilize IPMC materials. The energy-harvesting ability of IPMCs relates to their sensing capabilities, wherein deformation of the material generates an electric potential or short-circuit current that may be utilized for charging or powering low-load devices. These devices are typically designed to leverage passive deformations of an IPMC due to vibrations of a structure or movements of the surrounding medium [69, 101]. The principles of energy transduction have been modeled and tested extensively for samples of cantilever IPMC devices, demonstrating the utility of the material for such applications due to its low bending stiffness and compliance [102–104].

Cantilever IPMCs show variation in their harvesting performance based upon their surface electrode materials (platinum, gold, silver, etc.) [104], as well as on their loading conditions (bending, extension, shear) [105]. These devices have been demonstrated as energy harvesters when slammed against a stagnant fluid surface [106]. This approach utilized the IPMCs as surfaces of a wedge and leveraged the deformation of these compliant surfaces due to the impact forces to convert the mechanical energy into an electrical response; the authors also demonstrated the IPMCs capability to sense these types of impact loads. Similarly, rectangular IPMCs may be used to harvest energy from the deformation of plate type surfaces, as demonstrated in [107]. The authors therein extensively modeled the deformation of an underwater plate-like flag with an IPMC embedded into the surface, including hydrodynamic couplings key to an accurate analysis. Their work highlights stability regions of the flag as a function of the mass, stiffness, and tensioning effects, and provides an in-depth reporting of their experimental results.

Other works place emphasis on the base-excitation energy-harvesting capabilities of IPMCs, where inputs of base displacement are related to the electrical output response of the IPMC through some of the various modeling approaches as discussed in the previous sections. Aureli et al. [108] provide a model combining a

Kirchhoff-Love plate with the PNP equations for describing the mechanoelectrical transduction of a base-excited cantilever IPMC. Their modeling results are well supported by experimental results presented in the paper and the results emphasize the importance of impedance matching and driving the device at resonance frequency for maximization of harvesting efficiency. Similarly, Lim, Lei, and Tan [72] used a dynamic Euler-Bernoulli beam with viscous and strain-rate dependent damping to describe the mechanics of a cantilever IPMC. This beam model was coupled to a reduced and linearized electrochemistry model that describes the charge density and accounts for anionic redistribution due to volumetric change. The charge density is further related to the stress in the beam as is common in other sensing models. The resulting transfer function of electrical current from base deformation is reduced to a finite-order function and their modeling results match very well with experimental data.

Lastly, two examples of non-rectangular IPMC energy harvesters are presented. While rectangular-shaped cantilever IPMCs are the most ubiquitous, alternative shapes have been demonstrated to fit other application needs or constraints. Using annular and disc-shaped IPMCs for energy harvesting has been presented in [109, 110]. In a similar setting as the base-excited cantilevers, the authors in [109] used an annular shaped IPMC to harvest energy from underwater vibrations. Their model utilized an axisymmetric thin plate with transverse vibrations and limited the hydrodynamic interactions to added mass effects. Their parameter-identified model demonstrated great agreement with experiments and can hence elucidate some of the underlying physics for such energy-transduction phenomena. In a related way, a disc-shaped IPMC energy harvester was developed by the authors in [110]. This approach utilized radial extension of the disc as the driving deformation and an electrochemical model found in [73]. Their study focused on investigating the effects of the disc dimensions on the energy-harvesting performance and demonstrated the device's capability for mechanical to electrical transduction and charge generation.

6.3.3 IPMCs as Sensors and Artificial Skin

While there has been extensive research into the development of underwater soft robotic IPMC actuators, there has been concurrent interest in utilizing the IPMC's sensing abilities in related studies. The sensing mechanisms of IPMCs are primarily dependent on actively measuring the output response of the sensor after a mechanical input causes a redistribution of cations within the IPMC structure. Sensing studies have been considered more niche in comparison to the enthusiasm researchers have exhibited for IPMC actuation applications. This bias can be due in part to the practical limitations of IPMCs as sensors, specifically regarding its minimal voltage output. Sensor devices will thus often require sensitive equipment capable of detecting these small signals or signal-conditioning components such as instrumentation amplifiers to minimize external noise, in contrast to actuation, wherein easily constructed input signals on the order of a single volt provide large amplitude deformations. That said, an exploration of several applications involving IPMCs as sensors is provided in this section.

Segmented electrodes are an appealing technique to increase the DOF and maneuverability of actuated IPMCs. This concept has been expanded upon in the development of what are known as self-sensing actuators. Built upon the unique two-way transductional nature of IPMCs, researchers have experimented with coupling actuation and sensing in a single device [111]. Kruusamäe and colleagues developed a patterned-electrode IPMC that was segmented into three discrete sections [112]. A common challenge that arises in the development of self-sensing IPMC is the issue of cross-talk due the driving voltages being orders of magnitude larger than the detectable sensing signal. In an attempt to eliminate cross-talk, a segmented section functioned as shielding between the actuator and sensor elements [112]. Such a system can minimize the ultimate footprint of an IPMC-based device, as self-contained feedback eliminates the need for an external laser system. An alternative approach to self-sensing IPMCs is presented in [113], which relies on information gained through passive sensing, a sensing method that is more explicitly related to the measurement changes in material electrical properties after mechanical deformation. This method would thereby remove the need for discrete segmented electrodes, but instead requires an additional power supply and measurement circuit to detect these variations [114].

The inspiration of utilizing the characteristics found in biology has not been limited to solely replicating swimming modes of fish or other aquatic species. Researchers have also been interested in mimicking the various biocomponents relating to naturally occurring sensing components within aquatic environments. This aspect is especially apparent in studies inspired by the lateral line of fish and the manufacturing of artificial skin. The lateral line serves as a system of sensors used to detect pressure gradients, vibrations, and movement within the fish's local environment. The ability for an IPMC to capture this essential mechanism in biology has brought several devices mimicking this phenomenon to the forefront. In [115], an artificial lateral line prototype was constructed with IPMCs serving as the sensory hair cells found within the neuromast component of the lateral line. Two stimuli, comprising a vibrating sphere and oscillating fin, were used to determine the effectiveness of the sensor for underwater source localization with success up to multiple body lengths of the sensor array structure. This work has been furthered in [116] by encompassing the IPMC sensor within a 3D-printed structure imitating the lateral line cupula component. In a similar vein, Zhu et al. recently developed a pressure sensor that is inspired by the natural tactile and auditory sensing mechanisms in biology [117]. What these devices have in common is a desire to recreate the biological systems that allow animals to react and understand external stimuli generated by pressure waves and vibrations. In biology, we typically find this sensing capability in or near the skin layer of the animal and hence all these devices can be seen as a kind of artificial skin that may be used by robots to sense and understand their surroundings.

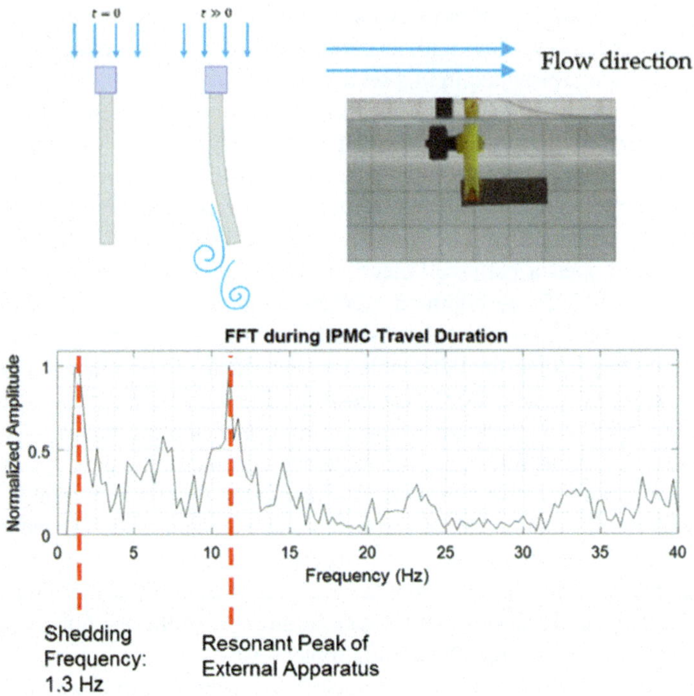

Fig. 6.3 Example frequency response of IPMC sensor in a towed vortex-shedding experiment

One of the more prominent uses of IPMC sensors in underwater applications has been in the development of IPMC-based flow sensor devices.[2] There have been two approaches in this endeavor, one of which involves interpretations of the characteristic response directly and the other requiring additional signal processing within the frequency domain. The former is dependent on the fluid dynamics phenomenon of Kármán vortex streets and requires the sensor to detect the frequency of vortex shedding either from the sensor itself or from an obstructing object positioned before the IPMC sensor. Using the Strouhal number and Reynolds number relationship, the vortex-shedding frequency is found to directly relate to the velocity of the surrounding fluid during the wide Reynolds number range where Strouhal number is considered constant (Fig. 6.3). Examples of such a device are provided in [118, 119].

Accompanying underwater platforms, IPMCs can serve to provide real-time flow-field measurements of the surrounding environment through studies of the wave amplitude and frequency of the IPMC sensing response [69]. IPMCs also exhibit a degree of scalability that can permit interesting bioinspired innovations such as artificial cilia [120–122]. The fabrication and application of such cilia are

[2]We thank Promode R. Bandyopadhyay for suggesting this experiment.

still very active fields of research, and implementation of these devices has not been fully explored. In principle, artificial cilia could be used as a sensor array for monitoring the flow field across a particular region. Using this sensing signal, or even that of another IPMC-based sensor, an array of cilia could feasibly be used to modulate the flow across a surface and provide enhanced efficiency and hydrodynamic performance of devices such as hydrofoils. This area of research will be particularly interesting to monitor over the coming years.

6.4 Conclusion

Here we have presented a broad review of the materials, fabrication, physical mechanisms, and applications associated with the highly functional IPMC electroactive polymer. The complex nature of these composite materials requires analysis from various fields of science and engineering. Despite this complexity, over the last few decades, researchers from around the world have developed a more comprehensive understanding of the interconnected multi-physics phenomena behind the two-way transductional nature of these important materials. As progress develops, new and exciting applications for IPMCs are sure to emerge and expand on our current perceptions about the fields of robotics, soft robotics, and biomimetics.

Acknowledgements The authors gratefully acknowledge support from the Office of Naval Research, Grant Number N000141612356. Also, KJK acknowledges the help from Drs. Viljar Palmre, Kam Keang, and Tyler Stalbaum conducting various underwater experiments using IPMCs.

References

1. Y. Bar-Cohen, Electroactive polymers as artificial muscles: A review. J. Spacecr. Rocket. **39**(6), 822–827 (2002)
2. L. Hines, K. Petersen, G.Z. Lum, M. Sitti, Soft actuators for small-scale robotics. Adv. Mater. **29**(13) (2017)
3. M. Shahinpoor, Y. Bar-Cohen, T. Xue, J.S. Harrison, J. Smith, Ionic polymer-metal composites as biomimetic sensors and actuators-artificial muscles. ACS Symp. Ser. **726**, 25–50 (1999)
4. K.J. Kim, W. Yim, J.W. Paquette, D. Kim, Ionic polymer-metal composites for underwater operation. J. Intell. Mater. Syst. Struct. **18**(2), 123–131 (2007)
5. M. Shahinpoor, Conceptual design, kinematics and dynamics of swimming robotic structures using ionic polymeric gel muscles. Smart Mater. Struct. **1**(1), 91–94 (1992)
6. W. Yim, J. Lee, K.J. Kim, An artificial muscle actuator for biomimetic underwater propulsors. Bioinspir. Biomim. **2**(2), S31–S41 (2007)
7. Oguro, K., Takenaka, H., Kawami, Y. Actuator Element, United States, (5,268,082) (1993)
8. M. Shahinpoor, K.J. Kim, Ionic polymer-metal composites: I. Fundamentals. Smart Mater. Struct. **10**(4), 819–833 (2001)

9. K.J. Kim, M. Shahinpoor, Ionic polymer-metal composites: II. Manufacturing techniques. Smart Mater. Struct. **12**(1), 65–79 (2003)
10. M. Shahinpoor, K.J. Kim, Ionic polymer-metal composites: III. Modeling and simulation as biomimetic sensors, actuators, transducers, and artificial muscles. Smart Mater. Struct. **13**(6), 1362–1388 (2004)
11. M. Shahinpoor, K.J. Kim, Ionic polymer-metal composites: IV. Industrial and medical applications. Smart Mater. Struct. **14**(1), 197–214 (2005)
12. K.J. Kim, Ionic polymer-metal composites as a new actuator and transducer material, in *Electroactive Polymers for Robotics Applications*, ed. by K. J. Kim, S. Tadokoro, (Springer, London, 2007), pp. 153–164
13. Y. Bahramzadeh, M. Shahinpoor, A review of ionic polymeric soft actuators and sensors. Soft Robot. **1**(1), 38–52 (2014)
14. J.Y. Li, S. Nemat-Nasser, Micromechanical analysis of ionic clustering in nafion perfluorinated membrane. Mech. Mater. **32**(5), 303–314 (2000)
15. W.Y. Hsu, T.D. Gierke, T.D. Gierke, Elastic theory for ionic clustering in perfluorinated ionomers. Macromolecules **15**(1), 101–105 (1982)
16. V.K. Datye, P.L. Taylor, A.J. Hopfinger, Simple model for clustering and ionic transport in ionomer membranes. Macromolecules **17**(9), 1704–1708 (1984)
17. A. Eisenberg, Clustering of ions in organic polymers. A theoretical approach. Macromolecules **3**(2), 147–154 (1970)
18. S. Trabia, Z.J. Olsen, K.J. Kim, Searching for a new Ionomer for 3D printable ionic polymer-metal composites: Aquivion as a candidate. Smart Mater. Struct. **26**(11) (2017)
19. S. Trabia, K. Choi, Z.J. Olsen, T. Hwang, J.-D. Do Nam, K.J. Kim, Understanding the thermal properties of precursor-ionomers to optimize fabrication processes for ionic polymer-metal composites (IPMCs). Materials (Basel). **11**(5), 1–12 (2018)
20. K.A. Mauritz, R.B. Moore, State of understanding of Nafion. Chem. Rev. **104**(10), 4535–4585 (2004)
21. S. Trabia, T. Hwang, K.J. Kim, A fabrication method of unique Nafion shapes by painting for ionic polymer-metal composites. Smart Mater. Struct. **25**(8), 1–11 (2016)
22. J.D. Carrico, N.W. Traeden, M. Aureli, K.K. Leang, Fused filament 3D printing of ionic polymer-metal composites (IPMCs). Smart Mater. Struct. **24**(12), 125021 (2015)
23. J.D. Carrico, K.K. Leang, Fused filament 3D printing of ionic polymer-metal composites for soft robotics, in *Electroactive Polymer Actuators and Devices (EAPAD) 2017*, ed. by Y. Bar-Cohen, (SPIE, 2017), p. 101630I
24. J.D. Carrico, K.J. Kim, K.K. Leang, 3D-Printed ionic polymer-metal composite soft crawling robot, in *Proc.—IEEE Int. Conf. Robot. Autom*, (2017), pp. 4313–4320
25. J.H. Jeon, S.W. Yeom, I.K. Oh, Fabrication and actuation of ionic polymer metal composites patterned by combining electroplating with electroless plating. Compos. Part A Appl. Sci. Manuf. **39**(4), 588–596 (2008)
26. M.A. Tsugawa, V. Palmre, J.D. Carrico, K.J. Kim, K.K. Leang, Slender tube-shaped and square rod-shaped IPMC actuators with integrated sensing for soft mechatronics. Meccanica **50**(11), 2781–2795 (2015)
27. Z. Chen, X. Tan, Monolithic fabrication of ionic polymer-metal composite actuators capable of complex deformation. Sensors Actuators A Phys. **157**(2), 246–257 (2010)
28. B.J. Akle, M.D. Bennett, D.J. Leo, K.B. Wiles, J.E. McGrath, Direct assembly process: A novel fabrication technique for large strain ionic polymer transducers. J. Mater. Sci. **42**(16), 7031–7041 (2007)
29. B.J. Akle, M.D. Bennett, D.J. Leo, High-strain ionomeric-ionic liquid electroactive actuators. Sensors Actuators A Phys. **126**(1), 173–181 (2006)
30. J.W. Lee, Y.T. Yoo, Preparation and performance of IPMC actuators with electrospun Nafion®-MWNT composite electrodes. Sensors Actuators B Chem. **159**(1), 103–111 (2011)
31. B.J. Akle, D.J. Leo, Single-walled carbon nanotubes—Ionic polymer electroactive hybrid transducers. J. Intell. Mater. Syst. Struct. **19**(8), 905–915 (2008)

32. V. Palmre, D. Brandell, U. Mäeorg, J. Torop, O. Volobujeva, A. Punning, U. Johanson, M. Kruusmaa, A. Aabloo, Nanoporous carbon-based electrodes for high strain ionomeric bending actuators. Smart Mater. Struct. **18**(9), 095028 (2009)
33. C.K. Chung, P.K. Fung, Y.Z. Hong, M.S. Ju, C.C.K. Lin, T.C. Wu, A novel fabrication of ionic polymer-metal composites (IPMC) actuator with silver Nano-powders. Sensors Actuators B Chem. **117**(2), 367–375 (2006)
34. J. Kim, J.-H. Jeon, H.-J. Kim, H. Lim, I.-K. Oh, Durable and water-floatable ionic polymer actuator with hydrophobic and asymmetrically laser-scribed reduced graphene oxide paper electrodes. ACS Nano **8**(3), 2986–2997 (2014)
35. J.W. Lee, J.H. Kim, Y.S. Chun, Y.T. Yoo, S.M. Hong, The performance of Nafion-based IPMC actuators containing Polypyrrole/alumina composite fillers. Macromol. Res. **17**(12), 1032–1038 (2009)
36. G. Di Pasquale, S. Graziani, F.G. Messina, A. Pollicino, R. Puglisi, E. Umana, An investigation of the structure–property relationships in ionic polymer polymer composites (IP 2 Cs) manufactured by polymerization in situ of PEDOT/PSS on Nafion ® 117. Smart Mater. Struct. **23**(3), 035018 (2014)
37. F. Cellini, A. Grillo, M. Porfiri, Ionic polymer metal composites with polypyrrole-silver electrodes. Appl. Phys. Lett. **106**(13), 131902 (2015)
38. S. Nemat-Nasser, S. Zamani, Y. Tor, Effect of solvents on the chemical and physical properties of ionic polymer-metal composites. J. Appl. Phys. **99**(10) (2006)
39. S. Nemat-Nasser, Y. Wu, Tailoring actuation of ionic polymer-metal composites through cation combination. Smart Struct. Mater. **5051**, 245–253 (2003)
40. S. Nemat-Nasser, S. Zamani, Modeling of electrochemomechanical response of ionic polymer-metal composites with various solvents. J. Appl. Phys. **100**(6) (2006)
41. S. Nemat-Nasser, Y. Wu, Comparative experimental study of ionic polymer-metal composites with different backbone ionomers and in various cation forms. J. Appl. Phys. **93**(9), 5255–5267 (2003)
42. J.M. Byun, T. Hwang, K.J. Kim, Formation of a gold nanoparticle layer for the electrodes of ionic polymer–metal composites by electroless deposition process. Appl. Surf. Sci. **470**, 8–12 (2019)
43. Z. Zhu, H. Chen, L. Chang, B. Li, Dynamic model of ion and water transport in ionic polymer-metal composites. AIP Adv. **1**(4) (2011)
44. M. Shahinpoor, Y. Bar-Cohen, J.O. Simpson, J. Smith, Ionic polymer-metal composites (IPMCs) as biomimetic sensors, actuators and artificial muscles—A review. Smart Mater. Struct. **7**(6), R15–R30 (1998)
45. J.D.W. Madden, N.A. Vandesteeg, P.A. Anquetil, P.G.A. Madden, A. Takshi, R.Z. Pytel, S.R. Lafontaine, P.A. Wieringa, I.W. Hunter, Artificial muscle technology: Physical principles and naval prospects. IEEE J. Ocean. Eng. **29**(3), 706–728 (2004)
46. S. Nemat-Nasser, J.Y. Li, Electromechanical response of ionic polymer-metal composites. J. Appl. Phys. **87**(7), 3321–3331 (2000)
47. M. Porfiri, An electromechanical model for sensing and actuation of ionic polymer metal composites. Smart Mater. Struct. **18**(1), 015016 (2009)
48. M. Aureli, M. Porfiri, Nonlinear sensing of ionic polymer metal composites. Contin. Mech. Thermodyn. **25**(2–4), 273–310 (2013)
49. P.G. de Gennes, K. Okumura, M. Shahinpoor, K.J. Kim, Mechanoelectric effects in ionic gels. Europhys. Lett. **50**(4), 513–518 (2000)
50. D. Schicker, T. Wallmersperger, Modeling and simulation of the chemo electro mechanical behavior of ionic polymer-metal composites. J. Appl. Phys. **114**(16), 163709 (2013)
51. Z. Zhu, K. Asaka, L. Chang, K. Takagi, H. Chen, Multiphysics of ionic polymer-metal composite actuator. J. Appl. Phys. **114**(8) (2013)
52. J.W. Paquette, K.J. Kim, J.-D. Nam, Y.S. Tak, An equivalent circuit model for ionic polymer-metal composites and their performance improvement by a clay-based polymer nano-composite technique. J. Intell. Mater. Syst. Struct. **14**(10), 633–642 (2003)

53. Y. Cha, M. Aureli, M. Porfiri, A physics-based model of the electrical impedance of ionic polymer metal composites. J. Appl. Phys. **111**(12) (2012)
54. D. Pugal, K.J. Kim, A. Aabloo, An explicit physics-based model of ionic polymer-metal composite actuators. J. Appl. Phys. **110**(8) (2011)
55. T. Wallmersperger, B. Kröplin, R.W. Gülch, Coupled chemo-electro-mechanical formulation for ionic polymer gels—Numerical and experimental investigations. Mech. Mater. **36**(5–6), 411–420 (2004)
56. Z.J. Olsen, K.J. Kim, A reduced dimensional mapping approach for modeling IPMCs with computational efficiency and rapid design development applications. Smart Mater. Struct. **27**(12) (2018)
57. T. Wallmersperger, A. Horstmann, B. Kroplin, D.J. Leo, Thermodynamical modeling of the electromechanical behavior of ionic polymer metal composites. J. Intell. Mater. Syst. Struct. **20**(6), 741–750 (2009)
58. P. Nardinocchi, M. Pezzulla, L. Placidi, Thermodynamically based multiphysic modeling of ionic polymer metal composites. J. Intell. Mater. Syst. Struct. **22**(16), 1887–1897 (2011)
59. S. Galante, A. Lucantonio, P. Nardinocchi, The multiplicative decomposition of the deformation gradient in the multiphysics modeling of ionic polymers. Int. J. Non. Linear. Mech. **51**, 112–120 (2013)
60. M. Rossi, T. Wallmersperger, Thermodynamically consistent three-dimensional electrochemical model for polymeric membranes. Electrochim. Acta **283**, 1323–1338 (2018)
61. Y. Cha, M. Porfiri, Mechanics and electrochemistry of ionic polymer metal composites. J. Mech. Phys. Solids **71**(1), 156–178 (2014)
62. M. Porfiri, Charge dynamics in ionic polymer metal composites. J. Appl. Phys. **104**(10) (2008)
63. M. Porfiri, H. Sharghi, P. Zhang, Modeling Back-relaxation in ionic polymer metal composites: The role of steric effects and composite layers. J. Appl. Phys. **123**(1), 014901 (2018)
64. M. Porfiri, Influence of electrode surface roughness and steric effects on the nonlinear electromechanical behavior of ionic polymer metal composites. Phys. Rev. E Stat. Nonlinear, Soft Matter Phys. **79**(4), 1–13 (2009)
65. P.H.R. Alijó, F.W. Tavares, E.C. Biscaia, A.R. Secchi, Effects of electrostatic correlations on ion dynamics in alternating current voltages. Electrochim. Acta **152**, 84–92 (2015)
66. M.S. Kilic, M.Z. Bazant, A. Ajdari, Steric effects in the dynamics of electrolytes at large applied voltages. I. Double-layer charging. Phys. Rev. E Stat. Nonlinear, Soft Matter Phys. **75**(2), 1–16 (2007)
67. M.S. Kilic, M.Z. Bazant, A. Ajdari, Steric effects in the dynamics of electrolytes at large applied voltages. II. Modified Poisson-Nernst-Planck equations. Phys. Rev. E Stat. Nonlinear, Soft Matter Phys. **75**(2), 1–11 (2007)
68. M. Porfiri, H. Sharghi, P. Zhang, Modeling Back-relaxation in ionic polymer metal composites: The role of steric effects and composite layers. J. Appl. Phys. **123**(1), 014901 (2018)
69. Q. Shen, T. Wang, K.J. Kim, A biomimetic underwater vehicle actuated by waves with ionic polymer-metal composite soft sensors. Bioinspir. Biomim. **10**(5), 055007 (2015)
70. H. Kim, Y. Cha, M. Porfiri, Voltage attenuation along the electrodes of ionic polymer metal composites. J. Intell. Mater. Syst. Struct. **27**(17), 2426–2430 (2016)
71. Y. Cha, F. Cellini, M. Porfiri, Electrical impedance controls mechanical sensing in ionic polymer metal composites. Phys. Rev. E Stat. Nonlinear, Soft Matter Phys. **88**(6), 1–12 (2013)
72. C. Lim, H. Lei, X. Tan, A dynamic physics-based model for base-excited IPMC sensors, in *Electroactive Polymer Actuators and Devices (EAPAD) 2012*, ed. by Y. Bar-Cohen, (SPIE, 2012)
73. Z. Chen, X. Tan, A. Will, C. Ziel, A dynamic model for ionic polymer-metal composite sensors. Smart Mater. Struct. **16**(4), 1477–1488 (2007)
74. K. Farinholt, D.J. Leo, Modeling of electromechanical charge sensing in ionic polymer transducers. Mech. Mater. **36**(5–6), 421–433 (2004)

75. B. Akle, W. Habchi, Finite element Modeling of the sensing and energy harvesting performance in ionic polymer metal composites, in *Electroactive Polymer Actuators and Devices (EAPAD)*, ed. by Y. Bar-Cohen, (SPIE, 2013), p. 86870G
76. V. Volpini, L. Bardella, A. Rodella, Y. Cha, M. Porfiri, Modelling compression sensing in ionic polymer metal composites. Smart Mater. Struct. **26**(3) (2017)
77. C. Zheng, X. Tan, A control-oriented and physics-based model for ionic polymer—metal composite actuators. IEEE/ASME Trans. Mechatronics **13**(5), 519–529 (2008)
78. A.J. McDaid, *Modelling and Control of IPMC Actuators for Biomedical Robotics Applications* (The University of Auckland, 2011)
79. J.D. Carrico, M. Fleming, M.A. Tsugawa, K.K. Leang, Precision feedback and feedforward control of ionic polymer metal composite actuators, in *Ionic Polymer Metal Composites (IPMCs): Smart Multi-Functional Materials and Artificial Muscles*, ed. by M. Shahinpoor, vol. 1, (Royal Society of Chemistry, 2015)
80. M.A. Rosly, H. Yussof, M.F. Shaari, Z. Samad, D. Kamaruzaman, A.R. Omar, Speed control mechanism for IPMC based biomimetic flapping Thruster, in *Proceedings—2017 IEEE 5th International Symposium on Robotics and Intelligent Sensors, IRIS 2017*, (2018), pp. 218–223
81. R.C. Richardson, M.C. Levesley, M.D. Brown, J.A. Hawkes, K. Watterson, P.G. Walker, Control of ionic polymer metal composites. IEEE/ASME Transactions on Mechatronics **8**(2), 245–253 (2003)
82. M.J. Fleming, K.J. Kim, K.K. Leang, Mitigating IPMC Back relaxation through feedforward and feedback control of patterned electrodes. Smart Mater. Struct. **21**(8) (2012)
83. K. Mallavarapu, K.M. Newbury, D.J. Leo, Feedback control of the bending response of ionic polymer-metal composite actuators, in *Smart Struct. Mater. 2001 Electroact. Polym. Actuators Devices, July 2001*, vol. **4329**, (Proc. SPIE, 2001), p. 301
84. S. Kang, J. Shin, S.J. Kim, H.J. Kim, Y.H. Kim, Robust control of ionic polymer—Metal composites. Smart Mater. Struct. **16** (2007)
85. H. Khadivi, B.S. Aghazadeh, C. Lucas, Fuzzy control of ionic polymer-metal composites. Conf. Proc. IEEE Eng. Med. Biol. Soc. **2007**, 4198–4201 (2007)
86. Chen, Z., Tan, X., Shahinpoor, M., 2005, Quasi-static positioning of ionic polymer-metal composite (IPMC) actuators Int. Conf. Adv. Intel. Mechatronics, pp. 24–28
87. Z. Chen, D.R. Hedgepeth, X. Tan, A nonlinear, control-oriented model for ionic polymer-metal composite actuators, in *Proc. IEEE Conf. Decis. Control*, (2008), pp. 1851–1856
88. Z. Chen, T.I. Um, H. Bart-Smith, A novel fabrication of ionic polymer-metal composite membrane actuator capable of 3-dimensional kinematic motions. Sensors Actuators A Phys. **168**(1), 131–139 (2011)
89. V. Palmre, J.J. Hubbard, M. Fleming, D. Pugal, S. Kim, K.J. Kim, K.K. Leang, An IPMC-enabled bio-inspired bending/twisting fin for underwater applications. Smart Mater. Struct. **22**(1) (2013)
90. C. Zheng, S. Shatara, X. Tan, Modeling of biomimetic robotic fish propelled by an ionic polymer–metal composite caudal fin. IEEE/ASME Trans. Mechatronics **15**(3), 448–459 (2010)
91. M. Aureli, V. Kopman, M. Porfiri, Free locomotion of underwater vehicles actuated by ionic polymer metal composites. IEEE/ASME Trans. Mechatronics **15**(4), 603–614 (2010)
92. Yang, T., Chen, Z., 2015, Development of 2D Maneuverable robotic fish propelled by multiple ionic polymer-metal composite artificial fins 2015 IEEE Int. Conf. Robot. Biomimetics, IEEE-ROBIO 2015, 1(2), pp. 255–260
93. T. Xie, Tunable polymer multi-shape memory effect. Nature **464**(7286), 267–270 (2010)
94. J. Rossiter, K. Takashima, T. Mukai, Shape memory properties of ionic polymer-metal composites. Smart Mater. Struct. **21**(11) (2012)
95. Q. Shen, S. Trabia, T. Stalbaum, V. Palmre, K. Kim, I.-K. Oh, A multiple-shape memory polymer-metal composite actuator capable of programmable control, creating complex 3D motion of bending, twisting, and oscillation. Sci. Rep. **6**, 24462 (2016)

96. Q. Shen, T. Stalbaum, N. Minaian, I. Oh, A robotic multiple-shape-memory ionic polymer—Metal composite (IPMC) actuator: Modeling approach. Smart Mater. Struct **28**(1), 015009 (2019)
97. Q. Shen, *Theoretical and Experimental Investigation on the Multiple Shape Memory Ionic Polymer-Metal Composite Actuator*,. Ph.D. Disseration (University of Nevada, Las Vegas, Las Vegas, 2017)
98. Q. Shen, Z. Olsen, T. Stalbaum, J. Lee, R. Hunt, K.J. Kim, J. Kim, I.-K. Oh, Basic design of a biomimetic underwater soft robot with switchable swimming modes and programmable artificial muscles. Smart Mater. Struct. (2020). In Print
99. S.-W. Yeom, I.-K. Oh, A biomimetic jellyfish robot based on ionic polymer metal composite actuators. Smart Mater. Struct. **18**(8), 085002 (2009)
100. Z.J. Olsen, K.J. Kim, Design and modeling of a new biomimetic soft robotic jellyfish using IPMC-based electroactive polymers. Front. Robot. AI **6**(112) (2019)
101. S.D. Peterson, M. Porfiri, Energy exchange between a vortex ring and an ionic polymer metal composite. Appl. Phys. Lett. **100**(11) (2012)
102. J. Brufau-Penella, M. Puig-Vidal, P. Giannone, S. Graziani, S. Strazzeri, Characterization of the harvesting capabilities of an ionic polymer metal composite device. Smart Mater. Struct. **17**(1) (2008)
103. B.R. Martin, *Energy Harvesting Applications of Ionic Polymers* (Virginia Polytechnic Institute and State University, 2005)
104. S.V. Anand, K. Arvind, P. Bharath, D. Roy Mahapatra, Energy harvesting using ionic electro-active polymer thin films with Ag-based electrodes. Smart Mater. Struct. **19**(4) (2010)
105. R. Tiwari, K.J. Kim, S.-M. Kim, Ionic polymer-metal composite as energy harvesters. Smart Struct. Syst. **4**(5), 549–563 (2008)
106. Y. Cha, C.N. Phan, M. Porfiri, Energy exchange during slamming impact of an ionic polymer metal composite. Appl. Phys. Lett. **101**(9) (2012)
107. A. Giacomello, M. Porfiri, Underwater energy harvesting from a heavy flag hosting ionic polymer metal composites. J. Appl. Phys. **109**(8) (2011)
108. M. Aureli, C. Prince, M. Porfiri, S.D. Peterson, Energy harvesting from base excitation of ionic polymer metal composites in fluid environments. Smart Mater. Struct. **19**(1) (2010)
109. Y. Cha, S. Abdolhamidi, M. Porfiri, Energy harvesting from underwater vibration of an annular ionic polymer metal composite. Meccanica **50**(11), 2675–2690 (2015)
110. Tiwari, R., Kim, K. J., 2010, "Disc-shaped ionic polymer metal composites for use in MECHANO-electrical applications," Smart Mater. Struct., 19(6), p. 065016
111. M. Konyo, Y. Konishi, S. Tadokoro, T. Kishima, Development of velocity sensor using ionic polymer-metal composites, in *Smart Structures and Materials 2004: Electroactive Polymer Actuators and Devices (EAPAD)*, ed. by Y. Bar-Cohen, (SPIE, 2004), p. 307
112. K. Kruusamäe, P. Brunetto, S. Graziani, A. Punning, G. Di Pasquale, A. Aabloo, Self-sensing ionic polymer-metal composite actuating device with patterned surface electrodes. Polym. Int. **59**(3), 300–304 (2010)
113. A. Punning, M. Kruusmaa, A. Aabloo, A self-sensing ion conducting polymer metal composite (IPMC) actuator. Sensors Actuators A Phys. **136**(2), 656–664 (2007)
114. W. MohdIsa, A. Hunt, S.H. HosseinNia, Sensing and self-sensing actuation methods for ionic polymer-metal composite (IPMC): A review. Sensors (Basel). **19**(18) (2019)
115. A.T. Abdulsadda, X. Tan, Underwater source localization using an IPMC-based artificial lateral line, in *Proc.—IEEE Int. Conf. Robot. Autom*, (2011), pp. 2719–2724
116. M.A. Sharif, X. Tan, A pressure difference sensor inspired by fish canal lateral line. Bioinspir. Biomim. **14**(5), 055003 (2019)
117. Z. Zhu, X. He, Q. He, X. Fang, Q. Hu, H. Chen, Ionic polymer pressure sensor with gradient shape based on ion migration. J. Appl. Phys. **125**(2) (2019)
118. G. Di Pasquale, S. Graziani, A. Pollicino, S. Strazzeri, A vortex-shedding Flowmeter based on IPMCs. Smart Mater. Struct. **25**(1) (2015)
119. M.A.A. Sharif, X. Tan, IPMC flow sensor exploiting self-generated vortices. Electroact. Polym. Actuators Devices **XX**, 44 (2018)

120. H. Lei, W. Li, X. Tan, Microfabrication of IPMC cilia for bio-inspired flow sensing, in *Electroactive Polymer Actuators and Devices (EAPAD) 2012*, ed. by Y. Bar-Cohen, (SPIE, 2012), p. 83401A
121. K.J. Kim, V. Palmre, T. Stalbaum, T. Kwang, Q. Shen, S. Trabia, Promising developments in marine applications with artificial muscles: electrodeless artificial cilia microfibers. Mar. Technol. Soc. J. **50**(5), 24–34 (2016)
122. S. Sareh, A.T. Conn, J.M. Rossiter, I. Ieropoulos, P. Walters, Optimization of bio-inspired multi-segment IPMC cilia, in *Proc. SPIE 7642 Electroactive Polymer Actuators and Devices (EAPAD) 2010*, ed. by Y. Bar-Cohen, (SPIE, 2010), p. 76421S

Chapter 7
Design and Analysis of Electrohydraulic Systems for Underwater Systems Utilizing Fluidic Artificial Muscle Actuators

Edward M. Chapman and Matthew Bryant

7.1 Introduction

Fluidic artificial muscles are lightweight, flexible, easy-to-construct actuators [1]. They are inexpensive and can often be used without precision alignment [2]. They have been used as the actuator for climbing and walking robots [3], and though they were originally created for use as an orthotic for polio victims [4], they have a long history of use in rehabilitation [5] and prosthetic devices [6]. Originally, FAMs were only actuated pneumatically, but in recent studies [7] they have been shown to have increased efficiency when actuated hydraulically due to the lower compressibility of liquid working fluids compared to gasses. FAMs can be built in a variety of ways [5, 8–11], most commonly by placing an elastomeric bladder or bladder material inside a braided sheath that is typically made of nylon or Kevlar [1]. This simple construction process allows FAMs to be built cheaply and quickly and to be used in a myriad of systems. These actuators are highly biomimetic, acting to convert energy in the form of pressurized fluid into a contraction force in much the same way that muscles convert stored chemical energy to cause contraction. The similarities to biological muscle are more evident when FAMs are analyzed as part of a fully coupled electrohydraulic system. As a mammalian muscle is limited by its size and length in its ability to move a load at a given speed [12–16], so too are FAMs [1, 17–19]. FAMs are further limited to act within a system by the ability of the system to replenish or maintain its pressure source.

E. M. Chapman
Department of Physics, United States Naval Academy, Annapolis, MD, USA
e-mail: echapman@usna.edu

M. Bryant (✉)
Department of Mechanical and Aerospace Engineering, North Carolina State University, Raleigh, NC, USA
e-mail: mbryant@ncsu.edu

D. A. Paley, N. M. Wereley (eds.), *Bioinspired Sensing, Actuation, and Control in Underwater Soft Robotic Systems*, https://doi.org/10.1007/978-3-030-50476-2_7

Recently, FAMs have been used in underwater systems as pumps and energy harvesters [20]. They have shown a great deal of promise in this burgeoning field and bear further exploration as actuators [21] as well. In order to successfully integrate FAMs into underwater electrohydraulic systems, it is necessary to understand and analyze the interactions and couplings between the FAMs, the hydraulic pressure source, the electro-mechanical power converter, and the electrical power source. It has been shown that a great deal of information in system design can be gained by fully modeling an electrohydraulic system using FAMs as linear actuators [22].

The analysis presented in this chapter explores the simple special case of a single degree of freedom underwater robot arm and can serve as a template for system design of FAMs in other underwater robotics applications. The analysis proceeds by modeling each subsystem and defining the coupling terms that interconnect the various component models to produce a system model. We then present a series of analyses to determine the operating envelopes of the system for a variety of loads and predict the system efficiency for several cases within the operating envelope. The primary purpose of this chapter is to provide an example case that illustrates methods for predicting the coupled interactions and system performance implications inherent to a soft robotic actuation system driven by an electrohydraulic power system. The approach and results will be of use to design engineers seeking to size and select components for both the soft actuators and the electrohydraulic power source in order to satisfy system-level operational efficacy requirements.

7.2 Load, Actuator, and Electrohydraulic Subsystem Models

Each subsystem will be modeled mathematically beginning from first principles. These systems are the FAMs, the specific load condition, the hydraulic subsystem, and the electrical subsystem. Figure 7.1 illustrates the subsystems' interactions; as shown, they are integrated in a way that cannot be fully decoupled, so exploring their mathematical relationships is critical to system design and construction.

This exploration will begin with the load case considered for this underwater system, continue with the fluidic artificial muscle force and volume model, and conclude with the hydraulic and electrical system models.

7.2.1 Load Condition Model

The system considered here is shown in Fig. 7.2. It is a single degree of freedom robotic arm with initial setpoint values shown in Table 7.1. Since this preliminary case study focuses on determining the system interaction, we assume an open-loop controller that generates the prescribed path and no external system inputs (i.e., zero drag, neutral buoyancy, etc.) for a slow-speed motion of a slender, massless rod with a point mass load. The FAM characteristics will be described in Sect. 7.2.2.

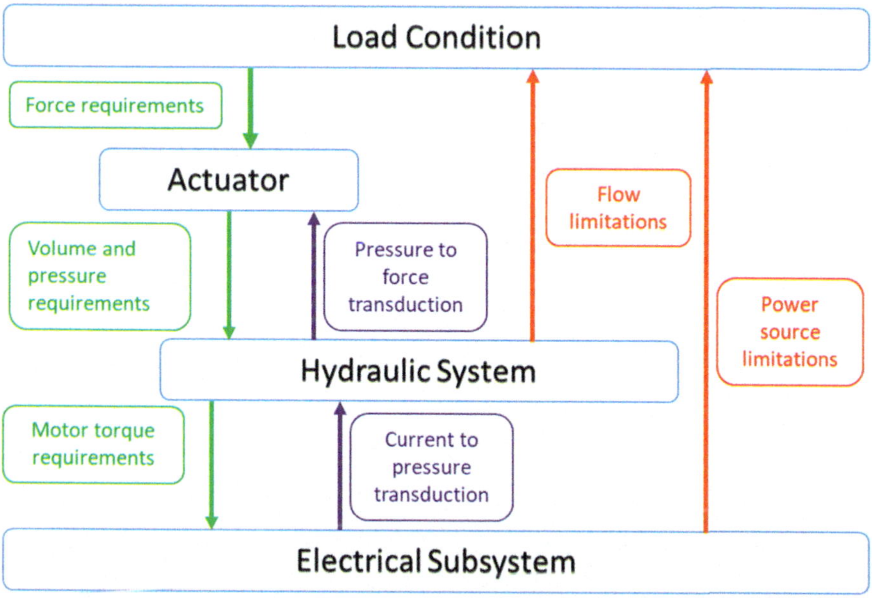

Fig. 7.1 The interactions between the load condition, actuator, hydraulic system, and electrical subsystem are strongly coupled

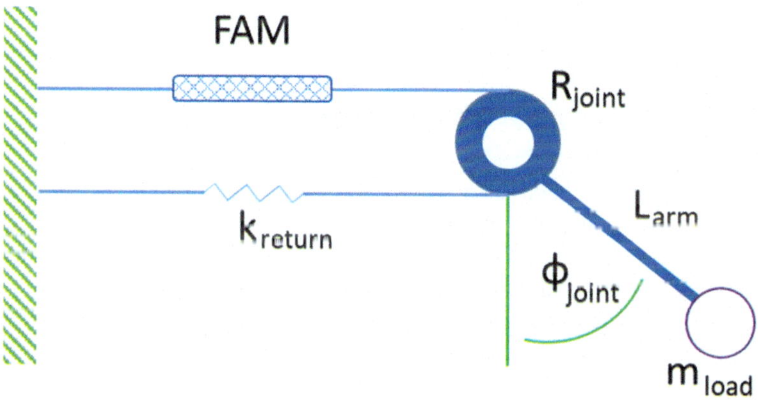

Fig. 7.2 The load condition explored in this study consists of a one degree of freedom underwater robot arm of length L_{arm} and mass m_{load} actuated by a FAM and returned with a return spring of stiffness k_{spring}

Beginning with Newton's second law in a rotational coordinate system, we can develop the following:

$$\ddot{\Phi}_{\text{joint}} = \frac{R_{\text{joint}} F_c - R_{\text{joint}}{}^2 \Phi k_{\text{return}} - L_{\text{arm}} \text{mgsin}(\Phi)}{J_{\text{load}}}, \tag{7.1}$$

Table 7.1 Dimensions of single DOF underwater robotic arm

Dimension (unit)	Value
R_{joint} (m)	0.05
L_{arm} (m)	1
m_{load} (kg)	5
k_{return} (N/m)	100

where J_{load} is the moment of inertia given by

$$J_{load} = {L_{arm}}^2 m_{load} \tag{7.2}$$

and

$$\Phi_{joint}(t) = \Phi_{max}\cos(\omega t) + \frac{\Phi_{max}}{2}, \tag{7.3}$$

where Φ_{max} is the maximum desired angle, $\omega{=}2\pi/T$ is the angular frequency, and T is period in seconds. These two values can be adjusted to find the maximum usable range and frequency for which the system is viable. The angle range will be limited by a combination of system operating pressure, load, and FAM characteristics, whereas the frequency range will be limited by both the force output limitation and the system's ability to restore pressure. This will be discussed in more detail in Sect. 7.3.

7.2.2 Fluidic Artificial Muscle Model

FAMs have been modeled in a variety of ways [1, 7, 8, 17, 18, 23–28]. The most common starting point for all of these models is the virtual work model [1]. Because each mathematical FAM model is highly specific to the FAM construction methodology, for this study the Tondu-cylindrical model [1] using only the virtual work principle is used. In doing so, we allow for system-specific design alterations that may occur depending on the construction of the FAMs selected [2]. This model [1] begins with

$$(\Delta P)\,dV = F\mathrm{dx} \tag{7.4}$$

and adapts the system using the geometry of the braid to result in the FAM contraction force,

$$F_c = \Delta P\pi {R_0}^2\left[a(1-\varepsilon)^2 - b\right], \tag{7.5}$$

where R_0 is the initial radius of the braid of the FAM and $\varepsilon = \frac{\Delta L}{L_0}$, where L is the length of the FAM, L_0 is its initial length, and ΔL is its change in length during pressurization and contraction. Additionally,

$$\Delta P = P_{\text{int}} - P_{\text{ext}}, \tag{7.6}$$

$$a = \frac{3}{\tan^2 \alpha_0}, \tag{7.7}$$

and

$$b = \frac{1}{\sin^2 \alpha_0}, \tag{7.8}$$

where P_{int} and P_{ext} are the internal pressure applied to the FAM and the external ambient underwater pressure, respectively, and α_0 is the unpressurized sheath's braid angle. Since the coupling between the FAM and the hydraulic system is via the volume consumption of the FAM during contraction, it is important to have a viable model for FAM volume as a function of contraction. Derived from the geometric relationships governing cylindrical FAM contraction [8], we arrive at

$$V_c = \pi L_0 {R_0}^2 \left[\frac{1 - \varepsilon - (1 - \varepsilon)^3 \cos^2 \alpha_0}{\sin^2 \alpha_0} \right], \tag{7.9}$$

where V_c is the consumed volume of the FAM as it contracts. As shown in previous work [3], using the chain rule to relate the volume consumption to the volumetric flowrate $\dot{V}_c$ into the FAM yields

$$\dot{V}_c = \pi L_0 {R_0}^2 \left(\frac{-1}{\sin^2 \alpha_0} + \frac{3(1 - \epsilon)^2}{\tan^2 \alpha_0} \right) \dot{\epsilon}, \tag{7.10}$$

where $\dot{\epsilon}$ is the time-rate of change of strain of the FAM due to pressurization.

7.2.3 *Hydraulic Subsystem Model*

The simple hydraulic system considered is shown in Fig. 7.3 and is made up of a positive displacement pump, a gas-charged accumulator, a bypass-and-relief valve, and the throttle control valve for the FAM. The bypass-and-pressure relief valve serves to prevent the gauge pressure of the accumulator from reaching more than the maximum operating pressure of the system. By modeling a rotary positive displacement pump, we can keep the low-pressure reservoir at or near ambient water pressure and define the pressure difference across the pump as the gauge

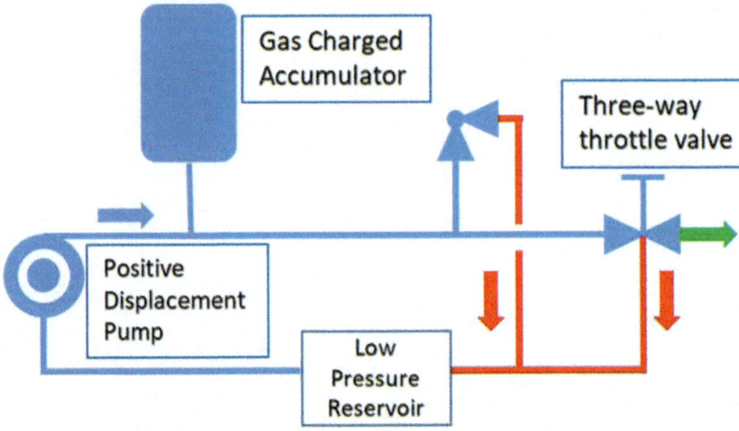

Fig. 7.3 Hydraulic system schematic, where blue represents the outgoing flow, red represents return fluid flow, and green represents the flow to the FAM

pressure of the system. The internal system pressure, P_{int}, which is moderated by the accumulator, is governed by the ideal gas law,

$$P_{\text{int}} = \frac{P_{\text{int},o} V_{\text{int},0}}{V_{\text{int},0} + V_c - V_{\text{pump}}}, \tag{7.11}$$

where V_{pump} is the volume of working fluid delivered to the accumulator from the pump, and $P_{\text{int},0}$ and $V_{\text{int},0}$ are the initial values of the accumulator pressure and gas-volume, respectively.

7.2.4 Electrical Subsystem Model

Beginning with a series RL motor circuit containing an applied voltage source and using Kirchhoff's voltage and current laws and Euler's second law of motion for a motor [29], we arrive at

$$\ddot{\theta}_{\text{motor}} = \frac{k_e I_{\text{circuit}} - B_{\text{motor}} \dot{\theta}_{\text{motor}} - \tau_{\text{pump}}}{J_{\text{motor}}}, \tag{7.12}$$

where I_{circuit} is the current in the circuit, B_{motor} is the motor frictional damping coefficient, J_{motor} is the rotational moment of inertia of the motor, and k_e is the motor's torque constant. The time-rate change of current in the circuit, $\dot{I}_{\text{circuit}}$, is described as

Table 7.2 Values of motor and pump operating characteristics

Characteristic (unit)	Value
V_{motor} (V)	18
Motor no load speed (RPM)	9970
Motor no load current (A)	0.0104
$k_{b,motor}$ (NM/A)	17.1×10^{-3}
$k_{e,motor}$ (Vs/radian)	0.0171
B_{motor} (kNM/rad-s)	1.74×10^{-10}
J_{motor} (kgm^2)	4.2×10^{-7}
R_{motor} (kΩ)	0.138
L_{motor} (mH)	0.89
D_{pump} (cm^3/rev)	1.274×10^{-3}

$$\dot{I}_{circuit} = \frac{V_{motor} - k_b - I_{circuit}R_{motor}}{L_{motor}}, \tag{7.13}$$

where V_{motor} is the circuit's applied voltage, k_b is the motor's electromotive force constant, and L is the inductance of the motor. The torque experienced by the motor-pump shaft, τ_{pump}, is described by

$$\tau_{pump} = \frac{P_{int}D_{pump}}{2\pi}, \tag{7.14}$$

where D_{pump} is the pump's displacement per revolution. The motor and pump characteristics selected for this case study [30] are described in Table 7.2.

7.3 System Response Results

The analyses performed on this system were threefold. The first determines the hydraulic and electrical system responses for a given load movement pattern. This was performed by prescribing the movement of the point mass through Eq. (7.1) and calculating its effect through the fully coupled series of equations in Sect. 7.2.

Figure 7.4 shows the movement of the robot arm and the resulting fluid volume consumption, accumulator pressure, and current variance. It is evident that these outputs are related to one another and that the ability of the system to operate is limited by the system pressure as well as the capacity of the motor-pump combination to restore this pressure for prolonged operation. In the case shown, the system operating pressure is always sufficient to meet the pressure required for the FAM to contract and meet the operating range.

As shown, in the case here ($m_{load} = 10$ kg, $\Phi_{max} = 10°$), the pump is able to replace all of the volume consumed by the FAM during contraction ($V_{pump,\,final} > V_{c,final}$). Also seen in Fig. 7.4 is that throughout contraction, the accumulator internal pressure, P_{int}, is greater than the required pressure of maximal

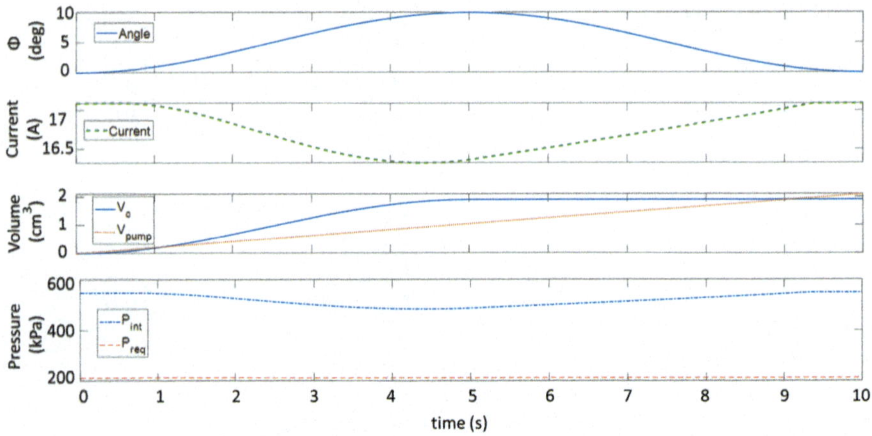

Fig. 7.4 Relationship between arm operation (shown here as angle change) with motor current draw, fluid volume consumption/supply, and internal system pressure as a result of the arm operation. The relatively slow robotic arm motion shown here allows the researchers to observe the relationships between contraction, pressure, and current-draw effectively

contraction. In some load-cases, typically for higher frequencies and larger ranges of motion, the FAM is able to successfully sweep the robot arm through the entire range of required/desired motion but the electrohydraulic system is unable to fully restore the operating pressure to its original value. In this case, the system pressure will slowly decay until the operating pressure is insufficient to actuate fully. In the event that this pressure decay occurs, we call the system frequency limited.

It has been shown [22] that, by considering only steady-state pump operation (i.e., by neglecting the effect of pressure transients on the electrical subsystem), a reasonable predictor of system operating limits can be found. These operating limits can be used to narrow design criteria prior to the more robust simulations performed in Sect. 7.2. In making this steady-state assumption, we determine the relationship between maximum volume consumption, V_c, of the FAM and maximum volume provided by the pump, V_p, where the volume provided by the pump during one period of operation, $V_{p,\mathrm{ss}}$, can be shown to be

$$V_{p,ss} = \frac{T D \dot{\theta}_{\mathrm{pump,SS}}}{2\pi} \tag{7.15}$$

and the volume consumed by the FAM, V_c, is described by Eq. (7.9). In order for the pump to recover the volume consumed by the FAM, the period of operation must be at least

$$T_{\min} = \frac{2\pi^2 L_0 {R_0}^2 \left[1 - \frac{\Phi_{\max} R_{\mathrm{joint}}}{L_o} - \left(1 - \frac{\Phi_{\max} R_{\mathrm{joint}}}{L_o} \right)^3 \cos^2 \alpha_0 \right]}{\sin^2 \alpha_0 D \dot{\theta}_{\mathrm{pump,SS}}}. \tag{7.16}$$

Additionally, this period is related to the maximum angular frequency of oscillation as

$$\omega_{\max} = \frac{\sin^2 \alpha_0 D\dot{\theta}_{\text{pump,SS}}}{2\pi L_0 R_0{}^2 \left[1 - \frac{\Phi_{\max} R_{\text{joint}}}{L_o} - \left(1 - \frac{\Phi_{\max} R_{\text{joint}}}{L_o}\right)^3 \cos^2 \alpha_0 \right]}, \tag{7.17}$$

where $\dot{\theta}_{\text{pump,SS}}$ is the steady-state operating speed of the pump at the initial internal gauge pressure, $P_{\text{int,0}}$, found using Eqs. (7.12) and (7.14).

As shown in Fig. 7.5, if an increased angle of motion for the robot arm is desired, the period of time required for the specified system to complete the motion will increase. This closed-form solution can be used to determine the limits of the system prior to detailed numerical analysis. In the event that the system pressure is insufficient to meet the load and range requirement, we call the system pressure limited. In order to find the steady-state pressure limitation, combine Eqs. (7.1) and (7.5) by setting the FAM force production, F_c, in Eq. (7.5) equal to the muscle force application, F_c, in Eq. (7.1) for a given load mass, m_{load}, and desired maximum position, $\Phi_{\max}$. This calculation yields a minimum required operating pressure, P_{req} of

$$P_{\text{req}} = \frac{\ddot{\Phi}_{\max} J_{\text{load}} + R_{\text{joint}}{}^2 \Phi_{\max} k_{\text{return}} + L_{\text{arm}} \text{mgsin}\left(\Phi_{\max}\right)}{R_{\text{joint}} \pi R_0{}^2 \left[a\left(1 - \frac{\Phi_{\max} R_{\text{joint}}}{L_o}\right)^2 - b \right]}, \tag{7.18}$$

where

$$\ddot{\Phi}_{\max} = -\Phi_{\max} \omega^2 \cos\left(\pi\right) = \Phi_{\max} \omega^2 \tag{7.19}$$

Figure 7.6 shows an example pressure limitation plot for a variety of loads under the assumption that the period is sufficiently long to allow for system recovery, as described by Eqs. (7.16) and (7.17).

As shown in Fig. 7.6, as the desired maximum angle increases, so too does the required system internal operating pressure. Also of note is the difference between Fig. 7.5 and Fig. 7.6 in that the minimum operating period is load independent (as muscle fluid volume consumption is a function only of contraction), whereas minimum operating pressure is both load and angle dependent. These operating performance limits are critical to prolonged system usage in soft-robotic systems, specifically those for underwater usage.

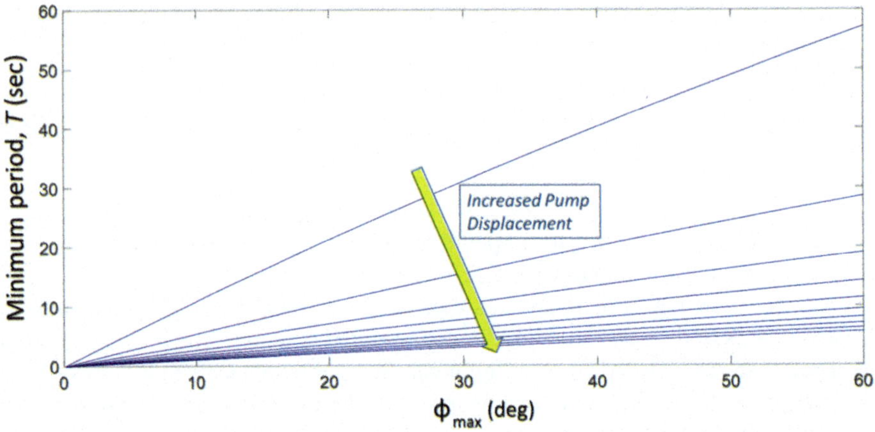

Fig. 7.5 Minimum operating period T with respect to desired arm angle for a 5 kg mass, where $T = 2\pi/\omega$, for increasing values of pump displacement

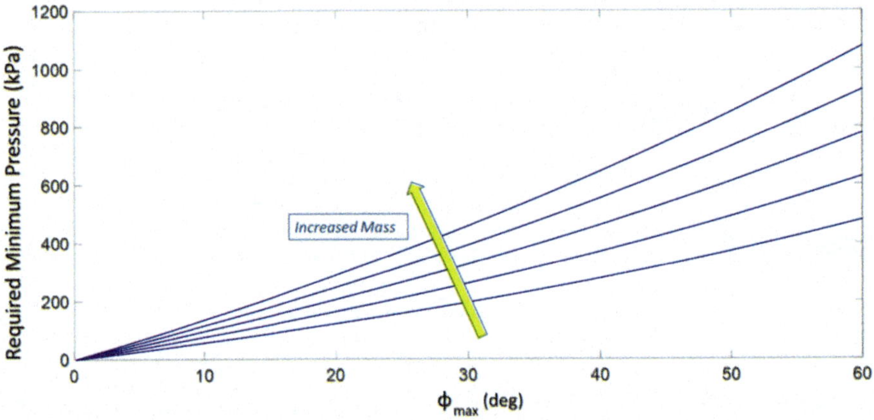

Fig. 7.6 Minimum pressure required for operating system to various maximum robot-arm angles for a 1 kg load (bottom) to a 5 kg load (top). As load mass increases, so does the required system pressure

7.4 Discussion

The aim of this study is to provide a framework for modeling a FAM-actuated electrohydraulic robotic arm by modeling each of its constituent subsystems and exploring their interactions. Its results allow researchers to generate the limits of such a system imposed by the output force of the actuator as well as the hydraulic limitations imposed by the pressurization system. It also, by way of steady-state solutions for maximum operating frequency and minimum operating pressure as functions of system parameters and desired motion kinematics, provides researchers a way to predict performance limitations in support of system design prior to in-depth analysis and construction. This case study provides guidance on a process for fully integrated system analysis for soft robotic systems using fluidic artificial muscles in a variety of operating environments. The approaches presented here can be applied to other examples of soft, fluidically powered actuators and robotic system for terrestrial or underwater applications. This type of coupled actuator and power transduction analysis captures the total dynamics and interdependencies of the system and thus provides a method to predict, analyze, and optimize system-level performance metrics.

References

1. B. Tondu, Modelling of the McKibben artificial muscle: A review. J. Intell. Mater. Syst. Struct. **23**(3), 225–253 (2012)
2. B. Tondu, P. Lopez, Modeling and control of McKibben artificial muscle robot actuators. IEEE Control. Syst. Mag. **20**(2), 15–38 (2000)
3. E.M. Chapman, T.E. Jenkins, M. Bryant, Modeling and analysis of a Meso-hydraulic climbing robot with artificial muscle actuation. Bioinspir. Biomim. **12**(6), 066010 (2017)
4. Anonymous, *More Help for Polio Victims* (Buckingham Post, 1958), p. 3
5. A.S. Camp, E.M. Chapman, P. Jaramillo Cienfuegos, Modeling and analysis of hydraulic piston actuation of McKibben fluidic artificial muscles for hand rehabilitation. Int J Robot Res (2019). Online First
6. G.S. Sawicki, D.P. Ferris, Mechanics and control of a knee ankle foot Orthosis (KAFO) powered with artificial pneumatic muscles. J. Biomech. **39**, S109–S109 (2006)
7. R. Tiwari, M.A. Meller, K.B. Wajcs, Hydraulic Artificial Muscles. J. Intell. Mater. Syst. Struct. **23**(3), 301–312 (2012)
8. M. Bryant, M.A. Meller, E. Garcia, Variable recruitment fluidic artificial muscles: Modeling and experiments. Smart Mater. Struct. **23**(7), 1–12 (2014)
9. F. Daerden, D. Lefeber, Pneumatic artificial muscles: Actuators for robotics and automation. Eur. J. Mech. Environ. Eng. **47**(1), 11–21 (2002)
10. M. De Volder, A.J.M. Moers, D. Reynaerts, Fabrication and control of miniature McKibben actuators. Sensor Actuat A-Phys **166**(1), 111–116 (2011)
11. T. Noritsugu, T. Tanaka, Application of rubber artificial muscle manipulator as a rehabilitation robot. IEEE/ASME Trans Mechantr **2**(4), 259–267 (1997)
12. A.V. Hill, P. Kupalov, Anaerobic and aerobic activity in isolated muscle. Proc R Soc Lond Ser B Biol Sci **105**(737), 313–322 (1929)
13. A.A. Biewener, *Animal locomotion* (Oxford University Press, Oxford, 2003)
14. E. Henneman, G. Somjen, D.O. Carpenter, Functional significance of cell size in spinal Motoneurons. J. Neurophysiol. **28**(3), 560–580 (1965)
15. E. Henneman, G. Somjen, D.O. Carpenter, Excitability and Inhibitability of Motoneurons of different sizes. J. Neurophysiol. **28**(3), 599–620 (1965)
16. E. Henneman, C.B. Olson, Relations between structure and function in the Design of Skeletal Muscles. J. Neurophysiol. **28**, 581–598 (1965)
17. M.A. Meller, M. Bryant, E. Garcia, Reconsidering the McKibben muscle: Energetics, operating fluid, and bladder material. J. Intell. Mater. Syst. Struct. **25**(18), 2276–2293 (2014)
18. M. Meller, J. Chipka, A. Volkov, Improving actuation efficiency through variable recruitment hydraulic McKibben muscles: Modeling, orderly recruitment control, and experiments. Bioinspir. Biomim. **11**(6), 065004 (2016)

19. C. Kothera, M. Jangid, J. Sirohi, Experimental characterization and static modeling of McKibben actuators. J. Mech. Des. **131**(9), 091010.1–10 (2009)
20. M. Philen, C. Squibb, L. Groo, Wave energy conversion using fluidic flexible matrix composite power take-off pumps. Energy Convers. Manag. **171**, 1773–1786 (2018)
21. F. Carpi, G. Frediani, C. Gerboni, Enabling variable-stiffness hand rehabilitation Orthoses with dielectric elastomer transducers. Med. Eng. Phys. **36**(2), 205–211 (2014)
22. E.M. Chapman, T. Jenkins, M. Bryant, Design and analysis of electrohydraulic pressure Systems for Variable Recruitment in fluidic artificial muscles. Smart Mater. Struct. **27**(10), 105024 (2018)
23. E. Ball, E. Garcia, Effects of bladder geometry in pneumatic artificial muscles. J Med Dev **10**(4), 041001 (2016)
24. C. Chou, B. Hannaford, Static and dynamic characteristics of McKibben pneumatic artificial muscles, in *Proceedings. IEEE International Conference on Robotics and Automation*, (1994), pp. 281–286
25. T.E. Jenkins, E.M. Chapman, M. Bryant, Control approach development for variable recruitment artificial muscles, in *SPIE Smart Structures and Materials Nondestructive Evaluation and Health Monitoring*, (International Society for Optics and Photonics, 2016), p. 979911-979911-10
26. T. Pillsbury, C. Kothera, N. Wereley, Effect of Bladder Wall thickness on miniature pneumatic artificial muscle performance. Bioinspir. Biomim. **10**(5), 055006 (2015)
27. R.M. Robinson, C.S. Kothera, N.M. Wereley, Variable recruitment testing of pneumatic artificial muscles for robotic manipulators. IEEE/ASME Trans Mechatr **20**(4), 1642–1652 (2015)
28. D. Sangian, S. Naficy, G.M. Spinks, The effect of geometry and material properties on the performance of a small hydraulic McKibben muscle system. Sensors Actuators A Phys. **234**(Supplement C), 150–157 (2015)
29. J.J. Craig, *Introduction to Robotics : Mechanics and Control* (Pearson Prentice Hall, Upper Saddle River, MJ, 2005)
30. Maxon Motor, 2015, Maxon Motor Company, 2015

Chapter 8
A Soft Robotic Model to Study the Effects of Stiffness on Fish-Like Undulatory Swimming

Zane Wolf and George V. Lauder

8.1 Introduction

A desire to understand how aquatic animals navigate their environment has accompanied humanity's sojourns into aquatic realms. We seek to learn how to maneuver and swim as efficiently as tuna or jellyfish. While the study of aquatic locomotion has a long and varied history, it has been rejuvenated in recent years by the application of modern methodologies. High-speed videography, particle image velocimetry (PIV), and robotic analogs, among others, have allowed researchers to revisit and verify old ideas as well as push into new territories, even studying the hydrodynamics of animals in situ [5, 6, 10, 12, 15–17, 42]. In particular, the use of robotics to study biological systems serves as a demarcation point, signifying not just advancing technological capabilities but also the burgeoning age of interdisciplinary science.

Sometimes the best way to study a fish, however, is to simply remove the fish from the equation. The field of robotics has facilitated the ability to create controlled experiments, using fish-like analogs to focus on a particular aspect of aquatic locomotion. These analogs can take the form of flat, two-dimensional, fish-like representations, or realistic, three-dimensional, autonomous robotic fish (e.g., [19, 35]). Similarly, the control of these robotic analogs can vary, often depending upon the complexity of the model and the question at hand. There are myriad advantages and disadvantages associated with each these systems as we discuss in more detail below. However, a complexity disparity has emerged in this field. Often, the models are simplistic to a fault or too complicated to serve as an accessible experimental model that can be easily built or altered.

Z. Wolf (✉) · G. V. Lauder
Harvard University, Department of Organismic and Evolutionary Biology, Cambridge, MA, USA
e-mail: rzwolf@g.harvard.edu; glauder@oeb.harvard.edu

D. A. Paley, N. M. Wereley (eds.), *Bioinspired Sensing, Actuation, and Control in Underwater Soft Robotic Systems*, https://doi.org/10.1007/978-3-030-50476-2_8

The experimental platforms presented in this chapter are aimed at bridging this complexity gap. By modifying a passive flexible 2D model of a fish by adding muscle-like soft robotic pneumatic actuators (PneuNets, [31]), we turn a passive, externally-activated experimental analog into an actively swimming model. PneuNets are activated with digital pressure regulators, which, when compared to previous control methods, introduce a variety of control features that can be explored and characterized in order to better understand how to swim and how to use this soft robotic model, or pneufish, efficiently. This chapter presents two versions of the pneufish, the duo- and quad-pneufishes, named for the number of pneumatic actuators. For both models we explore the available control parameter space and interpret the results with one eye towards the implications of this model as a robotic system and another towards the insights into aquatic locomotion in fish.

8.2 Background

8.2.1 How Fish Swim

One of the central concepts in fish swimming biomechanics is that fish can be categorized by their primary mode of swimming. Broadly, fish fall into two main categories: median and/or paired fin swimmers (MPF, use of fins to propel the fish) and body-caudal fin (BCF, use of the full body) [17, 18]. More specifically, BCF swimmers pass a wave of muscle activation along the body from head to tail. Two offset waves, one on each side of the body, produce the canonical body undulatory swimming patterns. The consequences of this motion are body movements in the lateral plane, starting with the head and increasing in amplitude down the body and into the tail. Typically, the tail-tip oscillatory amplitude that results from the undulatory body wave averages 0.2 body lengths (L), whereas head oscillation amplitude is on the order of 0.05 L.

Fish body musculature generating the wave of bending is unique in that it is subdivided into complex, overlapping, three-dimensional blocks known as myomeres (Fig. 8.1c) [13]. Each myomere is activated sequentially down the length of the body, and mirrored activation of the myomeres on the other side of the body occurs with a variable phase difference (Fig. 8.1d, e) [13]. This alternating activation of left and right side musculature with the traveling electrical stimulus generates the classic undulatory fish-swimming body kinematics (Fig. 8.1a, b).

The undulatory body wave results in forces actively generated by body musculature pushing on the surrounding fluid environment to generate thrust [8, 9, 17, 34]. However, fish (and other aquatic organisms) may possess morphological features, such as flexible fins or body shapes, that instead allow them to capitalize on the surrounding media acting upon them. An ongoing area of investigation in aquatic locomotion focuses on the interaction of active and passive movement—not only the separate contributions of each to locomotion but also their combined effect [9, 22].

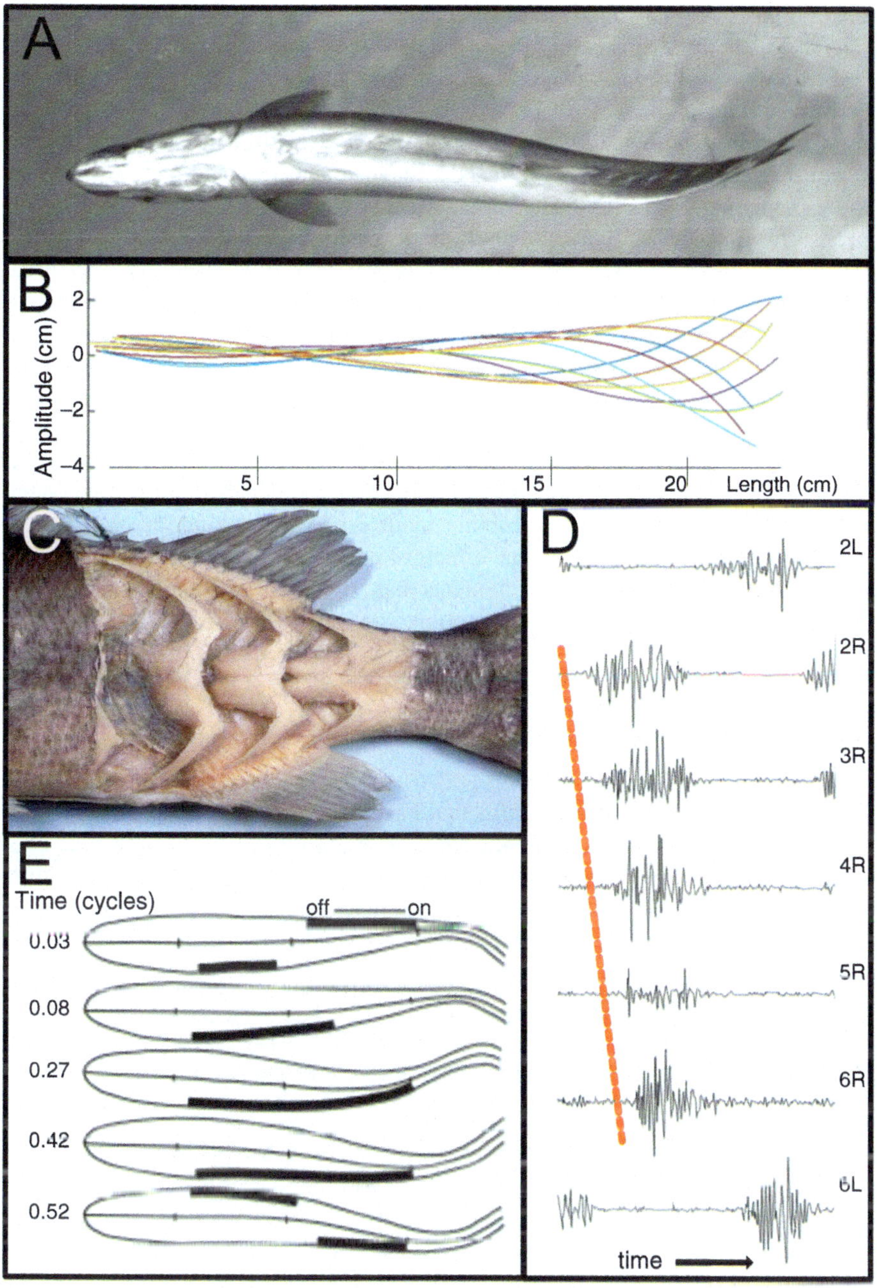

Fig. 8.1 (continued)

Fish propulsion is thus the result of both active and passive components and their interactions in three dimensions [9, 38].

Stiffness of fish tissues is a critical element that influences the effect that both active and passive movements have on propulsion, and stiffness can be controlled by fish for most of their locomotor structures. Passive stiffness is derived from an organism's composition—bones, muscle, tendons, etc. [8]. The mere presence of these components and their attachments to one another afford a certain amount of structural stiffness. Active stiffness is imparted from the interaction of those structural components with muscular contraction, as well as their shape. Fish can modulate their overall body stiffness by co-activating muscles on the left and right sides of their body [2, 24, 25]. Fish fins are also capable of active stiffness control as a result of the muscles attaching to each fin ray [1, 11]. The interaction of passive and active stiffness is not easily studied in live free-swimming fish, nor is the interaction easily studied using analogs, because modulating the overall stiffness of a robot once it has been manufactured often proves difficult. To change the stiffness of a robotic fish manufactured with rigid components, those components must be changed. Nonetheless, robotic models of various kinds have still been useful in teasing apart the individual effects of passive and active stiffness, and even models made of relatively stiff components (hard robots) have provided insights into stiffness modulation during fish locomotion.

8.2.2 Bio-Inspired Hard Robots

A diversity of hard robotic platforms that have been used to study fish propulsion is shown in Fig. 8.2c–f. Hard robotic systems are those in which stiffness is fixed due to the material properties of the component parts, although stiffness alterations certainly can be accomplished by exchanging platform components between experiments. As the components of each system are rigid, physical properties such as

Fig. 8.1 Summary of fish undulatory locomotion. (**a**) Image from high-speed video of a mackerel *(Scomber scombrus)* swimming in a laboratory flow tank, as seen in ventral view. (**b**) Body midlines of a mackerel swimming at 1 body length (BL) per second. (**c**) A largemouth bass has been dissected to showcase the complex three-dimensionality of myomeres [13]. (**d**) Jayne and Lauder [13] attached electrodes to five consecutive myomeres on the right side of a *Micropterus salmoides* and measured the time delay in activation as the fish swam at 2.4 BL/s. Orange dotted line emphasizes the anterior-to-posterior propagation of electrical activity (the electromyogram, EMG) in the myomeres. Right side activity is offset from the left side (top and bottom EMG lines). (**e**) Adapted from [13] summary of red axial muscle activity of *Micropterus salmoides* based on direct observation from 13 recording sites in one individual swimming at 2.4 L/s and with a cycle duration of 303 ms. Outlines of the fish were taken directly from digitized video images. The three reference marks on the midline of the fish indicate the posterior border of the skull, the joint between the trunk and the caudal vertebrae, and the most posterior intervertebral joint. The thick lines along the sides of the fish indicate locations of muscle activity

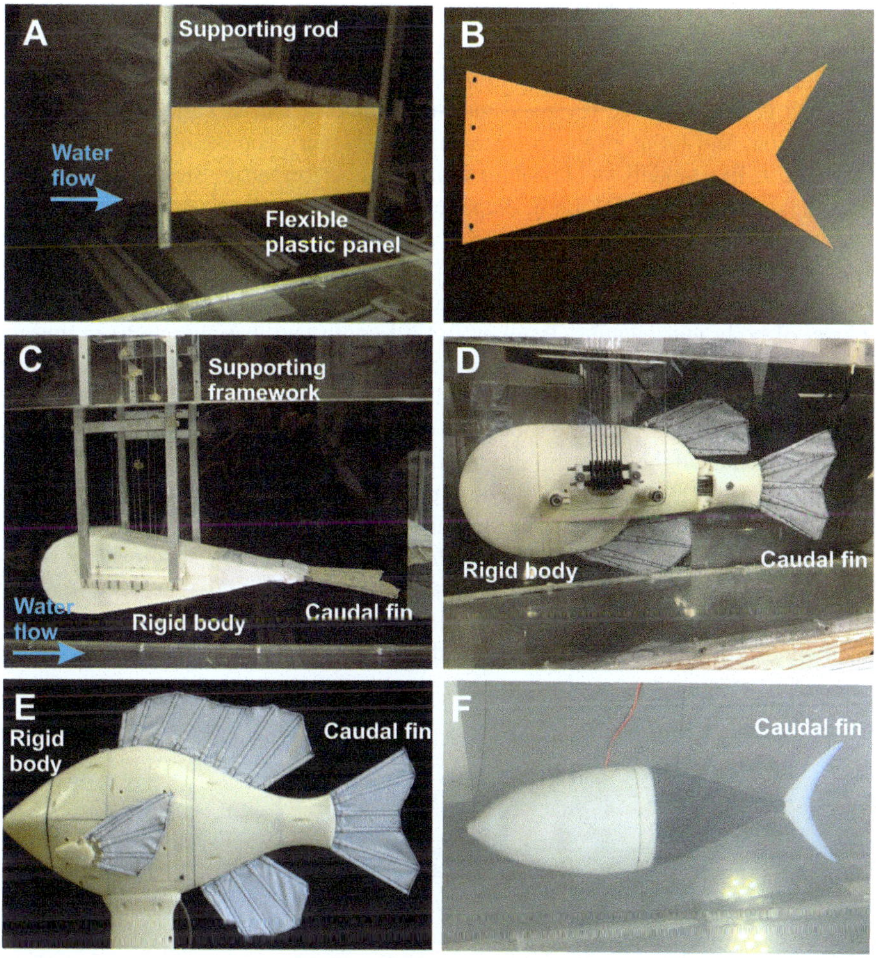

Fig. 8.2 A diversity of robotic platforms used to study fish propulsion. (**a**) and (**b**) Thin flexible propulsive panels attached to a robotic controller that produces a wave-like swimming motion [21]. (**c**) Caudal fin robot [7, 30], and (**d**) and (**e**) multi-fin robotic platforms [7]. (**f**) Semi-autonomous self-propelled tuna-like fish robot [42]). Power is provided externally, but the robot is otherwise unrestrained, and supported by two threads

Young's modulus of the body or fins determine the nature of the fluid-structure interaction. Previous research on fish propulsion using simple two-dimensional flapping panels such as those shown in Fig. 8.2a, b ([20, 21, 32, 36]) has been used to understand the fundamental dynamics of aquatic propulsion, to estimate scaling relationships, and quantify wake flow patterns under different movement programs. Locomotor movements of these passive foil models are typically generated by heave, pitch, or a combination of both at the leading edge, and the material properties of the flexible panels result in an undulatory wave passing down the panel length. Shelton et al. [35] used flexible panels to estimate fish body stiffness during swimming and arrived at a values for two different fish species that ranged from 3.3×10^{-5} Nm^2 for swimming knifefish to 9.9×10^{-4} Nm^2 for bluegill sunfish.

Long et al. [25] have reviewed the importance of stiffness modulation in fish locomotion, and Long et al. [26] used an evolutionary approach to alter stiffness among fish-like models and test their swimming performance. McHenry et al. [28] used a flexible molded fish body shape to estimate that in freely swimming sunfish (*Lepomis*) flexural stiffness of the body increases by a factor of two over resting body stiffness.

Finally, using a computational approach that generated body waveforms from a model of body musculature, Tytell et al. [39] confirmed that alterations in muscle activity patterns can change body stiffness, and suggested that fish are able to tune this stiffness to improve swimming speed and acceleration.

More realistic hard robotic models of individual fish fins (Fig. 8.2c) have been used to understand the effects of different fin movement patterns on thrust produced by single fins [7]. Multi-fin robotic platforms (Fig. 8.2d, e) allow the fluid dynamic interactions among fins to be studied, which is a key feature of fish propulsion [23, 37]. Finally, a deforming model of the fish body (Fig. 8.2f) allows more natural fish-like swimming behavior to be studied in a laboratory setting. To date, however, few such systems have allowed for active stiffness modulation during swimming, and their control mechanisms are complicated [43–45].

8.3 The Pneufish Model

The soft robotic apparatus known as the pneufish was originally designed and built by Jusufi et al. [14] and utilizes the same plastic shim foils as a midline backbone described in Sect. 8.2.2. Muscle-like actuators, or PneuNets, were created using 3D-printed molds and silicone epoxy [14, 31]. PneuNets are comprised of eleven chambers connected through a ventral channel. Upon inflation, the chambers swell and push against each other, resulting in a net curvature of the actuator (Fig. 8.3a). Two PneuNets were placed symmetrically on either side of the foil, and alternating activation of the PneuNets resulted in the curvature of the foil to one side or other. With repetitive activation, the motion resembled a fish-like swimming motion.

Jusufi et al. [14] initially characterized this soft robotic apparatus and its performance: amplitude decreased with frequency, net thrust decreased linearly with increasing flow speeds and increased with frequency. Roughly 4% of bilateral co-activation resulted in higher net thrust than no co-activation. This study laid the groundwork for establishing this apparatus as a model for studying fish-like undulatory motion.

When analyzing midlines from swimming motions generated by pneufish, the apparatus never exhibited what might be considered highly-flexible fish-like undulatory motion; i.e., the wavelength always exceeded the length of the pneufish. Additionally, this initial study did not examine the effect of pressure modulation or backbone stiffness. To address these issues and further solidify the usefulness of the apparatus as a model for undulatory motion, we redesigned the control system in hopes of offering more precise pressure actuation of the PneuNets as well as providing more control parameters overall.

Additionally, to make a fully undulatory apparatus, we extended the length of the backbone foil and added another pair of symmetrically placed PneuNets, creating the quad-pneufish (Fig. 8.3b). The duo-pneufish is the pneufish with a single pair of PneuNets. Each pneufish and associated air hoses are attached to a housing rod. Measured pneufish midlines during swimming show that the quad-pneufish model is capable of undulatory motion. Depending upon the control parameter combination, midline movements also could reproduce the flapping motion of the duo-pneufish (Fig. 8.3c). With both models, we investigated a larger control space than presented in [14], with a particular focus on the interaction between the passive and active aspects of stiffness.

We substituted the solenoid pressure valves used by Jusufi et al. [14] with digital pressure regulators (Model: ITV00050-2U(B)L, SMC Pneumatics, Yorba Linda, CA). These pressure regulators accept an input signal crafted using pulse-width modulation and a low-pass RC filter (100 nF capacitor and 16.1 kΩ resistor, resulting in a cut-off frequency of 98.8 Hz), which were provided by a micro-controller circuit board (Arduino, Uno, SmartProjects, Italy). The Arduino's output signal was set to a value between 0–5 V. Prior to experiments, the pressure regulators were characterized and their differences mapped using both a pressure gauge and radius of curvature analysis of the duo-pneufish during one-sided activation.

Unlike with the previous analog pressure valves, the input signal did not need to default to a null pressure of 0 kPa. Thus, the digital pressure regulators afforded the ability to dynamically increase the overall pressure (and hence stiffness) of the pneufish during activation. We used two control parameters: the maximum pressure in the activating PneuNet and the minimum pressure of the non-activating PneuNet, which provided opposition to the activating PneuNet. We retained the ability to control bilateral co-activation, or phasing, with four regulators: each PneuNet had its own maximum and minimum pressure.

Figure 8.3d diagrams the experimental setup. The pressure regulators were controlled using an Arduino Uno, where the activation frequency, maximum pressure, minimum pressure, inflation rate, and phasing were set. The pneufish were anchored in place in the flow tank and a 6-axis ATI force transducer, which measures force and torque, was attached to the housing rod. Amplitude and midline kinematics were collected from a ventral high-speed camera. To analyze these data, we applied linear modeling to determine the importance of each individual parameter, and combinations of parameter values, on performance data. Linear models were fit using R (glm, [33]), and analyzed and selected using Akaike Information Criterion (AIC, [40]) and the stepAIC function. Because our input parameters had various units and some are categorical in nature, we transformed all the variables to dimensionless variables. This had the benefit of allowing nonlinear trends to become apparent without needing to fit a nonlinear model.

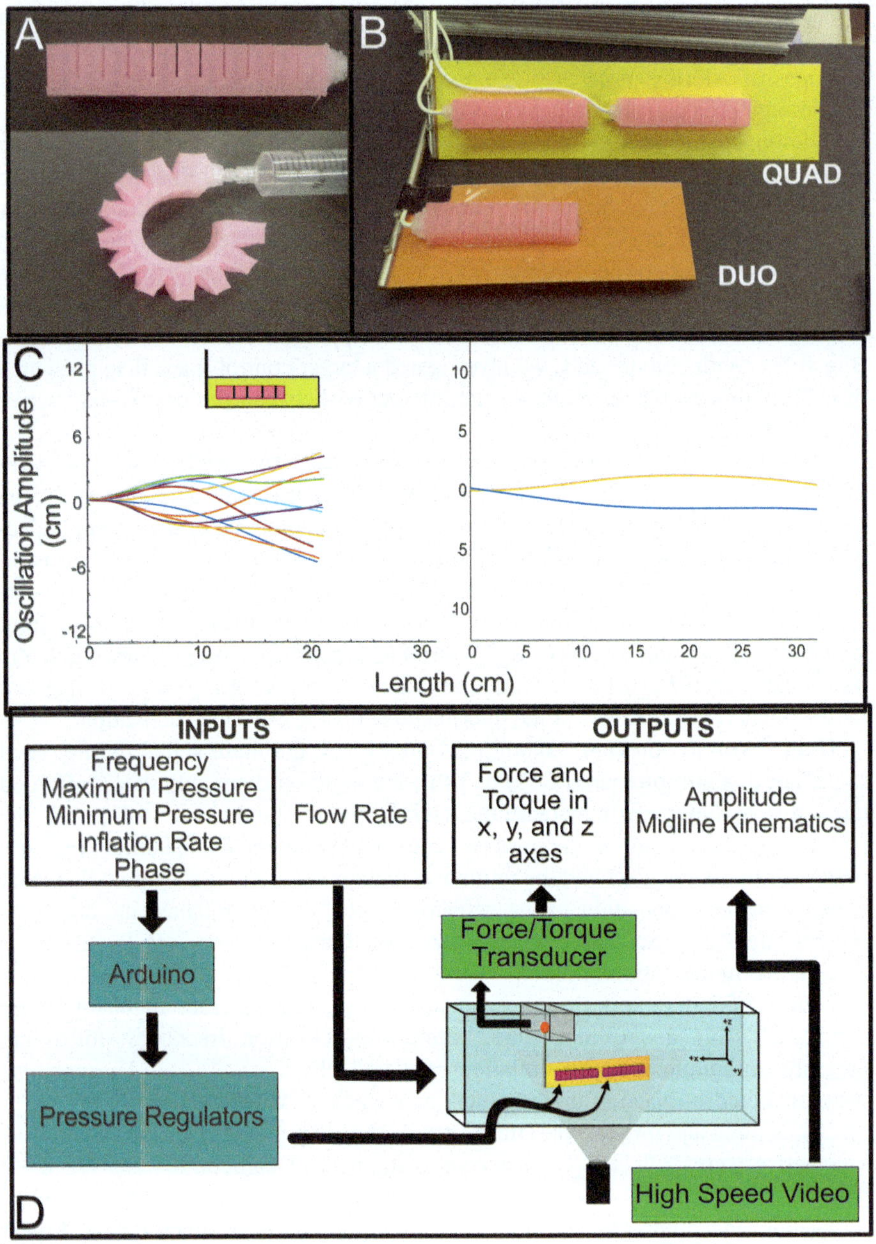

Fig. 8.3 Duo- and quad-pneufishes and experimental design. (**a**) *Top* shows a 2 × 2 × 10 cm uninflated PneuNet (made from DragonSkin 20 silicone), and *Bottom* shows an inflated PneuNet. The individual chambers expand and push against each other, resulting in a net curvature. (**b**) A duo- and quad-pneufish compared. The backbone foils are attached to a leading edge rod, and

8.3.1 The Duo-Pneufish

To more fully explore control of the pneufish model with regard to active and passive stiffness, the duo-pneufish was activated across a large parameter space, comprising 288 parameter combinations, with four replicates. One high-speed video was recorded for each parameter combination to obtain midline kinematics and amplitude data. The parameter values were as follows: Activation Frequency (0.25, 0.5, 0.75, and 1.0 Hz), Backbone Foil Stiffness (Flexible [8.1e−4 Nm^2], Medium [9.9e−4 Nm^2], and Stiff [1.1e−2 Nm^2]), Maximum Pressure (0.5, 0.75, and 1.0 kPa), Minimum Pressure (0, 5, 10, and 25%), and Flow Speed (0 and 5.3 cm/s).

Though not an input parameter, we initially predicted that tail-tip amplitude would serve as a good predictor of swimming performance (specifically, forward thrust) in the pneufish. We began the linear modeling process as simply as possible: Forward Thrust (Fx) $\sim$ Amplitude. A significant model was found: predicted net thrust was equal to $-2.7 + 1.2$(Amplitude) mN, where amplitude is measured in centimeters. The amplitude coefficient had a p-value of 1.1e−13 (computed within the glm function using a Wald test, where $p < 0.01$ is significant) and the R^2 value of this model was 0.17. Based on these results, amplitude is an important contributor to thrust, but it does not have the power to explain more than 17% of the variance in the data.

Our next step was to include flow speed as a second predictor, as we intuited that the presence or absence of drag, with an interaction effect with amplitude (the larger the tail-tip amplitude, the greater surface area of the pneufish body would be exposed to drag) would be the next simplest model we could make. A simple ANOVA comparing thrust data at 0 and 5.3 cm/s did in fact show that data taken at 5.3 cm/s had lower thrust values across the board (Fig. 8.4c). This model, Fx = $7.1 + 1.1 *$ (Amplitude) $- 8.4 *$ (Flowspeed) $- 0.79*$(Amplitude:Flow speed), had coefficient p-values that were <0.001, <0.01, and ≤ 0.001, respectively, and an R^2 value of 0.55. That is, amplitude and flow speed alone were able to account for half of the variation in our data set. While these two models provide some support for our initial ideas that the amount of water pushed by the pneufish and the amount of drag present were key factors in the performance of our model, half of the variation present in the data still needed to be explained.

Ultimately, we found the best-fit model, according to Akaike Information Criterion (AIC) analysis [40].

Fig. 8.3 (continued) inflation tubes (white) are attached to these rods. (**c**) Midlines of the duo-pneufish (left) and quad-pneufish (right) at 0.75 Hz. (**d**) The experimental setup: an Arduino Uno microprocessor was used to control the frequency, maximum pressure, minimum pressure, inflation rate of the PneuNets, and phasing (for quad only). The flow rate of water in the flow tank was controlled separately. A six-axis ATI force/torque transducer attached to the housing rod of the pneufishes collected force and torque data, while a ventral high-speed camera was used to extract amplitude and midline kinematic data

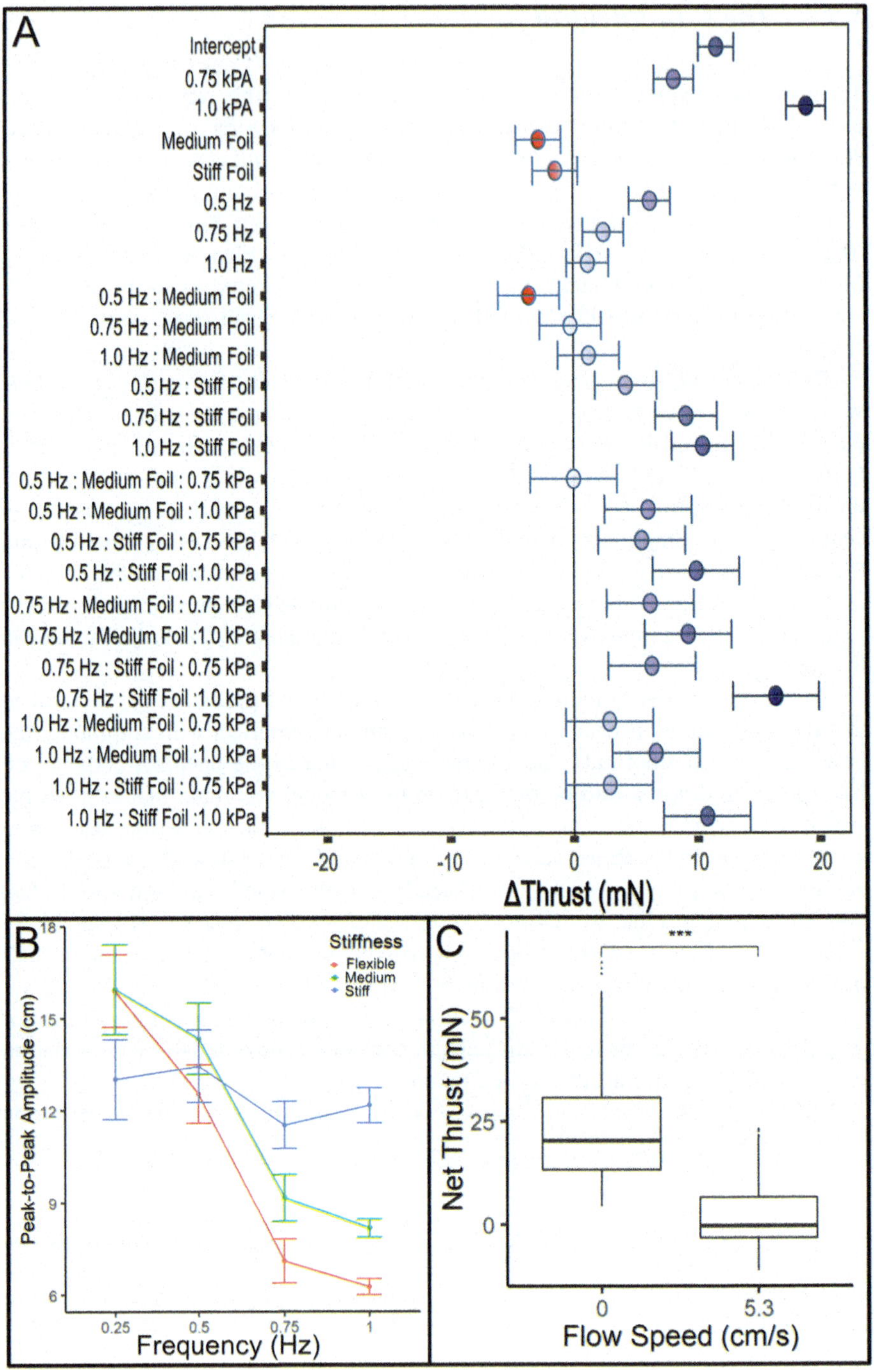

Fig. 8.4 (continued)

$$\begin{aligned}\text{Fx} \sim \text{Frequency}+\text{FlowSpeed}+\text{Foil Stiffness}+\text{Maximum Pressure}+\text{Minimum Pressure}+\\ \text{Frequency} * \text{FlowSpeed} + \text{Frequency} * \text{Foil Stiffness} + \text{Frequency} * \text{Maximum Pressure}+\\ \text{FlowSpeed} * \text{FoilStiffness} + \text{FlowSpeed} * \text{Minimum Pressure}+\\ \text{Foil Stiffness} * \text{Maximum Pressure}+\\ \text{Foil Stiffness} * \text{Minimum Pressure} * \text{Frequency}\end{aligned} \tag{8.1}$$

This final model and all coefficients are presented in Wolf and Lauder [41]. The coefficient values of each component in the model relate which parameters and parameter interactions (denoted with an asterisk) contribute the most to thrust performance in the pneufish. Parameters and interactions not included in the model (e.g., Maximum Pressure * Minimum Pressure) were removed during AIC analysis, as they did not explain variance in the data. For each parameter combination of interest (e.g., 0.5 Hz, 0 cm/s, 0.5 kPa Max. Pressure, 0% Min. Pressure, and interactions [0.5 Hz * 0.5kPa]), coefficients from the model are inserted into the equation to determine the thrust value generated by the pneufish at those control parameter values. As these are all categorical variables, rather than a mixture of continuous and categorical, there are no dimensional constants needed in this equation.

This full model had an adjusted R^2 value of 0.97 and 237 degrees of freedom. For brevity, we present coefficients of interest in Fig. 8.4a. By themselves, increasing foil stiffness decreased thrust production, and increasing frequency had diminishing returns on increasing thrust (i.e., the increase seen in thrust decreases with higher frequencies) (Fig. 8.4a). As frequency increased, the tail-tip amplitude, or peak-to-peak amplitude of the pneufish, decreased (Fig. 8.4b).

However, the combination of backbone stiffness, frequency, and maximum pressure suggests an interaction effect: high frequency motion is most effective with a stiffer body.

This interaction effect between active stiffness (maximum pressure), passive stiffness (backbone foil stiffness), and frequency has been postulated and observed in experiments conducted on dead fish. For example, McHenry et al. [28] conducted

Fig. 8.4 Selected Results from Duo-Pneufish Experiments. (**a**) Fitting a linear model to the data generates parameter coefficients presented here to show an interaction effect between frequency, stiffness, and maximum pressure. The *x*-axis value of each coefficient represents the change in thrust conferred by changing from the intercept parameter value to the new value (e.g., increasing from 0.5 kPa of maximum pressure to 0.75 kPa increases net thrust by roughly 8 mN). The color gradient represents the coefficients that increase thrust the most (dark blue) to those that decrease thrust the most (dark red). The intercept parameter combination is: flexible foil; 0.5 kPa max. pressure; 0.25 Hz; 0% min. pressure; and 0 cm/s flow speed. (**b**) Increasing activation frequency tends to decrease tail tip amplitude in the flexible and medium duo-pneufishes, but the stiffest pneufish remains nearly constant in tail-tip amplitude. (**c**) Increasing flow speed dramatically decreases net thrust experiments on models of sunfish, with various passive flexural stiffnesses, and investigated the effect of increasing activation frequency on performance parameters (e.g., wavelength, swimming speed, Froude efficiency). Ultimately, McHenry et al. posited that in order for fish to swim efficiently at higher frequencies, fish would need to actively increase their flexural stiffness at least twofold over their passive stiffness value [28].

The presence of this stiffness/frequency interaction effect in our pneufish model has two primary implications. First, this interaction appears to be an inherent feature of undulatory swimming mechanics [2–4, 16, 24, 25, 27–29]. Second, the actively swimming duo-pneufish model, despite being relatively simple compared to its hard robotic companions in the field, is able to replicate the findings of other established models and studies.

8.3.2 The Quad-Pneufish

One of the limitations of the duo-pneufish presented above is the lack of wave-like undulatory motion exhibited by the model. The midlines all demonstrate a "flapping" motion, but never is a wavelength greater than the length of the model (Fig. 8.3c). Therefore, one of our next goals was to make a model capable of true undulatory motion. Using two sets of PneuNets, the midlines of the quad-pneufish began resembling the midlines of swimming fishes (Figs. 8.1b and 8.3c).

With two sets of PneuNets to be activated, we decided to investigate the effect of differential pressure and co-activation in the front and rear PneuNet pairs on swimming performance. In particular, we were curious how these control parameters might affect the balance between forward and lateral force production. When swimming, a certain amount of energy is lost due to the lateral components of the momentum imparted from the swimming body to the water. Limiting the production of lateral forces while maximizing forward thrust would be ideal, theoretically.

Our parameter space for the quad-pneufish was as follows: Activation Frequency (0.25, 0.5, and 0.75 Hz), Backbone Foil Stiffness (Medium [9.9e−4 Nm^2]), and Co-activation between left and right PneuNets (0, 10, 20, and 30%). Additionally, we investigated front-back parameter pairs: Minimum pressure combinations of 0–0%, 0–30%, 30–0%, and 30–30%, and Maximum Pressure Combinations of every combination of 0.5, 0.75, and 1.0 kPa (7 total). Frequency and co-activation were not varied in such a manner during this experiment. In total, this parameter space had 336 parameter combinations, again performed in quadruplicate.

We fitted a linear model to both the net forward thrust and the peak-to-peak lateral forces with the same methodology described for the duo-pneufish models. Peak-to-peak, rather than mean, lateral force measurements were used because, as the movement and force generation should be symmetrical, then the average would be zero. The peak-to-peak values also reflect the side force oscillation that ultimately moves the center of mass, thus affecting propulsive efficiency. The AIC process to determine a best-fit model resulted in:

$$\begin{aligned}\text{Fx} \sim \text{Frequency} + \text{Maximum Pressure Front/Back Combo} + \text{Minimum Pressure Combo}+\\ \text{Co-Activation} + \text{Frequency} * \text{Max. Pressure Combo}+\\ \text{Frequency} * \text{Min. Pressure Combo} + \text{Frequency} * \text{Co-activation}\end{aligned} \tag{8.2}$$

This model had an R^2 of 0.97. Again, for brevity, we present only a selection of the parameters in Fig. 8.5a. The coefficients for the forward thrust model are presented in red, with the lateral force model coefficients represented in blue. Note that the lateral forces were measured to be an order of magnitude higher than the thrust forces. To decrease lateral force oscillation, it is best to set the active stiffness of the anterior PneuNets through minimum pressure, while leaving the rear PneuNets more flexible (Min. Pressure combo 30–0%, marked with asterisks).

Figure 8.5b shows the effect of having higher rear maximum pressure, higher front maximum pressure, and equal front/rear maximum pressure, while minimum pressure is kept zero, on net thrust production. Higher pressures generate more thrust. This implies that thrust is independent of frequency when rear pressure is higher. When front pressure is higher, thrust starts high, but transitions to a similar low pattern with higher frequencies. Oscillation amplitude could explain this. When moving slowly, a large anterior oscillation will result in an even larger tail-tip amplitude.

These results suggest that altering active stiffness can affect the balance of forward thrust and lateral forces, which can be a key consideration in how fish efficiently navigate their environment.

8.4 Conclusions and Future Directions

Designing mechanical systems that emulate the functional design of fishes is a substantial challenge. Both hard robotic systems with rigid components and soft mechanical systems with flexible elements offer insights into the dynamics of aquatic propulsion. And yet we still fall short of matching fish performance and understanding how fish modulate and control body surface deformation.

In particular, understanding how fish modulate body stiffness and the performance consequences of stiffness changes during swimming remain an active area of investigation. This chapter presents two soft robotic models that allowed us to alter stiffness and better understand how changing body stiffness affects kinematics and both thrust and lateral force production. We demonstrate an interaction between segmental body activation frequency and stiffness, and show that variable longi-

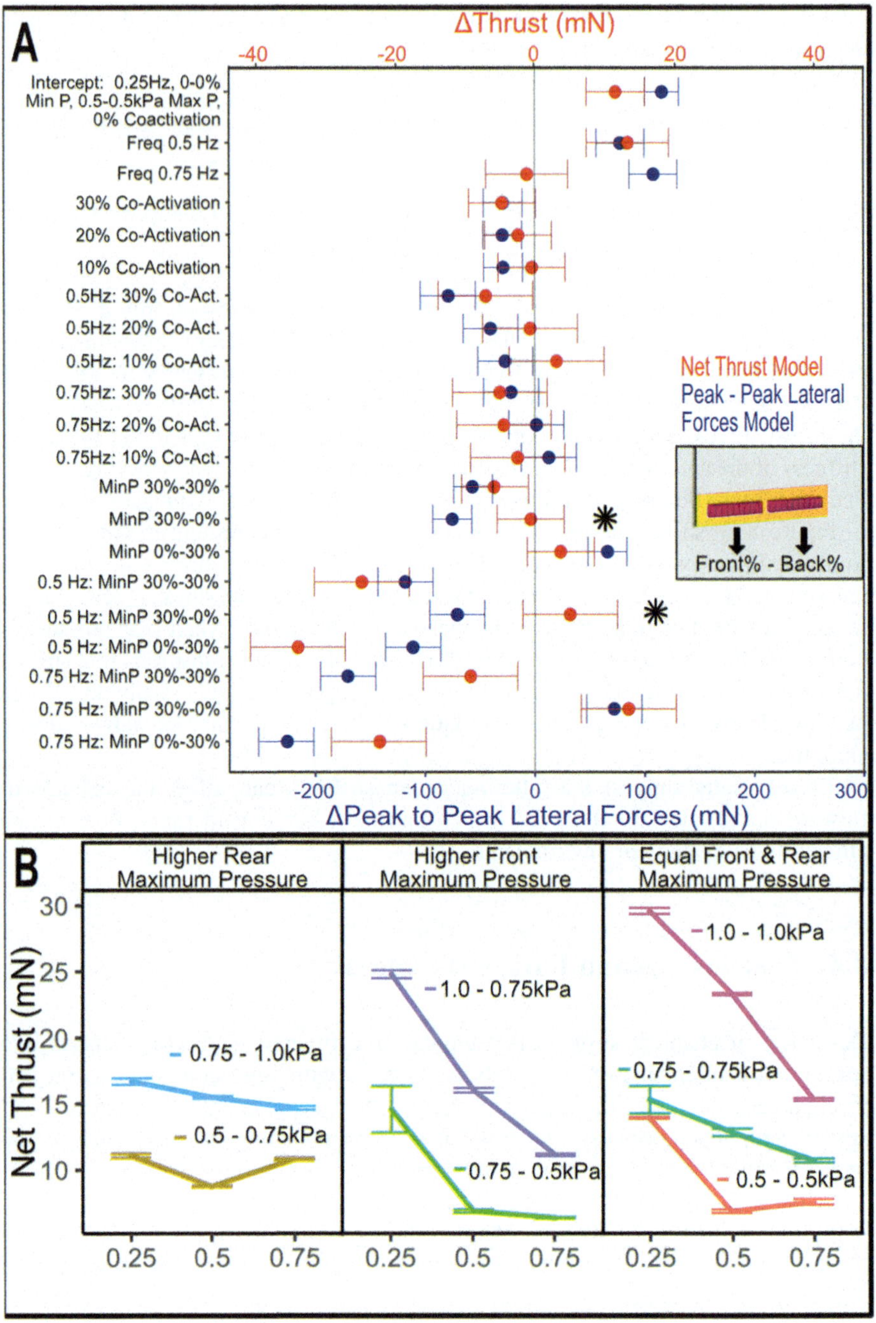

Fig. 8.5 (continued)

tudinal stiffness from the anterior to posterior of the pneufish can decrease lateral forces.

In the future, using pneumatic models with shorter PneuNets capable of generating more complex kinematics will permit us to focus more specifically on kinematic correlates of thrust and side forces, and to better mimic fish swimming kinematics with a greater extent of bilateral co-activation. Controlling this increased complexity with appropriate phasing and activation sequencing will involve its own challenges, but we believe that future soft robotic models nonetheless offer considerable opportunities as a platform for understanding fish locomotor dynamics.

Acknowledgments Many thanks to Ardian Jusifi, Daniel Vogt, and members of the Lauder Lab for assistance with this project and ongoing collaborative research on fish robotics and locomotion. Mackerel swimming kinematic data in Fig. 8.1 were obtained by Dr. Valentina Di Santo. This research was supported by the Department of Organismic and Evolutionary Biology at Harvard University, the Ashford Foundation, and the Office of Naval Research (Tom McKenna, Program Manager, ONR 341), grant number N00014-15-1-2234.

References

1. S. Alben, P.G. Madden, G.V. Lauder, The mechanics of active fin-shape control in ray-finned fishes. J. R. Soc. Interface **4**(13), 243–256 (2007). https://doi.org/10.1098/rsif.2006.0181
2. J.D. Altringham, D.J. Ellerby, Fish swimming: patterns in muscle function. J. Exp. Biol. **202**(23), 3397–3403 (1999)
3. D. Barrett, M.S. Triantafyllou, D. Yue, M. Grosenbaugh, M. Wolfgang, Drag reduction in fish-like locomotion. J. Fluid Mech. **392**, 183–212 (1999). https://doi.org/10.1017/S0022112099005455
4. J.E. Colgate, K.M. Lynch, Mechanics and control of swimming: a review. IEEE J. Ocean. Eng. **29**(3), 660–673 (2004). https://doi.org/10.1109/JOE.2004.833208
5. O.M. Curet, N.A. Patankar, G.V. Lauder, M.A. MacIver, Mechanical properties of a bio-inspired robotic knifefish with an undulatory propulsor. Bioinspir. Biomim. **6**(2) (2011). https://doi.org/10.1088/1748-3182/6/2/026004
6. E.G. Drucker, G.V. Lauder, Experimental hydrodynamics of fish locomotion: functional insights from wake visualization. Integr. Comp. Biol. **42**(2), 243–257 (2002). https://doi.org/10.1093/icb/42.2.243
7. C.J. Esposito, J.L. Tangorra, B.E. Flammang, G.V. Lauder, A robotic fish caudal fin: effects of stiffness and motor program on locomotor performance. J. Exp. Biol. **215**(1), 56–67 (2012). https://doi.org/10.1242/jeb.062711

Fig. 8.5 Selected results from quad-pneufish experiments. (**a**) Fitting a linear model to the data produces coefficients that show the combined effect of a parameter combination on both the net thrust (red) and lateral forces (blue). Note the scale difference between the net thrust coefficients (top axis) and the lateral force coefficients (bottom axis). Combinations that show the pattern of increasing net thrust while decreasing lateral forces are highlighted with a black asterisk. (**b**) The results of varying maximum pressure between the front and back PneuNet pairs on net thrust, plotted by frequency. The left panel shows maximum pressure pairs where the posterior PneuNet maximum pressure is higher than the front. The middle panel shows data where the anterior PneuNets' pressure is higher. The right panel shows data with equal front and back PneuNet pressures

8. F. Fish, G.V. Lauder, Passive and active flow control by swimming fishes and mammals. Annu. Rev. Fluid Mech. **38**(1), 193–224 (2006). https://doi.org/10.1146/annurev.fluid.38.050304.092201. http://www.annualreviews.org/doi/10.1146/annurev.fluid.38.050304.092201
9. F.E. Fish, G.V. Lauder, Control surfaces of aquatic vertebrates: active and passive design and function. J. Exp. Biol. **220**(23), 4351–4363 (2017). https://doi.org/10.1242/jeb.149617. http://jeb.biologists.org/lookup/doi/10.1242/jeb.149617
10. B. Flammang, G.V. Lauder, D.R. Troolin, T. Strand, Volumetric imaging of shark tail hydrodynamics reveals a three-dimensional dual-ring vortex wake structure. Proceedings. R. Soc. **278**(1725), 3670–8 (2011). http://rspb.royalsocietypublishing.org/content/early/2011/05/04/rspb.2011.0489.short
11. B.E. Flammang, S. Alben, P.G. Madden, G.V. Lauder, Functional morphology of the fin rays of teleost fishes. J. Morphol. **274**(9), 1044–1059 (2013). https://doi.org/10.1002/jmor.20161
12. N. Gravish, G.V. Lauder, Robotics-inspired biology. J. Exp. Biol. **221**(7), 1–8 (2018). https://doi.org/10.1242/jeb.138438. http://jeb.biologists.org/lookup/doi/10.1242/jeb.138438
13. B.C. Jayne, G.V. Lauder, Are muscle fibers within fish myotomes activated synchronously? Patterns of recruitment within deep myomeric musculature during swimming in largemouth bass. J. Exp. Biol. **198**(Pt 3), 805–15 (1995). http://www.ncbi.nlm.nih.gov/pubmed/9318580
14. A. Jusufi, D.M. Vogt, R.J. Wood, G.V. Lauder, Undulatory swimming performance and body stiffness modulation in a soft robotic fish-inspired physical model. Soft Rob. **4**(3), 202–210 (2017). https://doi.org/10.1089/soro.2016.0053
15. K. Katija, R.E. Sherlock, A.D. Sherman, B.H. Robison, New technology reveals the role of giant larvaceans in oceanic carbon cycling pp. 1–8 (2017)
16. G.V. Lauder, Swimming hydrodynamics: ten questions and the technical approaches needed to resolve them. Exp. Fluids **51**(1), 23–35 (2011). https://doi.org/10.1007/s00348-009-0765-8
17. G.V. Lauder, Fish locomotion: recent advances and new directions. Annu. Rev. Marine Sci. **7**(1), 521–545 (2015). https://doi.org/10.1146/annurev-marine-010814-015614. http://www.annualreviews.org/doi/abs/10.1146/annurev-marine-010814-015614
18. G.V. Lauder, P.G.A. Madden, Learning from fish: kinematics and experimental hydrodynamics for roboticists. Int. J. Autom. Comput. **3**(4), 325–335 (2006). https://doi.org/10.1007/s11633-006-0325-0
19. G.V. Lauder, E.J. Anderson, J. Tangorra, P.G. Madden, Fish biorobotics: kinematics and hydrodynamics of self-propulsion. J. Exp. Biol. **210**(16), 2767–2780 (2007). https://doi.org/10.1242/jeb.000265
20. G.V. Lauder, E.J. Anderson, J. Tangorra, P.G.A. Madden, Fish biorobotics: kinematics and hydrodynamics of self-propulsion. J. Exp. Biol. **210**(16), 2767–2780 (2007). https://doi.org/10.1242/jeb.000265
21. G.V. Lauder, B. Flammang, S. Alben, Passive robotic models of propulsion by the bodies and caudal fins of fish. Integr. Comp. Biol. **52**(5), 576–587 (2012). https://doi.org/10.1093/icb/ics096
22. J.C. Liao, Neuromuscular control of trout swimming in a vortex street: implications for energy economy during the Karman gait. J. Exp. Biol. **207**(20), 3495–3506 (2004). https://doi.org/10.1242/jeb.01125
23. G. Liu, Y. Ren, H. Dong, O. Akanyeti, J.C. Liao, G.V. Lauder, Computational analysis of vortex dynamics and performance enhancement due to body-fin and fin-fin interactions in fish-like locomotion. J. Fluid Mech. **829**, 65–88 (2017). https://doi.org/10.1017/jfm.2017.533
24. J.H. Long Jr., Muscles, elastic energy, and the dynamics of body stiffness in swimming Eels. Am. Zool. **38**(4), 771–792 (1998)
25. J.H. Long, K.S. Nipper, J.H. Long Jr., K.S. Nipper, The importance of body stiffness in undulatory propulsion. Am. Zool. **36**(6), 678–694 (1996). https://doi.org/10.1093/icb/36.6.678
26. J.A. Long, G.C. Young, T. Holland, T.J. Senden, E.M.G. Fitzgerald, An exceptional Devonian fish from Australia sheds light on tetrapod origins. Nature **444**(7116), 199–202 (2006). https://doi.org/10.1038/nature05243
27. M.A. MacIver, E. Fontaine, J.W. Burdick, Designing future underwater vehicles: principles and mechanisms of the weakly electric fish. IEEE J. Ocean. Eng. **29**(3), 651–659 (2004). https://doi.org/10.1109/JOE.2004.833210

28. M. McHenry, C. Pell, J.H. Long Jr., Mechanical control of swimming speed: stiffness and axial wave form in undulating fish models. J. Exp. Biol. **198**(Pt 11), 2293–305 (1995). http://www.ncbi.nlm.nih.gov/pubmed/9320209
29. K.M.W. Mcletchie, Drag reduction of an elastic fish model, in *Oceans 2003. Celebrating the Past... Teaming Toward the Future (IEEE Cat. No. 03CH37492)* vol. 5 (2003), pp. 2938–2944
30. A.P. Mignano, S. Kadapa, J.L. Tangorra, G.V. Lauder, Passing the wake: using multiple fins to shape forces for swimming. Biomimetics **4**(1), 1–23 (2019). https://doi.org/10.3390/biomimetics4010023
31. B. Mosadegh, P. Polygerinos, C. Keplinger, S. Wennstedt, R.F. Shepherd, U. Gupta, J. Shim, K. Bertoldi, C.J. Walsh, G.M. Whitesides, Pneumatic networks for soft robotics that actuate rapidly. Adv. Funct. Mater. **24**(15), 2163–2170 (2014). https://doi.org/10.1002/adfm.201303288
32. D.B. Quinn, G.V. Lauder, A.J. Smits, Scaling the propulsive performance of heaving flexible panels. J. Fluid Mech. **738**, 250–267 (2014). https://doi.org/10.1017/jfm.2013.597
33. R Core Team: R: a language and environment for statistical computing (2013). http://www.r-project.org/
34. R.E. Shadwick, S. Gemballa, Structure, kinematics, and muscle dynamics in undulatory swimming, in *Fish Physiology*, vol. 23 (Academic Press, San Diego, 2006), pp. 241–280. https://doi.org/10.1016/S1546-5098(05)23007-8
35. R.M. Shelton, P.J.M. Thornycroft, G.V. Lauder, Undulatory locomotion of flexible foils as biomimetic models for understanding fish propulsion. J. Exp. Biol. **217**, 2110–20 (2014). https://doi.org/10.1242/jeb.098046. http://www.ncbi.nlm.nih.gov/pubmed/24625649
36. A.J. Smits, Undulatory and oscillatory swimming. J. Fluid Mech. 1–70 (2019). https://doi.org/10.1017/jfm.2019.284
37. E.D. Tytell, Median fin function in bluegill sunfish Lepomis macrochirus: streamwise vortex structure during steady swimming. J. Exp. Biol. **209**(8), 1516–1534 (2006). https://doi.org/10.1242/jeb.02154
38. E.D. Tytell, E.M. Standen, G.V. Lauder, Escaping flatland: three-dimensional kinematics and hydrodynamics of median fins in fishes. J. Exp. Biol. **211**(2), 187–195 (2008). https://doi.org/10.1242/jeb.008128
39. E.D. Tytell, C.Y. Hsu, T.L. Williams, A.H. Cohen, L.J. Fauci, Interactions between internal forces, body stiffness, and fluid environment in a neuromechanical model of lamprey swimming. Proc. Natl. Acad. Sci. **107**(46), 19832–19837 (2010). https://doi.org/10.1073/pnas.1011564107
40. W. Venables, B. Ripley, *Modern Applied Statistics with S*, 4th edn. (Springer, Berlin, 2002)
41. Z. Wolf, A. Jusufi, D. Vogt, G.V. Lauder, Fish-like aquatic propulsion studied using a pneumatically-actuated soft-robotic model. Bioinspir. Biomim. **15**(4), 046008 (2020). https://doi.org/10.1088/1748-3190/ab8d0f
42. J. Zhu, C. White, D.K. Wainwright, V. Di Santo, G.V. Lauder, H. Bart-Smith, Tuna robotics: a high-frequency experimental platform exploring the performance space of swimming fishes. Sci. Robot. **4**(34), 1–12 (2019). https://doi.org/10.1126/scirobotics.aax4615
43. M. Ziegler, M. Hoffmann, J.P. Carbajal, R. Pfeifer, Varying body stiffness for aquatic locomotion, in *Proceedings – IEEE International Conference on Robotics and Automation*, (2011), pp. 2705–2712
44. Y.J. Park, T.M. Huh, D. Park, K.J. Cho, Design of a variable-stiffness flapping mechanism for maximizing the thrust of a bio-inspired underwater robot. Bioinspir. Biomim. **9**(3), 036002 (2014). https://doi.org/10.1088/1748-3182/9/3/036002
45. K. Li, H. Jiang, S. Wang, J. Yu, A soft robotic fish with variable-stiffness decoupled mechanisms. J. Bio. Eng. **15**(4), 599–609 (2018). https://doi.org/10.1007/s42235-018-0049-1

Chapter 9
A Biomimetic Robotic Jellyfish Based on Shape Memory Alloy Springs

Mohammad A. Kazemi-Lari, Anthony D. Dostine, Jiadi Zhang, Alan S. Wineman, and John A. Shaw

9.1 Introduction

The development of robotic underwater vehicles has grown continuously in recent decades thanks to the potential applications of these systems and advancements made in the sensors and actuation devices. These robots have been utilized in numerous applications, including water sampling, underwater ecological change monitoring, oil and gas explorations, and hydrographic surveys [7, 14]. Compared to the traditional methods of underwater sensing with human-operated submarines, underwater robots offer potential benefits in cost-effectiveness, compactness, range of operation, and extended lifetime. Since animal locomotion in nature tends to use the optimal motion for its living environment, biomimetic robots have become a particular research focus. In early underwater robots, locomotion was accomplished by traditional piston and motor actuators, usually involving complex linkage mechanisms. Despite improvements in later attempts, the size and weight were still critical issues and the lifetime was limited because of the excessive number of parts (see the review in [1]).

Jellyfish are among the simplest animals, containing only a few organs, with axisymmetric bodies. Jellyfish are known to be one of the most efficient ocean swimmers, spending less energy per distance than many other underwater creatures. This is accomplished by recapturing part of the expended energy of each stroke [6]. Jellyfish are classified in the subphylum Medusozoa, which comprises a major part of the phylum Cnidaria. Of the swimming Medusozoa found in nature, the

M. A. Kazemi-Lari · A. D. Dostine · J. Zhang · J. A. Shaw (✉)
Adaptive Materials and Structures Lab, Aerospace Engineering Department, University of Michigan, Ann Arbor, MI, USA
e-mail: mkazemi@umich.edu; jashaw@umich.edu

A. S. Wineman
Mechanical Engineering Department, University of Michigan, Ann Arbor, MI, USA

D. A. Paley, N. M. Wereley (eds.), *Bioinspired Sensing, Actuation, and Control in Underwater Soft Robotic Systems*, https://doi.org/10.1007/978-3-030-50476-2_9

two main classes are Scyphozoa (*true jellyfish*) and Cubozoa (*box jellyfish*). Both have hemi-ellipsoidal bells, but as shown in Fig. 9.1, the Scyphozoa's is *oblate* (saucer-like), whereas the Cubozoa's is *prolate* (bullet-like). True jellyfish swim through a combination of rowing and jet propulsion, whereas box jellyfish swim using jet propulsion only [3, 13]. In jet propulsion, water is squeezed out of the bell periodically to produce the required thrust. By contrast, rowing propulsion uses the weak subumbrellar muscles to row at the bell periphery to form vortex rings in the water [4]. True jellyfish generally swim quite slowly, whereas box jellyfish swim somewhat faster through a darting motion. The slow motion of the true jellyfish inspired us to mimic its locomotion through the use of shape memory alloys (SMAs) as artificial muscles.

Our objective is to demonstrate a simple, compact, and biomimetic soft underwater robot using smart materials. Soft robots, as opposed to conventional robots, may be more environmentally friendly as they reduce unintended harm to the aquatic ecosystem of underwater vegetation and marine animals [12]. Among the known smart materials, NiTi-based SMAs provide some of the highest energy densities available, with typical transformation strains near 5% at actuation stresses of 200 MPa or more. In addition, NiTi-based SMAs are corrosion-resistant, biocompatible, and have ultimate tensile strengths above 1 GPa. A well-known disadvantage of SMAs, however, is the slow actuation compared to other smart materials or conventional actuators. The response time of SMAs is dictated by heat-transfer considerations; in an actuation mode, the response time is most often limited by the cooling segment. An underwater environment, however, improves the cooling rate of the SMA and enables useful actuation cycle rates.

Similar past efforts include the following. Yang et al. [17] developed a jellyfish microrobot of a hemispherical body shape made of a rubber material that incorporated a circumferential SMA actuator and four ionic polymer–metal composite (IPMC) legs for swimming. The proposed design achieved a maximum speed of 6 mm/s at an operating frequency of 0.6 Hz. Villanueva et al. [15] designed a robotic jellyfish that incorporated a beam-shape composite actuator based on SMA wires as the active component of the muscle. The design was inspired by contraction of a jellyfish bell through a combination of rowing and jetting mechanisms for locomotion [16] and reached a maximum speed of 54 mm/s at 0.5 Hz. Also, Gao et al. [5] fabricated a biomimetic cuttlefish robot that swam at relatively high speed (87.6 mm/s) through a jetting propulsion mechanism. The propulsion was achieved by squeezing water out of the mantle cavity by heating of SMA wires embedded in a soft silica gel. Their design achieved a maximum average jet thrust of 0.14 N at a jetting frequency of 0.83 Hz. Furthermore, Kim et al. [11] developed a turtle-like swimming robot where propulsion was achieved through flapping motion of two soft actuators made using SMA wires that served as the turtle's flippers and could generate bending and twisting motions. The robot was able to reach a swimming speed of 22.5 mm/s.

Our ongoing work explores an alternative SMA actuation approach to create a simple, compact, high-performance, biomimetic jellyfish, with initial results first reported in [10]. The main contribution is the simplicity of design, requiring only a soft rubber matrix with a single embedded SMA spring acting as the artificial muscle. Large deformations are achieved without the need for extra mechanical parts and joints, actuation cycles are demonstrated at comparable rates to previous attempts (0.5 Hz), and the robot swims with a motion reminiscent of the actual jellyfish.

The chapter begins with an overview of the proposed robot design, materials, and fabrication method. A nonlinear finite-element simulation demonstrates the buckling behavior of the robot structure. The typical swimming performance of our prototype is reviewed, followed by an experimental parameter study of the sensitivities of important operational and design parameters, including new results beyond those in [10].

9.2 The Robot Jellyfish

9.2.1 Design and Operation

We chose to mimic the true jellyfish of the oblate form (Fig. 9.1a), which uses the hybrid rowing and jetting modes for propulsion. The robot design is a flat rubber slab ($h = 3.8$ mm thick) in a sunburst shape, consisting of a central circular ($D = 76$ mm nominal diameter) disk with up to eight radially protruding flaps (see Fig. 9.2a). The tip-to-tip diameter across the flaps is about 140 mm. An SMA tension spring, embedded near the periphery of the central disk in a circular loop, serves as the circumferential muscle of the robot (Fig. 9.2f). The alloy is chosen to have a characteristic transformation temperature well above room temperature, so the SMA phase is martensite at room temperature and austenite when sufficiently

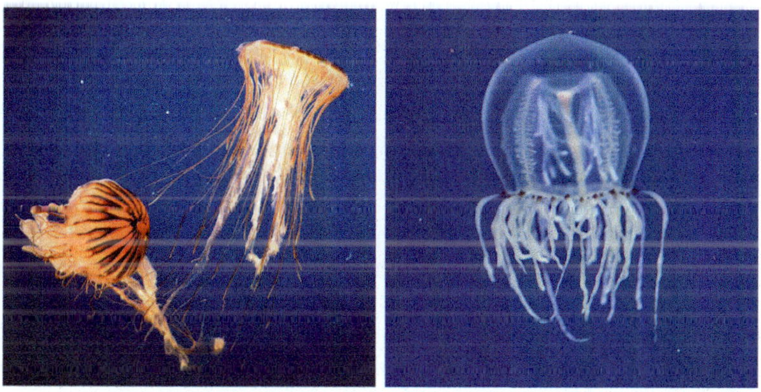

(a) (b)

Fig. 9.1 Photographs of the two main classes of swimming jellyfish. (**a**) Scyphozoa (True jellyfish) [8]. (**b**) Cubozoa (Box jellyfish) [9]

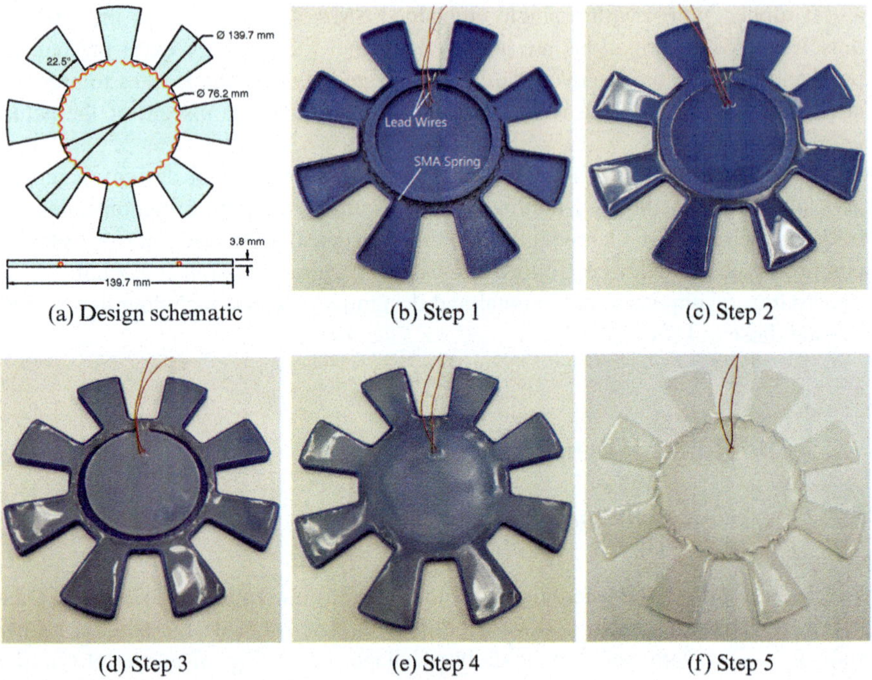

Fig. 9.2 Jellyfish robot design (**a**) and fabrication procedure (**b–f**); (**b**) the SMA spring loop and lead wires are placed in the primary mold; (**c**) liquid silicone rubber is poured in the primary mold; (**d**) after curing, the incomplete robot is removed and placed in the secondary mold; (**e**) additional rubber is poured to complete the robot body; (**f**) after curing, the completed robot jellyfish is removed

hot. The SMA spring is pre-stretched at room temperature such that the phase is oriented martensite and is cast in place within the rubber. The SMA spring is heated by running electrical current through two lead wires attached to the ends of the spring loop. The shape memory effect causes the spring to transform to the much stiffer austenite and contract in length, which creates a radial inward pressure to buckle the rubber disk into a dome shape. The flaps then follow the rotation of the disk edge, causing them to paddle downward and inward towards the axis of the dome.

The SMA spring cools back to the ambient temperature and reverts to the more compliant martensite phase once the electric current is turned off. The spring is sized such that the rubber matrix has sufficient stiffness to return the structure toward its pre-buckled, flat configuration. The locomotion of the robot results from the abrupt buckling of the disk that creates a propulsive jet along with the downward rowing of the flaps, followed by a relatively slow reset to the flatter configuration in order to minimize the opposing drag in the water. Each actuation cycle produces a net translation upward, and the robot is kept swimming by repeating the cycle.

9.2.2 *Materials*

The synthetic jellyfish structure consists of two components, an SMA spring loop and an elastomeric matrix. Close-coiled SMA tension springs ($d_s^o = 2.5$ mm outer diameter), made from $d_w = 0.38$ mm diameter 90C Flexinol® wire, were purchased from Dynalloy Inc. The spring index was $c = d_s/d_w = 5.67$, based on the nominal spring diameter $d_s = d_s^o - d_w$. The uniaxial thermo-electro-mechanical response of this wire was characterized previously in [2]. Differential scanning calorimetry measured the austenite and martensite finish temperatures near $A_f = 85$ °C and $M_f = 40$ °C, respectively.

The elastomeric matrix is made from the two-part Ecoflex 00-30 platinum-catalyzed silicone rubber from Smooth-On Inc. The as-cured rubber has a Shore hardness of 00-30 and a mass density of $\rho = 1070\,\text{kg/m}^3$, as indicated by the supplier. The silicone rubber is quite compliant, similar to a soft gel shoe insert, and can withstand large deformations without permanent damage.

9.2.3 *Fabrication*

The body of the prototype robot jellyfish is fabricated from silicone rubber using custom-made molds as described in [10]. Two molds were designed and 3D printed, and the body of the jellyfish was cast in a few steps. We found it important to precisely place the SMA spring loop in a concentric initial configuration, so the primary mold has a raised ring to provide a circular slot (73.7 mm inner diameter, 2.5 mm width) to hold it in place during casting. It also includes a radial slot to allow electrical lead wires to be attached to the spring loop and led radially toward the disk's center. The secondary mold had the same outer envelope as the primary mold, but without the raised ring, and it was used to complete the casting of the center disk of the jellyfish body.

About 38 coils of the SMA spring were excised, and the spring sample was heated above the austenite finish temperature until a nearly close-coiled spring was obtained ($L_0 = 27$ mm reference length). The SMA spring was then pre-stretched to fit the circumferential length of the primary mold's slot, giving a pre-stretch of about $\lambda_0 = L/L_0 \approx 16$ in the baseline case. At each end of the SMA spring, a short length of the SMA wire was straightened, crimped with a ferrule, and soldered to 30-gauge lead wires. This assembly was then placed in the primary mold (Fig. 9.2b).

After weighing and degassing, liquid Ecoflex 00-30 silicone rubber was poured over the SMA spring within the outer region of the primary mold (Fig. 9.2c) and then allowed to cure for at least 4 h at room temperature. After curing, the incomplete jellyfish was removed from the primary mold and placed into the secondary mold (Fig. 9.2d). Additional liquid rubber was then poured into the remaining central hole in the jellyfish body and cured for another 4 h (Fig. 9.2e). The jellyfish was removed from the mold and longer lead wires were soldered to the leads exiting the jellyfish body so that it could be connected to a power supply. Excluding lead wires, the entire robot jellyfish mass is about 40 g (Fig. 9.2f).

9.3 Modeling and Simulation

The buckling behavior of the prototype robot structure was examined in [10] through a combination of a bifurcation analysis at the onset of buckling and a nonlinear finite-element simulation of the post-buckling response. Here, the constitutive model of the rubber and the finite-element simulation are briefly reviewed.

9.3.1 Rubber Constitutive Model

The mechanical response of silicone rubber exhibits large recoverable deformations with very little hysteresis, so the behavior of the rubber matrix is assumed to be isotropic, incompressible, and nonlinearly elastic. Accordingly, the Mooney–Rivlin hyperelastic model has the strain energy density function W of the form

$$W(I_1, I_2) = C_1\,(I_1 - 3) + C_2\,(I_2 - 3)\,, \tag{9.1}$$

where C_1 and C_2 are two material constants and I_1 and I_2 are invariants of the left Cauchy–Green deformation tensor $\boldsymbol{B} = \boldsymbol{F}\boldsymbol{F}^{\top}$ in terms of the deformation gradient tensor $\boldsymbol{F}$. In terms of the principal stretch ratios $(\lambda_1, \lambda_2, \lambda_3)$, the first two principal invariants of $\boldsymbol{B}$ are $I_1 = \lambda_1^2 + \lambda_2^2 + \lambda_3^2$ and $I_2 = \lambda_1\lambda_2 + \lambda_2\lambda_3 + \lambda_3\lambda_1$. Since rubber is nearly incompressible and the deformation is isochoric, the third principal invariant is $I_3 = \lambda_1\lambda_2\lambda_3 = 1$. The Cauchy stress tensor $\boldsymbol{\sigma}$ is calculated by differentiating the strain energy density function with respect to its invariants, leading to

$$\boldsymbol{\sigma} = -q\boldsymbol{I} + 2\left[C_1\boldsymbol{B} - C_2\boldsymbol{B}^{-1}\right], \tag{9.2}$$

where q is a scalar field associated with the incompressibility constraint. Under plane stress conditions, q is found by setting the lateral stress $\sigma_3 = 0$. In the small strain limit, the Mooney–Rivlin model gives a shear modulus of $\mu = 2(C_1 + C_2)$.

The Mooney–Rivlin model shows good agreement with tensile test data of typical elastomeric materials in the moderate stretch range (say $1 < \lambda < 2$), but it fails to account for the hardening of the material at large stretches. The finite-element simulation of the jellyfish (below) confirms that the local stretch ratios are moderate, so the Mooney–Rivlin model is sufficient for our purposes.

Mooney–Rivlin parameters were determined by performing three constitutive tests on thin samples made of the silicone rubber [10]. The first test was pure tension of a dog-bone shaped specimen with a relatively long gauge section, stretched along the long direction ($\lambda = \lambda_1$). The second was a planar tension test of a short, wide specimen, gripped to constrain the width stretch ratio ($\lambda_2 = 1$) while stretching the short dimension ($\lambda = \lambda_1$). The third test was lateral air inflation of an initially flat circular membrane specimen clamped at its outer diameter. The equi-biaxial stretch ($\lambda = \lambda_1 = \lambda_2$) was measured at the center (crown) of the inflated configuration by a high-resolution camera and image analysis of markers on the sample. The normal pressure p_n was measured by a pressure transducer, and the curvature κ at the crown was measured by image analysis of a second high-resolution camera viewing the profile shape. In this way, the equi-biaxial stress at the crown was estimated using the membrane equilibrium relation $p_n = 2h_0\kappa\sigma/\lambda^2$. According to the Mooney–Rivlin model, the Cauchy stress–stretch relations for each case are

$$\sigma = \begin{cases} 2\left(C_1 + C_2\lambda^{-1}\right)\left(\lambda^2 - \lambda^{-1}\right) & \text{uni-axial tension} \\ 2\left(C_1 + C_2\right)\left(\lambda^2 - \lambda^{-2}\right) & \text{planar tension} \\ 2\left(C_1 + C_2\lambda^2\right)\left(\lambda^2 - \lambda^{-4}\right) & \text{equi-biaxial tension} \end{cases} \tag{9.3}$$

Since experimental measurements were performed with respect to their reference (initial) dimensions, the Cauchy stresses in Eq. (9.3) were converted to engineering (nominal) stresses by $\sigma_{eng} = \sigma/\lambda$. The material parameters of the Mooney–Rivlin model, $C_1 = 13$ kPa and $C_2 = 1.4$ kPa, were obtained by a simultaneous nonlinear least-squares fit of data from the three constitutive tests.

9.3.2 Finite-Element Simulation

To study the buckling and post-buckling behavior, a finite-element analysis (FEA) in COMSOL was performed to explore the nonlinear displacement and strain fields. The geometry of the entire robot jellyfish in its flat reference configuration was modeled (Fig. 9.3a), and a structured mesh was defined with a finer mesh near the periphery of the center disk (see [10] for more details). For simplicity, the SMA spring was modeled as a slender-ring inclusion, with a cross-sectional area of the SMA wire and an elastic modulus much larger than that of the rubber matrix. Contraction of the SMA element was simulated by prescribing a negative inelastic hoop strain (eigenstrain) β in the inclusion. This inelastic strain served as the loading parameter throughout the analysis. A zero axial (transverse) displacement $w = 0$ was imposed along the periphery of the inner disk. The inner disk was given a very small (1 Pa) lateral pressure as a loading imperfection to aid numerical convergence at the onset of buckling.

The 3-D deformed configurations at different stages during simulation are shown in Figs. 9.3b–d with contours of the normalized axial displacement, $\overline{w} = 2w/D$. A Von Karman plate analysis [10] of a radially loaded incompressible flat disk (of the aspect ratio used here) predicts a bifurcation point in the response at a hoop strain of $\varepsilon_c = -0.0028$ (see the unfilled circle in the inset of Fig. 9.3e).

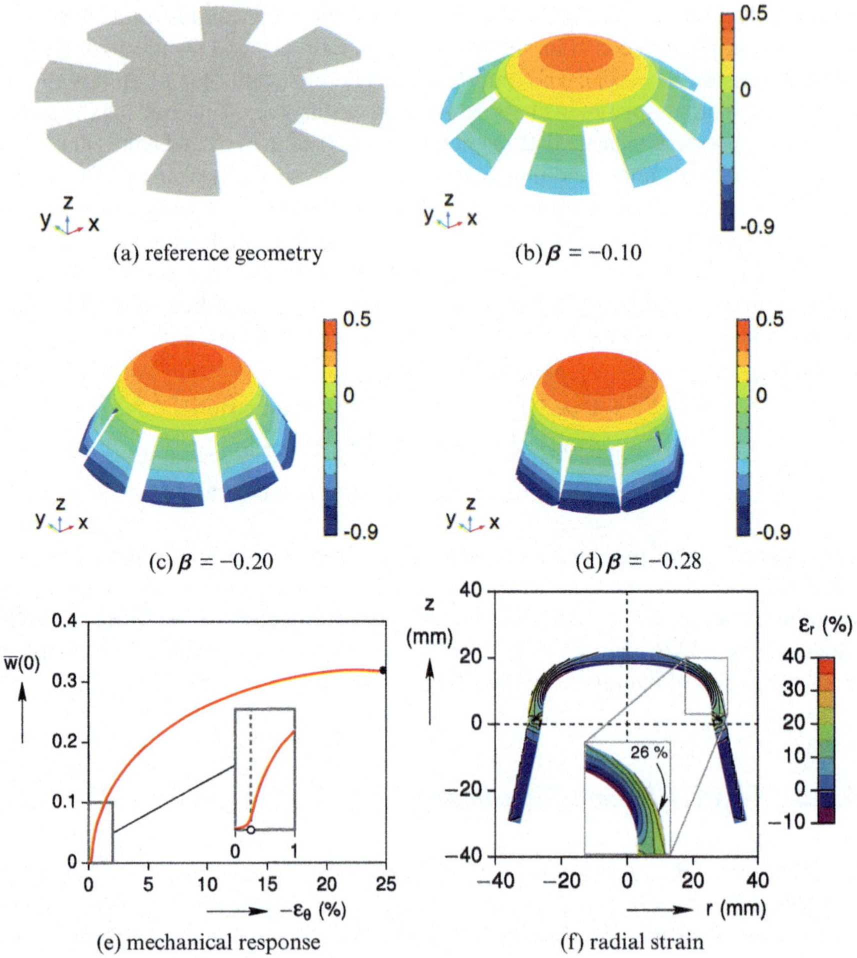

Fig. 9.3 Finite-element simulation results of the robot jellyfish deformation; (**a**) reference geometry, (**b–d**) contours of normalized axial displacement, (**e**) response of the body during simulation with a magnified view of the response near the onset of buckling, and (**d**) the radial Green-Lagrange strain contour inside the deformed robot body at the end of simulation

This is indeed where the lateral deflection begins to accelerate in the finite-element simulation as the central disk deforms into a dome configuration. Incidentally, this axisymmetric buckling of the central disk can only be achieved by designs having separation between the flaps. We found that alternative designs (not shown here) with insufficient gaps between the flaps, or a continuous disk of a larger diameter than that of the spring loop, led to a distorted, non-axisymmetric post-buckling deformation.

Further details of the deformation field within the jellyfish body at the end of the simulation ($\beta = -0.28$) are provided in Fig. 9.3f. Using the Green-Lagrange strain tensor $\boldsymbol{\varepsilon} = (\boldsymbol{F}^{\top}\boldsymbol{F} - \boldsymbol{I})/2$, the radial ($\varepsilon_r$) strain contours are shown in a section (θ=constant) of the body. Ignoring the strain concentration near the ring inclusion at the edge of the inner disk, the maximum bending strain ($\varepsilon_{r,\max} = 26\%$) occurs at the outer edge of the disk as indicated in Fig. 9.3f. This corresponds to a radial stretch ratio of $\lambda_{r,\max} = 1.3$, confirming that the strains were within the accurate range of the Mooney–Rivlin model.

9.4 Experimental Results

The swimming performance of a prototype robot jellyfish is presented in this final section. Deformation and force measurements in a special static setup are also presented to quantify the underwater performance and to explore the influence of operation and design parameters, such as the input actuation power and frequency, the pre-stretch ratio of the SMA spring, and the number of flaps of the robot.

9.4.1 Swimming Performance

The swimming performance of the submerged robot was evaluated based on its propulsive thrust, stroke, and the ability to reach steady-state swimming without losing performance over time. The robot had a slight negative buoyancy and was placed at the bottom of a water tank with electrical wires extending from the robot to an external power supply (see first photograph in Fig. 9.4a). Direct electrical current was applied periodically to resistively heat the SMA spring and cause it to contract within the robot.

Figure 9.4a provides a sequence of selected photographs, showing the prototype jellyfish swimming from the bottom to the top of the water tank during nine actuation cycles. The jellyfish traveled a vertical distance of about 245 mm in just over 16 s. The actuation period was 2 s (0.5 Hz actuation frequency), consisting of a constant voltage of 11.25 V (about 53.7 W) across the SMA spring for 0.25 s followed by no voltage for 1.75 s, which is equivalent to an energy consumption of about 13.4 J per cycle. A video (not shown here) of the swimming prototype showed within each actuation cycle a fast contraction followed by a slow relaxation, and the motion was reminiscent of an actual swimming jellyfish. The vertical position (z) and the upward stroke (Δz) per cycle plotted in Fig. 9.4b were measured from image analysis of the photographs. As seen in Fig. 9.4b, the stroke started small (under 10 mm) but the amplitude quickly grew after a few cycles to reach a nearly steady-state value of almost 40 mm/cycle by cycle 8, which corresponds to an average vertical speed of approximately 20 mm/s.

The stroke is initially small in the first few cycles, since both the rubber and SMA spring start at water temperature (cold) and the short duration of the input

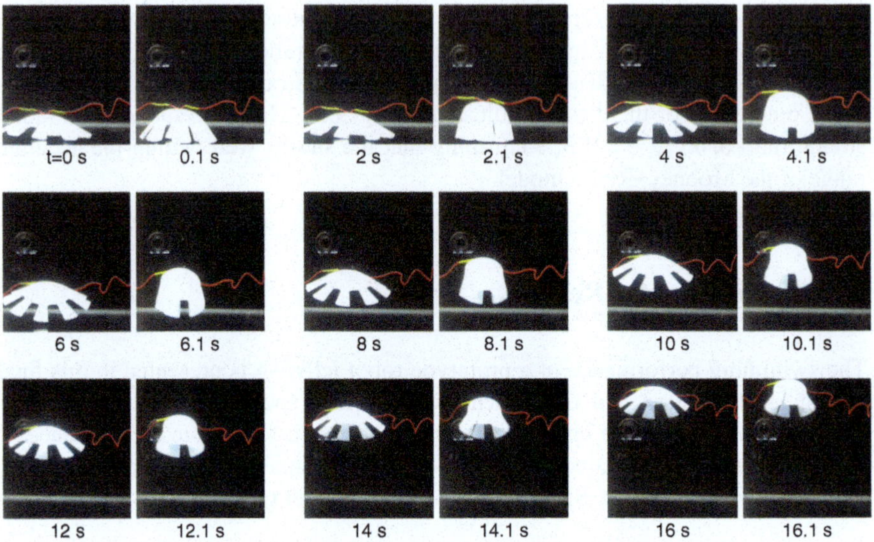

(a) Photographs of swimming robot

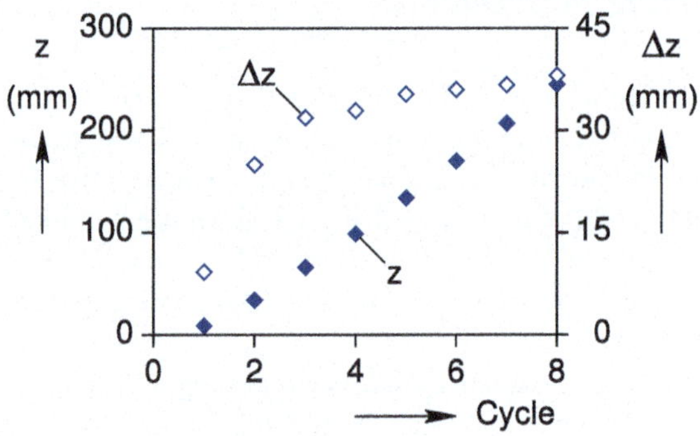

(b) Vertical position and stroke

Fig. 9.4 Swimming performance: (**a**) selected photographs of the prototype robot swimming upward (0.5 Hz actuation frequency), (**b**) vertical position z and stroke Δz at successive actuation cycles

power is insufficient to completely transform the SMA from tensile martensite (M^+) to austenite (A). Consequently, the SMA spring only partially contracts. Upon continued cycling, however, the average temperature of the body and the spring loop increases until the temperature oscillations span the $M^+ \rightarrow A$ and $A \rightarrow M^+$ transformation temperatures during steady-state operation. That is, the SMA spring requires a few transient cycles before it warms up to an optimal mean temperature to achieve full contraction and relaxation. Subsequent actuation cycles produce larger deformations of the robot's body and larger swimming strokes.

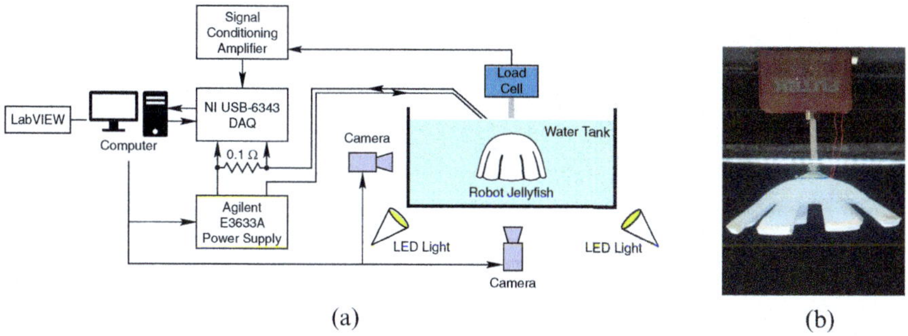

Fig. 9.5 Static test setup for deformation and thrust measurements. (**a**) Schematic of experimental setup. (**b**) Closeup photograph of robot

9.4.2 *Thrust Performance*

A static test rig (Fig. 9.5a) was developed to quantify the thrust performance and overall deformation of the jellyfish robot. The submerged prototype jellyfish was attached by a rigid rod to a load cell (Futek LRF400, 10 N capacity) above the water tank (Fig. 9.5b) to measure the thrust; synchronized images were taken by two CCD cameras to monitor the body deformation.

The submerged robot was actuated relatively fast, with approximately 12 V (about 73 W power) across the SMA spring for 0.2 s, followed by 1.8 s off for recovery, resulting an energy consumption of about 15 J in each actuation cycle. This corresponded to a total period of 2 s and a duty cycle of 10%. From analysis of the digital images, the radial displacements of the tip of the flaps were measured (u_f) and then normalized by the initial flap tip radius ($R_f = 69.2$ mm, the outer envelope of the entire robot) to obtain a strain-like measure ($\overline{u}_f = u_f/R_f$) of the overall body deformation. The time history of this normalized flap displacement over several cycles is shown in Fig. 9.6a, exhibiting contractions of near 60% after a few cycles. The gray line in Fig. 9.6a depicts the average flap displacement during each cycle, and the unfilled circles indicate the deflection at the end of each cycle. While the prototype reached an almost constant contraction after 9 cycles, it did not completely return to its initial state. The flap displacement at the end of each actuation cycle, or residual displacement ($\overline{u}_{f,\,res}$), is plotted at a magnified scale in Fig. 9.6b, showing an evolving residual displacement with cycling that grew to almost 30% by the end of the experiment. The cause of this ratcheting behavior was due to incomplete temperature recovery, as shown and verified in [10] by temperature measurements.

The thrust performance was measured by the load cell, as shown in the force (P_z) history of Fig. 9.6c. The peak thrust started near $P_{z,max} = 0.4$ N during early cycles and decayed gradually to about 0.3 N by the end of the experiment. Likewise, negative thrust values ($P_{z,min} < 0$) were also measured during each cycle which became smaller in magnitude after the first few cycles. We defined a net thrust as

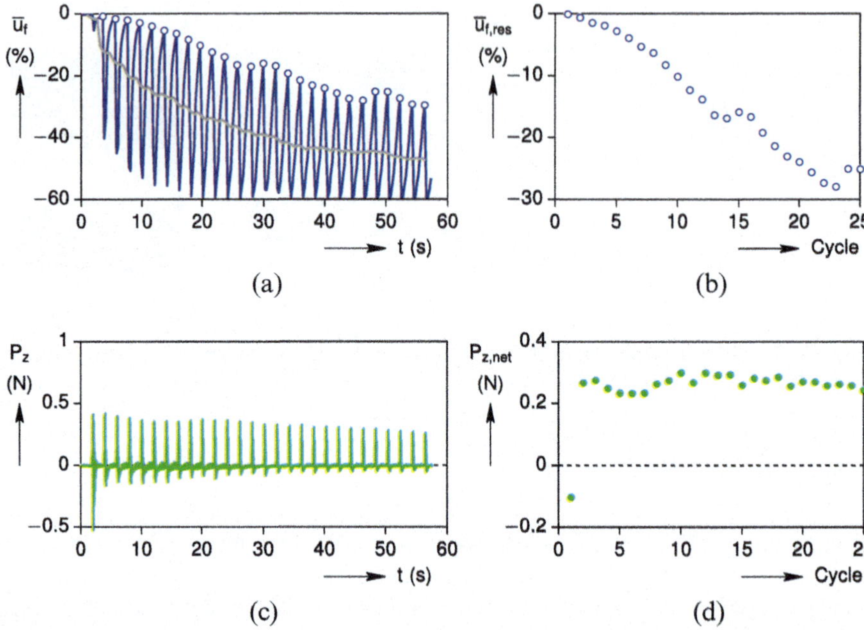

Fig. 9.6 Experimental results with 2 s actuation period at input energy of 15 J (10% duty cycle). (**a**) Flap displacement history. (**b**) Residual flap displacement. (**c**) Thrust history. (**d**) Net thrust

the simple sum $P_{z,\mathrm{net}} = P_{z,\max} + P_{z,\min}$, which is plotted in Fig. 9.6d. With sustained cycling, the net thrust quickly rose to a steady-state value of about $P_{z,\mathrm{net}} = 0.25$ N.

9.4.3 Operation and Design Study

Numerous additional experiments were performed to evaluate the influence of various design and operation parameters on thrust performance. The influence of two operation and two design parameters is presented in Fig. 9.7. In all cases presented here, the period of actuation was fixed at 2 s.

In the first parameter sweep (Fig. 9.7a, b), the influence of input energy was investigated on the thrust performance of the prototype robot. The duty cycle was fixed at 10% (electric current was provided for 0.2 s in each cycle), while nine experiments were performed at various input energies to the SMA spring, from 3 to 18 J (0.2 s at 15 W–90 W). The evolution of the net thrust at each energy level was measured during 28 actuation cycles (Fig. 9.7a). The net thrust remained negative at the four lowest energy levels (up to 9 J). A net positive thrust first appeared at 11 J after three actuation cycles, and it reached a sustained value of about 0.1 N thereafter. The performance improved at 13 J with a sustained net thrust of almost 0.2 N during late cycles. At 15 J and 16 J the thrust was higher yet during early

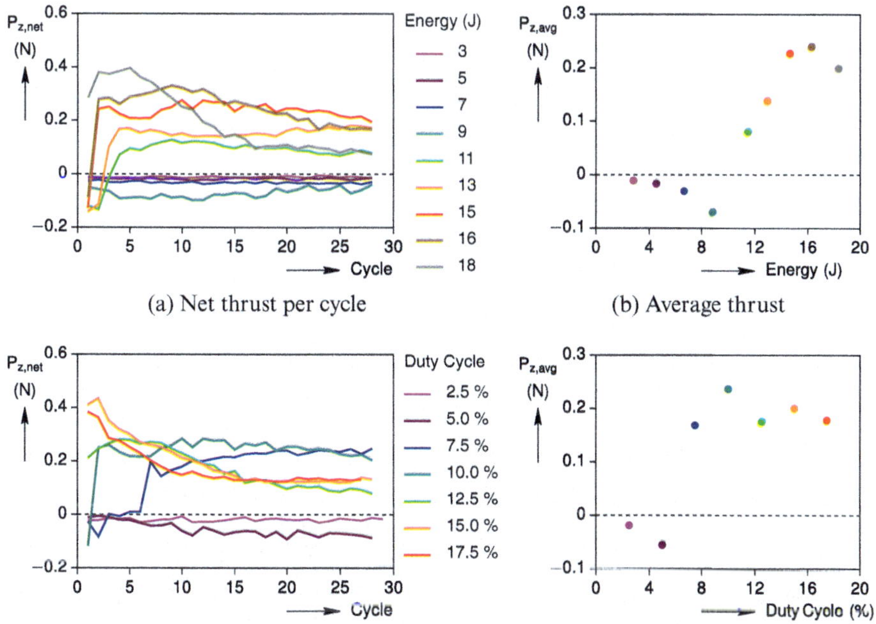

(a) Net thrust per cycle (b) Average thrust

(c) Net thrust per cycle (d) Average thrust

Fig. 9.7 The operational parameters, input energy, and duty cycle require threshold values to produce positive net thrust, and the thrust trends are non-monotonic with increasing values. The effects of input energy (**a, b**) are shown at a fixed 10% duty cycle and 0.5 Hz actuation frequency. The effects of duty cycle (**c, d**) are shown at a fixed input power of 73 W and 0.5 Hz actuation frequency

cycles, but it began to diminish during late cycles. At the highest energy level of 18 J, the initial thrust was about 0.4 N, but it quickly decayed with cycling and even underperformed the 13 J case by cycle 16. As shown in [10], the decrease in the propulsion performance during late cycles at high energy inputs was caused by the SMA spring becoming progressively hotter with cycling, such that it was unable to cool and relax sufficiently before the next actuation cycle. This contributed to incomplete deformation recovery. At the highest energy level, the deformation eventually reached a nearly static value with almost no oscillations (tetanic-like contraction).

To summarize the trends, the net thrust was averaged over the 28 cycles at each input energy level and plotted in Fig. 9.7b. Interestingly, the performance only got worse (more negative) as the energy was increased at the four lowest input energies. The plot clearly shows that a threshold energy was required to achieve a positive net thrust (somewhere between 9 J and 11 J). The average thrust improved with larger intermediate energy levels (11 J–16 J), but then dropped off at excessively high input energy (18 J). While the best average thrust in this series of experiments was obtained at 16 J, it should be recognized that this was only true for the duration chosen for these experiments (28 cycles). A lower energy level would be optimal for longer experiments that involve more actuation cycles, since the thrust histories at 15 J and 16 J were decaying by the end of their experiments. Alternatively, the jellyfish robot could be given a brief rest, allowing the SMA spring to cool and recover, and then the performance shown could be repeated for another 28 cycles.

In a second parameter sweep, the influence of the duty cycle was investigated (Fig. 9.7c, d). Here, the input power was held fixed at 73 W, and seven experiments were performed at duty cycles between 2.5% and 17.5%. Similar to the previous energy sweep, a non-monotonic trend in the thrust performance was observed with increasing duty cycle. No net thrust (Fig. 9.7c) was obtained at the two smallest duty cycles (2.5% and 5%). At the three largest duty cycles (12.5%, 15%, and 17.5%), the initial flap displacements (not shown here) and net thrusts were relatively large, but these decreased in later cycles. Again, this was due to excessive heating and the inability to cool and reset, thereby resulting in large residual displacements (see [10] for more details). Both cases at 15% and 17.5% duty cycle, resulted in a tetanic-like contractions by the experiments' end. Better performance was achieved at intermediate duty cycles, 7.5% and 10%, and they demonstrated a tradeoff between them. The case with 10% duty cycle reached a larger net thrust in fewer cycles than the 7.5% duty cycle case, but this came at the expense of a larger residual displacement near the end of the experiment. Thus, the 10% duty cycle would be the optimal choice (average thrust of 0.25 N) for the jellyfish robot to sprint quickly, but the 7.5% duty cycle would be the better choice (average thrust of 0.18 N) for long–distance swimming.

A new study was performed on two design parameters. The number of flaps was the first design parameter varied, where for each successive experiment the same jellyfish robot was modified by removing two flaps at a time to maintain symmetry of the robot (Fig. 9.8a, b). The input energy was 15 J and the duty cycle was 10%, and five experiments (28 actuation cycles each) were performed on robots with 0, 2, 4, 6, and 8 flaps. As shown in Fig. 9.8a the net thrust/cycle remained negative during all cycles for the robot with no flaps. A net positive thrust first appeared for the two–flap robot after eight actuation cycles, but its subsequent value remained quite small. With four flaps, the thrust remained positive after early cycles and, for six flaps and eight flaps, the performance improved further. The trend was monotonic with increasing flap number, shown more clearly by averaging the net thrust over all 28 cycles for each case (Fig. 9.8b). Thus, more flaps produced better performance, yet it is not clear if more than eight flaps (each narrower) would give further improvement.

Finally, the influence of the second design parameter, the pre-stretch ratio of the SMA spring (λ_0), was investigated (Fig. 9.8c, d) for robots with all eight flaps and a duty cycle of 7.5%. In this study, a different robot was fabricated for each experiment. The final spring length was fixed by the circumference of the robot's central disk, so the initial SMA spring length had to be decreased (fewer coils) as the pre-stretch was increased. A fixed input voltage would, therefore, produce a different temperature change due to the different mass of the SMA springs, which would obscure the effects of the pre-stretch alone. Consequently, the input energy density was held nearly constant at 3.44 J/kg, and five experiments were performed

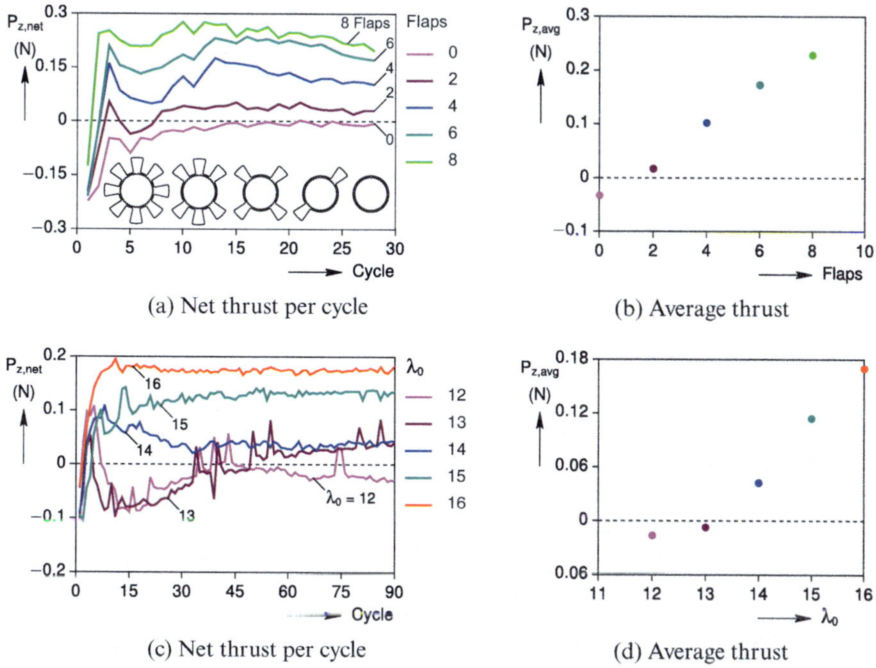

(a) Net thrust per cycle

(b) Average thrust

(c) Net thrust per cycle

(d) Average thrust

Fig. 9.8 The design parameters, number of flaps, and pre-stretch ratio (λ_0) of the SMA spring require threshold values to produce positive net thrust, and the thrust trends are monotonic with increasing values. The effects of flaps (**a, b**) are shown at 15 J input energy, 0.5 Hz actuation frequency, and 10% duty cycle. The effects of spring pre-stretch ratio (**c, d**) are shown at 3.44 J/kg input energy density, 0.5 Hz actuation frequency, and 7.5% duty cycle

for 90 actuation cycles at pre-stretch ratios of $\lambda_0 = 12$, 13, 14, 15, and 16. The corresponding energy inputs per cycle were 13.9 J, 12.8 J, 11.9 J, 11.1 J, and 10.4 J.

It can be seen that at $\lambda_0 = 12$ (Fig. 9.8c), the robot was unable to maintain a positive net thrust. Although not shown here, this was a result of a significant amount of unrecovered deformation at the end of each cycle. For a spring pre-stretch ratio of $\lambda_0 = 13$, the robot experienced a significant drop in the net thrust after the fifth cycle. The performance gradually improved, however, after 50 cycles to generate a small positive net thrust of about 0.03 N by the end of 90 cycles. Nevertheless, the average thrust over all cycles was still slightly negative (Fig. 9.8d). For $\lambda_0 = 14$, the robot sustained a net positive thrust, but its performance deteriorated after an initial peak near 0.1 N to an almost steady value of 0.04 N after 40 cycles. At $\lambda_0 = 15$, the net positive thrust generally grew and reached an asymptote near 0.13 N after about 40 cycles, which is more than a three-fold increase in steady-state thrust compared to the previous pre-stretch ratio. This improvement was attributed mostly to the faster cooling of the SMA spring as the distance between adjacent coils (the pitch) was larger at higher stretch ratios. The thrust performance improved further for $\lambda_0 = 16$, where the net positive thrust reached a steady-state value of 0.17 N after only 10 cycles. The averaged net thrusts in Fig. 9.8d clearly show a monotonic increase in the thrust performance, at least for the range of the pre-stretch ratios considered here.

Note that for the operational and design study on the number of flaps, the same previously cycled SMA spring was used in all the experiments. The design study on the pre-stretch ratio, however, started with a new (uncycled) spring at the smallest stretch ratio, and the same spring was trimmed to size and reused at each successively larger pre-stretch ratio.

9.5 Conclusion

We present a simple, compact, and high-performance biomimetic robot jellyfish that mimics the morphology and kinematics of the actual creature. The main contributions of the current study are the simplicity of design and the exploration of an alternative SMA actuation approach as the artificial contractile muscle. The design consists of only a soft rubber matrix with a single embedded SMA spring loop, which when Joule heated, achieves large deformations without the need for extra mechanical parts and joints. The buckling instability of the main body is harnessed to create an effective rowing mechanism of flaps to propel the robot through water.

A numerical finite-element simulation captures the post-buckling deformation of the robot and provides details of the deformation fields within it. This provides a useful stepping stone towards the development of design and simulation tools for the fully coupled electro-thermo-mechanical problem, which will be pursued in the future. Such modeling tools are essential to develop a design methodology for similar underwater robots to achieve a desired level of performance.

Our experimental results demonstrate actuation cycles at similar rates to previous attempts at biomimetic underwater robots. The robot swims from the bottom to the top of a water tank in a few actuation cycles via a pulsatile motion reminiscent of an actual jellyfish. It achieves swimming strokes of near 40 mm per cycle and a net steady-state propulsive thrust of 0.25 N when operated at 0.5 Hz frequency. An experimental parameter study reveals the influence of the input energy, duty cycle, number of flaps, and pre-stretch of the spring on the thrust performance of the robot. The performance is non-monotonic with the input energy and duty cycle, indicating the existence of a sweet spot that might be exploited in the robot's operation. It also reveals tradeoffs, depending on whether speed or endurance is preferred. In the range of parameters studied, the performance shows monotonic improvement with the number of flaps and pre-stretch of the SMA spring.

References

1. W.S. Chu, K.T. Lee, S.H. Song, M.W. Han, J.Y. Lee, H.S. Kim, M.S. Kim, Y.J. Park, K.J. Cho, S.H. Ahn, Int. J. Precis. Eng. Manuf. **13**(7), 1281 (2012)
2. C.B. Churchill, J.A. Shaw, in *Proceedings of the SPIE 15th Annual International Symposium on Smart Structures and Materials Behavior and Mechanics of Multifunctional and Composite Materials*, vol. 6929 (2008), p. 6929F
3. S.P. Colin, J.H. Costello, J. Exp. Biol. **205**(3), 427 (2002)
4. J.O. Dabiri, S.P. Colin, J.H. Costello, M. Gharib, J. Exp. Biol. **208**(7), 1257 (2005)
5. F. Gao, Z. Wang, Y. Wang, Y. Wang, J. Li, J. Bionic Eng. **11**(3), 412 (2014)
6. B.J. Gemmell, J.H. Costello, S.P. Colin, C.J. Stewart, J.O. Dabiri, D. Tafti, S. Priya, Proc. Natl. Acad. Sci. **110**(44), 17904 (2013)
7. G. Griffiths, *Technology and Applications of Autonomous Underwater Vehicles*, vol. 2 (CRC Press, Boca Raton, 2002)
8. https://live.staticflickr.com/2850/8843118889_46fa640013_b.jpg. Accessed 06 Jan 2019
9. https://dsvphotos.smugmug.com/Reference-Years/Z2013-2014-2015/i-KqQ4854/A. Accessed 06 Jan 2019
10. M.A. Kazemi-Lari, A.D. Dostine, J. Zhang, A.S. Wineman, J.A. Shaw, in *Proceedings of SPIE*, ed. by A.L. Raul, J. Martin-Palma, M. Knez. Bioinspiration, Biomimetics, and Bioreplication IX, vol. 10965 (SPIE, Denver, 2019), pp. 1096,504–1–1096,504–17
11. H.J. Kim, S.H. Song, S.H. Ahn, Smart Mater. Struct. **22**(1), 014007 (2012)
12. C. Majidi, Soft Robot. **1**(1), 5 (2014)
13. M.J. McHenry, J. Jed, J. Exp. Biol. **206**(22), 4125 (2003)
14. J. Nicholson, A. Healey, Mar. Technol. Soc. J. **42**(1), 44 (2008)
15. A. Villanueva, K. Joshi, J. Blottman, S. Priya, Smart Mater. Struct. **19**(2), 025013 (2010)
16. A. Villanueva, C. Smith, S. Priya, Bioinspir. Biomim. **6**(3), 036004 (2011)
17. Y. Yang, X. Ye, S. Guo, in *IEEE International Conference on Integration Technology, 2007. ICIT'07* (IEEE, Piscataway, 2007), pp. 673–678

Chapter 10
Control and Functionality of Octopus Arms and Suckers

Hosain Bagheri, Spring Berman, Matthew M. Peet, Daniel M. Aukes, Ximin He, Stephen C. Pratt, Rebecca E. Fisher, and Hamidreza Marvi

10.1 Introduction

The problem-solving skills of the octopus suggest the presence of a highly developed brain. They have shown great competence in a wide range of tasks such as navigating through mazes and unscrewing lids [25, 26, 40, 67, 74], differentiating the patterns, shapes, and colors of objects [33, 68, 93, 95, 96], tool usage (e.g., using bivalves and coconut shells as shelter) [24, 59], communication, and even

H. Bagheri · M. M. Peet · S. C. Pratt
Department of Mechanical and Aerospace Engineering, Arizona State University, Tempe, AZ, USA
e-mail: hbagheri@asu.edu; mpeet@asu.edu; stephen.pratt@asu.edu

S. Berman · H. Marvi (✉)
School for Engineering of Matter, Transport and Energy, Arizona State University, Tempe, AZ, USA
e-mail: spring.berman@asu.edu; hmarvi@asu.edu

D. M. Aukes
The Polytechnic School, Ira A. Fulton Schools of Engineering, Arizona State University, Mesa, AZ, USA
e-mail: danaukes@asu.edu

X. He
Department of Material Science and Engineering, University of California, Los Angeles, Los Angeles, CA, USA
e-mail: ximinhe@ucla.edu

R. E. Fisher
School of Life Science, Arizona State University, Tempe, AZ, USA

Department of Basic Medical Sciences, University of Arizona College of Medicine—Phoenix, Phoenix, AZ, USA
e-mail: rfisher@email.arizona.edu

D. A. Paley, N. M. Wereley (eds.), *Bioinspired Sensing, Actuation, and Control in Underwater Soft Robotic Systems*, https://doi.org/10.1007/978-3-030-50476-2_10

play [43, 53]. However, perhaps most remarkably, two-thirds of the total neurons are located in the arms [11], granting these appendages an incredible degree of autonomy [57]. This highly distributed nervous system, coupled with dexterous arms, provides endless solutions for locomotion, object manipulation, and exploration, positioning the octopus as an ideal model for bioinspired soft robotics. This chapter reviews octopus biology, highlighting structure–function relationships that can be used to inspire new soft robot designs, and summarizes the current state of the field of soft robotics, including examples of octopus-inspired soft robot arms and suckers.

10.2 Octopus Biology

10.2.1 Locomotion

The octopus has four primary modes of locomotion: jet propulsion, swimming, bipedal walking, and crawling. Jet propulsion and swimming are the faster forms of transportation. Jetting occurs with the arms curled or elongated and is accomplished by rapidly drawing water in and out of the mantle, causing the octopus to travel backwards with its mantle leading the way. During swimming, the octopus travels forward by expanding and then retracting its arms, leading with its head and eyes. A bipedal gait is less common among octopuses. *Octopus marginatus* and *Octopus aculeatus* use the posterior-most arms to perform a rolling bipedal gait. With the remaining six arms freed, the octopus can camouflage itself to appear as a piece of algae or a tumbling coconut shell, as it briskly walks backwards across the ocean floor [38, 39]. Crawling is the slowest, and most common, form of locomotion. Used both in and out of water, crawling is achieved by pairs of arms that alternate to push and pull the animal forward [58]. All of these locomotor modes require highly coordinated arm movements while the slower modes of transport, walking and crawling, also rely upon hundreds of suckers interacting with the substrate.

10.2.2 Distributed Nervous System

The nervous system of the octopus is one of the most sophisticated in the animal kingdom. The primary features include a central brain mass, two optic lobes, an axial nerve cord (ANC) in each arm, and an interbrachial commissure, connecting adjacent ANCs (Fig. 10.1). In the adult common octopus (*Octopus vulgaris*), the nervous system consists of 520 million neurons, 350 million of which reside in the arms, while the optical lobes and central brain contain 130 and 40 million neurons, respectively [11]. Cephalopod brains use a hierarchical structure of lower, intermediate, and higher motor centers [62]. The optic lobes interpret visual input, administer visual behavior, and trigger higher motor centers [11]. The ANCs include sensory and motor circuits that control many complex arm motions (i.e., extension, retraction, bending, and twisting) [57]. Adjacent ANCs are linked to one another

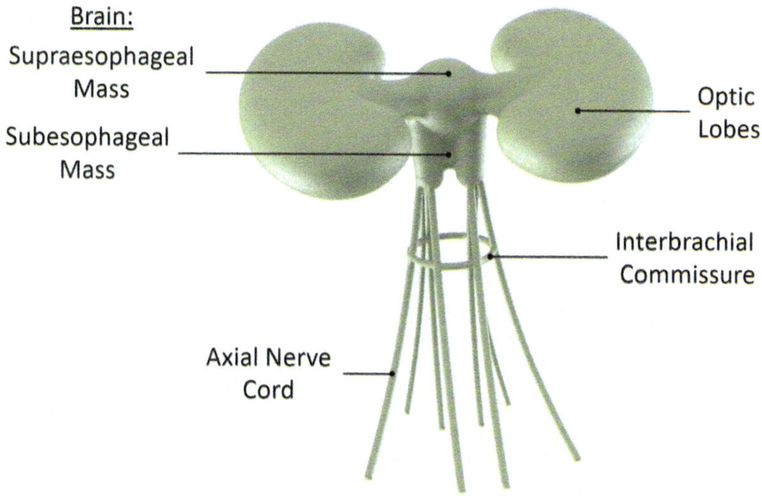

Fig. 10.1 Octopus nervous system, including the central brain mass (supra- and subesophageal mass), optic lobes, interbrachial commissure, and eight axial nerve cords

through the interbrachial commissure [61, 99]. The arms also feature sucker ganglia, positioned at the base of each sucker [61, 84]. The size of each ganglion correlates with the size of the sucker that it innervates and coordinates motor and sensory functions [99]. Thus, to some degree, each arm and each sucker has a brain, or control center, of its own.

When the arm is intact, the sucker ganglia can freely communicate with the ANC, while the ANC is able to communicate with the brain, via the brachial nerve, and adjacent arms, via the interbrachial commissure (Fig. 10.1). However, when the arm is amputated proximally, the sucker ganglia and ANC are disconnected from the interbrachial commissure and brain. It has previously been observed that removal of the supraesophageal mass from the octopus' brain results in sticky suckers that struggle to detach [76]. However, it has also been observed that when all eight arms are collectively amputated, the amputated arms are able to make coordinated movements similar to those of intact arms. Furthermore, in another study, when one of the eight arms was mechanically stimulated, adjacent arms reacted as well [32], alluding to the importance of the interbrachial commissure in coordinating arm movements [10, 99]. The suckers also play an important role in this regard. A study examining the prevention of *Octopus vulgaris* arm interference through sucker chemosensors, revealed that amputated arms suckers do not attach to arms with the skin intact, but instantly adhere to skinned arms. They further noticed that intact octopuses grab and adhere to amputated arms with the skin intact, thus overriding the typical reflex [68]. Another study observed that sucker attachment occurred in both intact and amputated arms of *Octopus bimaculoides*; detachment was achieved readily in intact arms and with much delay in amputated arms [3]. These results suggest that sucker attachment can occur in the absence of communication with the interbrachial commissure and brain, but one or both of these structures play a significant role in sucker detachment. These findings could be confirmed through neural recordings of the ANC and sucker ganglia during attachment and detachment when the arm is intact and amputated.

10.2.3 Structure and Functions of the Mantle

With an almost entirely soft body, the octopus is capable of squeezing through the smallest gaps and openings, limited only by the size of its beak. The octopus mantle is composed of radial, circular, and longitudinal muscles. The longitudinal muscles, located most superficially, sustain the mantle length, whereas the radial and circular muscles are used for respiration and jetting [44]. Contraction of the circular muscles thickens the mantle wall, decreasing the diameter of the mantle cavity and expelling water, while contraction of the radial muscles thins the mantle wall, increasing the diameter of the mantle cavity and allowing it to fill with water. Thus, alternating contractions of the circular and radial muscles achieves both respiration and jet propulsion [9, 44, 98].

10.2.4 Structure and Functions of the Arms

Octopuses use their eight arms for locomotion [38, 39, 58], object manipulation [13, 25], and prey capture and handling [44, 45]. The arms are muscular hydrostats, maintaining a constant volume to extend, shorten, bend, and twist. Octopus arms may be classified as right and left, and numbered one to four, from anterior to posterior (Fig. 10.2b). In previous studies, arm preference has been detected in some

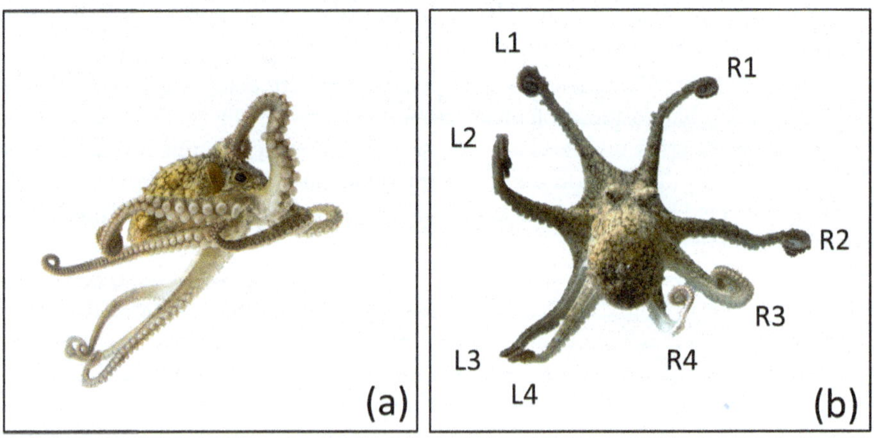

Fig. 10.2 *Octopus bimaculoides* (**a**) side view and (**b**) aboral view with right (R) and left (L) arm conventional labeling

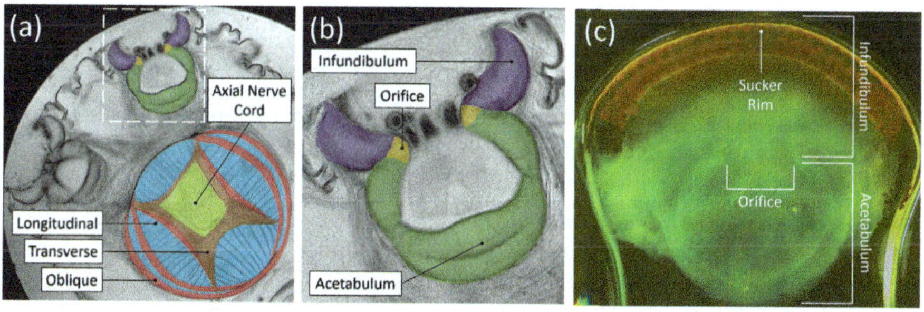

Fig. 10.3 MRI of the L2 arm of *Octopus bimaculoides* at its midpoint, highlighting: (**a**) the primary muscle fiber orientations (longitudinal, transverse, and oblique) and the axial nerve cord, and (**b**) the features of a sucker (infundibulum, orifice, and acetabulum). (**c**) Microscopic section of a sucker

octopus species. *Octopus marginatus* and *Octopus aculeatus* use their posterior-most arms to perform an alternating rolling bipedal gait [38, 39]. *Octopus vulgaris* appears to use its posterior arms more frequently during crawling movements [58], whereas the anterior arms more frequently engage in exploration and prey capture [13]. Furthermore, a lateral bias has also been observed in *Octopus vulgaris* and has been linked with right versus left eye preference (e.g., right eye with right arm 1 (R1), left eye with left arm 2 (L2) [12, 13]). However, these findings are based on a limited number of studies, in just a few species, so additional data are needed as octopus species exhibit different arm characteristics (i.e., lengths, diameters, arm and sucker strength), which may reflect their use and function. For example, in *Octopus vulgaris* the middle arm pairs (L2, L3, R2, R3) are slightly longer than the first and fourth, whereas in *Octopus macropus*, the anterior arms are the longest and most robust [13]. *Octopus bimaculoides* showed no correlation between arm identity (i.e., anterior versus posterior, right versus left) and sucker pull-off force, which could be used as a means of understanding arm functionality [3].

Octopus arms consist of muscle fibers that are oriented in three primary directions: transverse, longitudinal, and oblique [9, 44, 48, 61] (Fig. 10.3a). The transverse muscle fibers surround the axial nerve cord and are oriented perpendicular to the arm's longitudinal axis. Peripheral to the transverse muscle, the internal longitudinal fibers are organized into four main quadrants (i.e., oral, aboral, and lateral internal longitudinal). In addition, a thin layer of external longitudinal fibers is located more peripherally. All of these longitudinal muscle fibers are orientated parallel to the arm's longitudinal axis [45]. Finally, the oblique muscle fibers spiral around the circumference of the arm, forming three distinct layers [9, 61]. The internal oblique layer is located between the transverse and lateral internal longitudinal fibers. The median oblique layer is situated between the lateral internal longitudinal and external longitudinal fibers, whereas the external oblique layer is located most superficially, peripheral to the external longitudinal fibers [44, 45].

Since the functions of these muscle layers have not been confirmed via experimental studies, but based upon the muscular hydrostat model, the following actions are hypothesized. Contraction of the transverse muscle fibers decreases the arm diameter and elongates the arm, with slight decreases in arm diameter leading to rapid and large increases in arm length [44]. In contrast, contraction of the longitudinal fibers increases the arm diameter and shortens the arm length [99]. Contraction of the longitudinal fibers on one side of the arm, along with the contraction of all transverse fibers, induces arm bending. The longitudinal muscles, situated more peripherally, generate more torque, thus a bend is generated through their contraction, whereas the transverse fibers preserve the arm length, preventing the shortening that would result with contraction of the longitudinal fibers [44, 48]. Finally, contraction of the oblique layers twists the arm. The oblique muscles vary in their handedness; the external and internal obliques share a similar handedness that is opposite to the median oblique. Thus, if a left-handed or right-handed helix is activated, the arm twists counterclockwise or clockwise, respectively [44, 45, 48, 61]. Depending upon which muscle fiber or fibers are contracted, the arm will elongate, shorten, bend, or twist.

10.2.5 Structure and Functions of the Suckers

In order to achieve many of its daily tasks, the octopus employs not only arm movements but also the hundreds of suckers that are located on the oral side of each arm. Suckers are used for chemotactile sensing, object manipulation and transfer, prey capture, body anchoring, maneuvering, and locomotion such as crawling [47, 61, 85]. Octopus suckers are incredibly flexible, enabling them to adhere to objects of vastly different dimensions and textures [92]. The stalk, connecting the sucker to the arm, allows the sucker to elongate to twice its resting length and rotate and bend in any direction. The sucker itself consists of three primary regions: the acetabulum, orifice, and infundibulum (Fig. 10.3b). The acetabulum is a spheroidal cavity with smooth inner walls and a rough fibrillar protuberance at its base, which totals about 80% of the acetabular exposed surface. The orifice, which also features rough fibrillar walls, is torus-shaped and joins the acetabulum to the infundibulum. The infundibulum, another spheroidal cavity, is lined with radial and concentric ridges, starting at the surface of the sucker rim and continuing down to the orifice [89, 91, 92]. The walls of these three regions feature radial, circular, and meridional muscle fibers. The radial muscles are uniformly distributed throughout the walls of the sucker, whereas the circular muscles circumscribe the acetabular and infundibular cavities. Meridional muscles are perpendicularly aligned with respect to the radial and circular muscles, and occupy all regions apart from the sucker rim [44, 47, 89, 91].

To date, Tramacere et al. [89] have proposed the most thorough hypotheses of sucker attachment and detachment mechanisms. In brief, the radial and meridional muscles are activated for attachment, whereas the circular muscles are activated for detachment. The attachment mechanism consists of four phases. (1) While the acetabulum remains in its resting configuration, the infundibulum establishes a seal on the substrate as it transforms from a hollow cone to a disk. This is achieved through the contraction of the infundibular radial muscles, which thin the infundibulum and increase the surface contact area. At this point, there exists no pressure differential between the infundibulum, the acetabulum, and the pressure external to the sucker. (2) The acetabular meridional muscles begin to contract, pressurizing the water (resisting expansion) in the acetabular cavity. As suction is created, the pressure in the sucker cavities reduces with respect to the external pressure. (3) The meridional muscles continue to contract, causing the acetabular fibrillar protuberance to make contact with the fibrillar walls of the orifice, creating a toroidal water chamber. At this point the two internal chambers (i.e., the acetabular torus and the infundibulum) are of equal pressure, but still less than the external pressure. (4) The acetabular radial muscles cease contraction, while the cohesive forces of the water inside the infundibular chamber and the interlocking fibrillar surfaces of the acetabular base and orifice passively preserve the configuration. This allows the sucker to adhere for an extended period of time without muscle contraction. The pressure inside the infundibular chamber is conserved, but less than that of the acetabular chamber and the external pressure. In contrast, detachment is achieved by two concurrent muscle contractions: (1) infundibular circular muscles release the seal between the infundibulum and the substrate; and (2) acetabular circular muscles increase the pressure inside the acetabulum and release the torus water chamber by disengaging the fibrillar surfaces of the acetabular and orifice walls. Despite their compliance and size, the attachment and detachment mechanism of suckers are remarkably effective and efficient.

In addition, it has been postulated that other passive mechanisms may contribute to sucker attachment. Octopuses are known for frequently shedding and regenerating the lining (i.e., chitinous cuticle) that coats the walls of the infundibulum and acetabulum [46, 47]. Packard et al. reasoned this shedding was "to maintain the efficiency of the suckers as adhesive organs" [69]. Others have also hypothesized that the micro protuberances on the chitinous cuticle assist in creating a tight seal to achieve sucker attachment [3, 47]. Through a series of sucker pull-off experiments where the acetabular walls were punctured (compromising suction) in amputated *Octopus bimaculoides* arms, it was shown that adhesion still persisted to a minimal degree. Furthermore, during their preliminary parameterization studies, they observed that consecutive pull-off tests upon the same sucker revealed a great decrease in pull-off force after the first trial. Therefore, for their experimental procedure they performed one trial per sucker to avoid damaging the chitinous cuticle. Lastly, some have observed mucus surrounding the suckers [1, 3, 47]. Accogli et al. speculated the mucus assisted with attachment. Therefore, while suction is the primary mechanism of attachment, additional adhesive mechanisms likely contribute, making it worthwhile to further investigate these mechanisms.

10.3 Soft Actuators and Octopus-Inspired Robotic Arms

It has become evident that the octopus is a superb model for the field of soft robotics. Numerous inspirations can be derived and directed towards the design and development of soft robots that can bend, deform, configure, attach/detach, swim, and adapt to their surroundings—tasks that are either limited or impossible with rigid bodies. Rigid-bodied robots, indeed, are mainly adequate for single-tasked operations and pose hazards for human interaction (e.g., factory assembly line) and operations (e.g., endoscopy, cardiac surgery) [22, 77]. Currently, there are four main types of mechanisms for soft actuations: cable, dielectric electroactive polymers (EAPs), shape memory alloys (SMA), and fluidic. Using these forms of soft actuation, diverse octopus-inspired robots have been developed. In the following section, each actuation system is discussed, including its advantages, disadvantages, and examples of specific applications. A summary is presented in Table 10.1.

10.3.1 Cable Actuators

The lengthening and retracting of cables can be used for actuation of soft robots (Fig. 10.4a) [15, 22, 73, 78]. With minimal mechanical losses, the process is efficient, yet a complex design is required for achieving versatile deformations. Laschi et al. prototyped a silicone octopus-inspired arm, composed of a hollow superellipse, with concave sides, buttressed by four cylinders, oriented tangential to the concave sides of the superellipse [54]. Each of the four cylinders acts as a longitudinal muscle, and consisted of a cable running down its length with embedded circular plates for anchoring. The superellipse acts as the transverse muscle and consisted of eight radially oriented cables with embedded anchor plates. Laschi et al. validated that pulling the cables associated with the superellipse (i.e., the transverse muscle) decreases the arm diameter and increases its length. Likewise, they confirmed that by pulling one cylinder (i.e., the longitudinal muscle) the arm bends [54]. Calisti et al. designed and built a simple crawling robot platform to examine the crawling mechanism of the octopus through the shortening and bending of a conical silicone cable-driven arm. Their experimentation revealed that aside from object grasping, bending can be utilized for crawling [15]. The group went on to design a more extensive cable-driven arm consisting of 12 cables to perform dynamic experimentation with a soft octopus-like arm underwater [14, 18, 73].

Table 10.1 Summary of octopus-inspired robotic arms

Actuation mechanism	Implementation of octopus inspirations	Measured metrics	References
Cable	Arm density and dimension proportionally. Arm elongation, shortening, twisting, and bending motion	30% shortening (longitudinal cables), 43° bend (longitudinal cables), 60% elongation with 20% arm diameter reduction (transverse cables)	[14, 18, 54, 73]
		3.99 mm/s crawling speed (octopus: 200 mm/s)	[15]
Pneumatic	Lamina unidirectional muscle fiber structure. Elongation, shortening, twisting, and bending depend on fiber direction	23% elongation (90° lamina at 1.5 kPa). 4% elongation and 46° twist (45° lamina at 1 kPa). 8% shortening and 29° twist (90 and 45° lamina at 2.9 kPa). 40° bend (90° and 0° lamina at 3.8 kPa)	[50]
	Swimming gait (forward thrust), body steering and rotation (symmetric and asymmetric arm utilization). Multidirectional arm bending	9 cm/s swimming velocity (~0.477 BL/s). 174° arm bending at 356 kPa	[29, 30]
	Multidirectional arm bending	54 and 96% elongation for one and two arm segments, respectively, at 80 kPa. 98 and 52° arm bending with one and two arm segments, respectively, at 80 kPa	[31]
	Axial radial, and oblique muscle placement. Arm shortening, bending, and twisting motion	14.1% shortening and 72.1° bending at 300 kPa. 200° twisting at 500 kPa	[23]
	Arm elongation and bending motion	40% elongation and 53° bending at 200 kPa	[35]
	Arm bending, twisting, and grasping motion. Sucker end effector	50 m^{-1} curvature at 24 kPa	[60]
	Arm elongation, bending, and grasping motion	66% distal elongation and 380° twist at 413 kPa. 890 and 250 N lift capacity for proximal axial and transverse, respectively	[65, 66, 94]

(continued)

Table 10.1 (continued)

Actuation mechanism	Implementation of octopus inspirations	Measured metrics	References
Continuum motion	Crawling gait; pushing and pulling of compliant arms	Crawling: 0.3 BL/s and 1.8 COT (Octopus avg and max: 0.62 and 1.94 BL/s, respectively [38])	[16, 20]
	Body steering, crawling, and swimming (forward and backward thrust) gait. Collective arm grasping. Arm geometry (with and without connective webbing) with inactive suckers	Swimming without webbing: 0.26 BL/s and 1.42 COT	[81]
		Swimming with webbing: 0.5 BL/s and 0.62 COT (Octopus avg and max: 1.27 and 3.06 BL/s, respectively [38])	[80]
SMA	Arm density and dimension proportionally. Sinusoidal arrangements of nerve cord during arm's relax state inspired cord arrangement, allowing for large arm elongation	70% elongation with 20% (23% in octopus) diameter reduction. 64° bend (transverse SMA and longitudinal cables). 20% shortening and 43° bending (longitudinal cables)	[19, 27, 55, 63]
EAP	Bending motion	90 and 35° bending for 2DOF and 3DOF arm, respectively	[2, 70]

BL/s: Body Length per second, *COT*: Cost of Transport, *DOF*: Degrees of Freedom

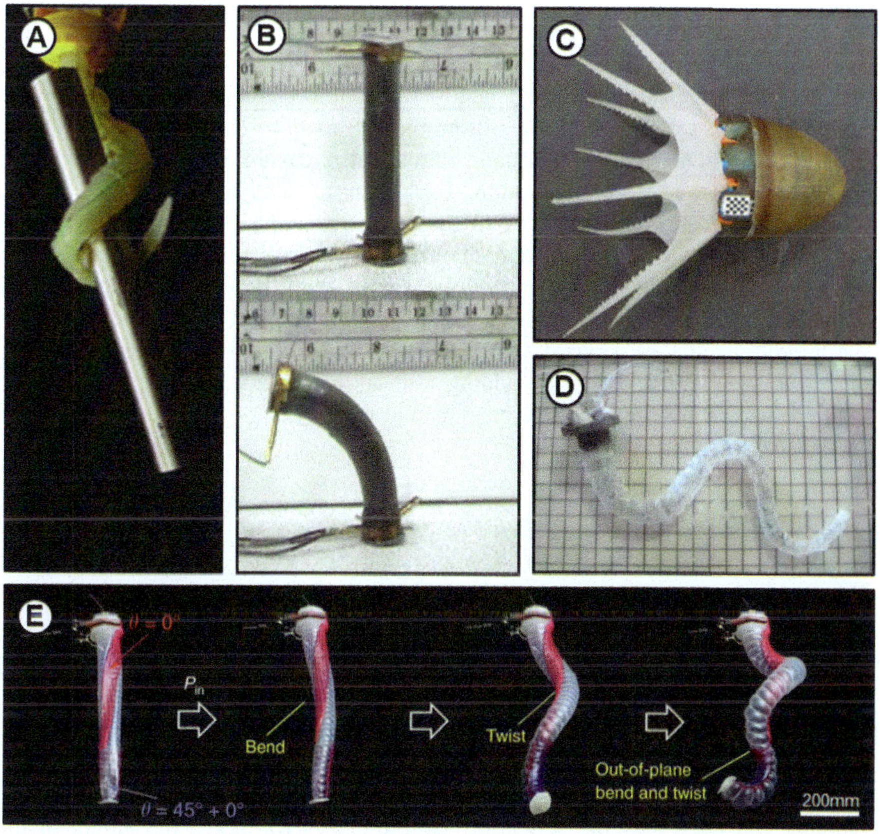

Fig. 10.4 Examples of octopus-inspired robotic arms actuated through (**a**) cables, (**b**) dielectric electroactive polymers, (**c**) rigid bodies with soft motions, (**d**) shape memory alloy, and (**e**) fluidic. (**a**) Adapted by CC 4.4 [64]. Copyright 2019, John Wiley and Sons, Inc, (**b**) Adapted with permission [70]. Copyright 2004, IOP Publishing Ltd. (**c**) Adapted with permission [80]. Copyright 2014, IEEE. (**d**) Adapted with permission [19]. Copyright 2012, IEEE, (**e**) Adapted by CC 4.0 [50]. Copyright 2019, Nature Research

10.3.2 Dielectric ElectroActive Polymers

Dielectric electroactive polymers (EAPs) utilize electrical potential fields to actuate (Fig. 10.4b) [34, 52]. While they can provide large deformations at a relatively fast response rate, a high voltage is required to do so. Furthermore, a rigid structural backbone is essential to pre-strain the elastomer, otherwise it yields low stress. MERbot is a six-legged walking robot inspired by the octopus, with each leg (Fig. 10.4b) fabricated from a spring core wrapped by pre-strained electroelastomer films. By bending the legs sequentially, the robot can locomote forward [2, 70].

10.3.3 Rigid Bodies with Soft Motions

As observed from the previously mentioned examples, some soft robots may be more rigid than soft, but they can still mimic soft motions (Fig. 10.4c) [21, 36, 71, 78, 100]. PosaiDRONE is an aquatic robot that is capable of buoyancy control, jet propulsion, and arm locomotion. Each arm features four three-bar linkage mechanisms, each connected with a compliant tapered silicone arm. Therefore, cranking the bar linkage initiates a sculling motion, propelling the robot forward when the silicone arms are in and out of contact with the ground. The robot has been used to explore how center of mass and buoyancy influence reaction forces and the hydrodynamics of forward and backward locomotion [16]. The group further enhanced their initial design by adding four additional arms, where two of them could be used actively to shorten, bend, and grasp [20].

10.3.4 Shape Memory Alloys

Shape memory alloys (SMA) operate based on a change in temperature [7, 51, 75]. While they provide a high mass-specific force, the process is inefficient when surrounded with alternating thermal environments. Moreover, excessive heating and straining of the SMA wire may lead to irreversible damage. Mazzolai et al. utilized cables and SMA coils to actuate an octopus-inspired soft robotic arm. The arm consists of a thin silicone shell with an underlying conical meshed structure, which achieves multiple functionalities such as passive movement, structural support, and housing and anchoring the actuators. Twelve cables run down the length of the arm, acting as longitudinal muscles, and when all are actuated the arm shortens by 20%. In addition to bending in four distinct directions, the arm is also capable of bending at single or multiple locations, depending on the series of cables actuated. Nine SMA membranes were distributed perpendicular to the arm (Fig. 10.4d), where each membrane consisted of a disk center with eight helical SMA coils propagating radially and anchoring the mesh housing by polyethylene wires. While all of the SMA membranes were connected in series, each could be actuated independently. Through actuation of the SMA membranes, the diameter of the arm was decreased by 20%, extending the arm by 90%. The arm has the ability to shorten, elongate, and bend [19, 27, 55, 63].

10.3.5 Fluidic Actuators

Pressurized fluid or pneumatic air can be used to inflate elastomer channels to achieve high displacement and force. Pneumatic actuators provide advantages as they are lightweight, high yielding, and can achieve varied movements similar to the octopus arm (e.g., extension, retraction, twisting, bending). However, they also have many disadvantages, such as nonlinear rubber properties leading to open-loop control algorithms, high-strain requirements for actuation, rupture vulnerability and failure upon inflation, a convoluted fabrication process, low repeatability in production, and heavy and complex components such as compressors and valves [41, 42, 49, 72, 77]. OctArm is an octopus-inspired continuum arm [83], consisting of four long sections partitioned by five base plates, where three pneumatic McKibben artificial muscles are mounted. Supplied air actuates the muscles through flexing and/or extending the arm tubing. Cable encoders measure the degree of deformation of each muscle along the arm, and the arm has the ability to extend, shorten, curl, and grasp objects of various sizes [65, 66, 94]. Similar efforts have been made by Guglielmino and colleagues [35].

Gong et al. used a pneumatically actuated soft arm to study the open-loop kinematics of a pneumatic soft system, finding the relationship between input pressure and end-effector location [31]. The design consists of a silicone exterior, polydimethylsiloxane (PDMS) core, and three arced chambers surrounding the PDMS core that run along the arm length [60]. Depending on the actuated chamber(s), the arm can bend and twist in the desired direction. Doi et al. developed a synthetic octopus arm that took into consideration the longitudinal, transverse, and oblique muscles. They manufactured thin McKibben muscles, using four for each longitudinal bundle, and weaved one between the arm's rubber core and each longitudinal bundle to make the transverse muscle. In addition, they braided two clockwise and two counterclockwise McKibben muscles on the outermost edge of the arm to represent the oblique muscles. Actuating these artificial pneumatic muscles shortens, stiffens, bends, and twists the arm [23]. Similar motions were accomplished through the implementation of unidirectional fibers into pneumatic structures, inspired by the muscle lamina structure (Fig. 10.4e) [50]. Fras et al. designed a pneumatically driven soft octopus-inspired swimmer with eight arms. Each arm consists of two internal chambers. Depending on which chambers were inflated, the arm bends, enabling the robot to rotate and turn about its primary axis. When all chambers inflated, the arms provide a swimming stroke, propelling the robot forward [29, 30]. Sfakiotakis et al. implemented a similar design, although the arms are less elastic, being composed of polyurethane, and are actuated through a servo motor at the base of each arm (Fig. 10.4c) [80, 81].

10.4 Synthetic Suckers

Like the octopus arm, inspiration can be gained from the structure and functions of octopus suckers. If controlled attachment is essential for functionality [6], synthetic suckers provide an elegant solution for manipulating, grasping, and transferring objects of different dimensions, densities, and textures. Potential applications include the agricultural industry, where they can be used for picking and transferring fruits and vegetables, to the automobile industry, where they can augment assembly lines. In addition, synthetic suckers are useful in various experimental conditions such as dry, wet, hot, and cold. Lastly, they also have potential medical applications, forming and/or securing bandages and serving as therapeutic drug delivery patches. Presented below are some instances where suckers have inspired the design of synthetic attachment/detachment mechanisms using existing soft actuators: passive, shape memory alloys (SMA), fluidic, and dielectric electroactive polymers (EAPs). A summary is presented in Table 10.2.

10.4.1 Passive Suction Cups

The soft and deformable characteristics of suckers have inspired the development of passive micro suction cups. Miniaturized suction cups (mSCs) are dry adhesive medical patches inspired by octopus suckers for monitoring vital signs and delivering therapeutic drugs [17, 56]. The patch consists of a silicone base and polydimethylsiloxane (PDMS) mSCs array interface. The interface layer houses cardiac sensing electrodes, while strain gauges and temperature sensors are located between the interface and the base. Negative pressure and van der Waals forces initiate and sustain the patch's attachment. Aside from functioning under deformations (i.e., tension and compression), it is reusable for approximately 8 cycles [17]. Preliminary prototyping of the patch consisted of a PDMS base and hydrogel interface, where adhesion is temperature dependent. In the presence of heat, attachment is achieved through the negative pressure that is established in the mSC, and in the absence of heat the patch is detached [56]. Another group has developed a similar design for medical and industrial applications, such as silicon wafer manufacturing [4, 5, 97]. They tested the performance of their octopus-inspired architectures (OIAs) (Fig. 10.5a) against three other designs, including perforated cylinders, cylindrical pillars, and cylindrical holes, on silicon wafers. The patch is made of a polyurethane-acrylate-based polymer (s-PUA), and the attachment mechanism is dependent on the applied normal load. Through an applied load, a pressure differential is created between the suction cup cavity and the external environment, resulting in attachment. The OIAs take inspiration from the octopus sucker's acetabular fibrillar protuberance and its functionality during the attachment process [89]. This design was tested under dry, moist, oiled, and underwater conditions, taking advantage of van der Waals forces under dry conditions and capillary forces under wet conditions. The greatest attachment forces were attained by the OIAs under oiled and, secondly, underwater conditions [4, 5, 97].

10.4.2 Shape Memory Alloys

Hu and colleagues have designed a piston-like sucker with embedded SMA actuators [37]. Their design is composed of a circular silicone membrane connected

Table 10.2 Summary of octopus-inspired synthetic suckers

Actuation method	Mechanism	Implementation of octopus inspirations	Measured metrics	References
Passive	Dry solid surface: suction (negative pressure through compression) and van der Waals force. Wet solid surfaces: suction and capillary force	Architectural feature inspired by sucker rim, infundibulum, and acetabular protuberance. Wet and dry adhesion	26, 37, 41, and 154 kPa normal adhesion of octopus-inspired architecture (OIA) patch with 50 μm radii array under dry, moist, under water, and under oil conditions. Twice than cylindrical hole (26, 14, 14, and 61 kPa) and non-patterned patch (9, 10, 9, and 9 kPa) under similar conditions, respectively. 180 kPa max normal adhesion with 15 μm radii array OIA patch under oil	[4, 5]
	Hydrogel layer thermal response creating pressure differential suction	Radial and meridional muscle functionality during contraction to decrease cavity volume, and induce suction through pressure differential	94 kPa normal adhesion for patch with 1.0 μm diameter and 5.5 μm pitch arrays under temperature change of 20 °C (off) and 61 °C (on). 59.1 kPa normal adhesion under temperature change of 22 °C (off) and 35 °C (on), more than four times than non-patterned patch (14 kPa)	[56]
	Van der Waals force and suction through negative pressure	Suction cup geometry	1.5 kPa normal adhesion for patch with 0.5 μm diameter arrays. More than twice that of non-patterned patch (0.7 kPa)	[17]

(continued)

Table 10.2 (continued)

Actuation method	Mechanism	Implementation of octopus inspirations	Measured metrics	References
Fluidic	Vacuum suction induces pressure differential between membrane	Suction cup with flat and concave membrane. Assessment and enhancement of arm grasping	Grasping force increased by 1.4, 2.4, and 12.5 times in air (3.6 N), water (2.3 N), and oil (1.3 N), respectively	[64]
	Applied load and vacuum pressure	Elastic/rigidity of acetabulum, softness/pliability, and radial surface grooves of infundibulum	2.1 N loading capacity under −50 kPa	[79]
	Vacuum and pressure differential	Suction through pressure differential. Micro bumps on infundibulum sucker surface	4 N adhesive force with success gripping rate (20 trials) increasing from 50% (no micro bumps) to 100% (with micro bumps) when grasping 180 gf	[86, 87]
	Suction through syringe	Sucker morphology (size and proportionality), radial surface grooves of infundibulum	Adhesion (pull-off) improved through increased number and width of groves. 8 N pull-off force of on aluminum, acetal, and acrylic surface at 90 kPa, 98 kPa, and 98 kPa, respectively	[88, 92]
SMA	Negative pressure through sealing elastic cavity with current activation.	Adhesion through volume change and pressure reduction	Maximum negative pressure of 12 kPa (15 N) at current of 4 A, while holding rubber surface with 200 gf	[8, 37]
EAP	Negative pressure through straining elastic cavity with voltage activation	Pressure differential suction, radial surface grooves of infundibulum	8.34 N pull-off force underwater, through pressure differential of 3.63 kPa at 10.75 kV	[82]
	Pressure differential through activating electrostatic pressure in active membrane through applied voltage	Attachment mechanism, sucker morphology (size and proportionality), radial surface grooves of infundibulum, and acetabular radial muscles	Decreasing compliant dielectric layer from 5 to 3 increased pressure differential from 3.6 to 6.6 kPa, at internal volume of 0.56 cm^3	[28]

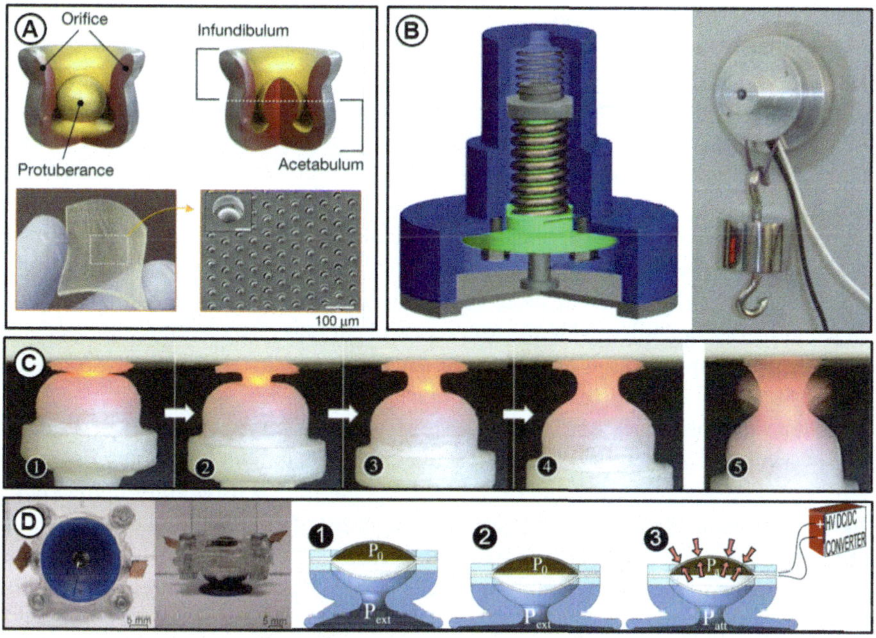

Fig. 10.5 Examples of octopus-inspired synthetic suckers actuated through (**a**) passive suction cups, (**b**) shape memory alloy, (**c**) fluidic, and (**d**) dielectric electroactive polymers. (**a**) Adapted with permission [4]. Copyright 2017, Nature Research. (**b**) Adapted by CC 3.0 [8]. Copyright 2009, SAGE Publishing. (**c**) Adapted with permission [79]. Copyright 2017, The Royal Society Publishing. (**d**) Adapted with permission [28]. Copyright 2014, IOP Publishing Ltd

to a rigid membrane consisting of three cavities, radially spaced from one another. Each cavity has a SMA coiled rod at its center, which is connected to the silicone membrane on one end and is free to move on the other end with respect to the main structural housing. Upon activation, the SMA wires elongate, pushing the rod up with respect to the main housing structure. In the process, the silicone membrane is stretched, creating suction through negative pressure. A bias spring in the middle of the main housing structure pushes down upon the rigid membrane, expediting detachment [37]. A miniaturized version consisting of a single cavity (Fig. 10.5b) was also manufactured [8].

10.4.3 Fluidic Actuators

Continuing their efforts to design an octopus-inspired silicone cable-driven arm [73], Renda et al. enhanced their design with pneumatically actuated silicone suckers (Fig. 10.4a) [64]. They implemented three suction cup designs with open, flat, and concave membrane interfaces. The open suction cup, directly connected to the fluidic channel, creates suction upon vacuum. The design improves attachment; however, it is prone to clogging by particles in the environment, thus decreasing suction. The flat membrane interface, similar to the open suction cup but featuring a thin flap that collapses upon vacuum, creates suction between the object and interface through negative pressure. By directing air onto the flat membrane, the suction cup expands into a dome, detaching itself from the object. The concave membrane interface features a thin membrane lining the internal cavity of the suction cup. Through pre-loading the base of the sucker (done so by the arm), passive attachment can be achieved, whereas detachment occurs through the introduction of air into the chamber. The performance of these sucker designs was tested with various objects in a confined space filled with air, water, and oil. It was observed that the synthetic suction cups improved the arm's attachment force and its ability to grasp objects [64].

Sareh and colleagues designed a sucker (Fig. 10.5c), fabricated out of silicone, with variable stiffness [79]. To hold vacuum, the acetabulum has a high degree of stiffness. The acetabular chamber houses an optical fiber sensor, emitting/receiving light to measure distance and contact. In contrast, the infundibular chamber is stiff to allow for greater conformation to create an efficient seal. Furthermore, radial grooves in the infundibulum interface create a tighter seal, strengthening the attachment mechanism, and even accelerating the detachment process [79]. Earlier prototyping showed that increasing the number of radial grooves, as well as their width and depth, increases attachment force [88, 92]. A different group, similarly inspired by octopus suckers, enhanced the contact surface of a granular jamming vacuuming gripper [86, 87]. The gripper consists of a silicone pouch housing small glass beads. When pressed against an object, the pouch deforms to conform around the object and, following the introduction of vacuum, the pouch is able to hold onto the object. The pouch's interaction interface was improved by adding arrays of flat membrane suction cups surrounded by micro bumps inspired by the fibrillar protuberance in the octopus sucker [90]. A porous filter is located behind each flat membrane suction cup, which upon vacuum causes the membrane to collapse creating suction through negative pressure. These added design features increased friction and attachment force between the gripper and objects [86, 87].

10.4.4 Dielectric ElectroActive Polymers

Sholl et al. created a soft sucker made of EAP that can be used underwater [82]. The design consists of a silicone cylinder containing EAP film wrapped around a rod connected to a silicone sucker. Through the activation of the EAP elements, a pressure differential is created, and thus suction. They implemented flat and radially grooved sucker designs, and the results show an increase in adhesion with the radial groove design [82]. For a proof-of-concept, another group took further inspiration from the octopus sucker by implementing an actuated acetabular dome and a passive silicone infundibulum (Fig. 10.5d). Through an applied pre-load the sucker attachment initiates, followed by the activation of the acetabular EAP, where a pressure differential between the sucker and the external environment ensures attachment [28].

10.5 Conclusion

As the aforementioned examples of the design and development of octopus-inspired robotic arms and synthetic suckers clearly demonstrate, a bioinspired interdisciplinary approach is quite fruitful. Through the integration of biology, physics, and engineering, elegant solutions to various multifaceted engineering problems can be found. In addition, reciprocal learning is established: through biology, solutions are discovered for engineering; and, through engineering, insights are gained for future biological studies. By focusing on biology and acquiring further knowledge of the control mechanisms and functions of octopus arms and suckers, more effective synthetic arms and suckers can be made. Preliminary research reveals that the control and functionality of octopus arms and suckers are more complex than what has been presented, necessitating further investigation. Additional studies are needed to document neural and muscular activity within the arm to conclusively determine arm movements (i.e., extension, shortening, bending, and twisting) and the mechanisms involved in sucker attachment and detachment. The ultimate goal of these future investigations is to develop dexterous soft robotic arms and suckers with distributed hierarchical control architectures for manipulation, attachment, and locomotion. Decentralized sensing and actuation can be used to minimize communication and power consumption, enabling a scalable design and adaptive autonomous system. Compared to existing rigid body systems, the development of soft robotic arms will provide vast improvements in versatility, flexibility, adaptability, functionality, and safety. Importantly, these robots will be able to perform unstructured manipulation and locomotion in uncertain, dynamic environments. Future bioinspired octopus robots are envisioned to have autonomous functional arms, capable of performing meticulous and complex tasks through independent and collective arm movements, and featuring efficient synthetic suckers for object manipulation.

References

1. G. Accogli, G. Scillitani, D. Mentino, S. Desantis, Characterization of the skin mucus in the common octopus *Octopus vulgaris* (Cuvier) reared paralarvae. Eur. J. Histochem. **61**(3) 2518, (2017)
2. I.A. Anderson, T.A. Gisby, T.G. McKay, B.M. O'Brien, E.P. Calius, Multi-functional dielectric elastomer artificial muscles for soft and smart machines. J. Appl. Phys. **112**(4), 041101 (2012)

3. H. Bagheri, A. Hu, S. Cummings, C. Roy, R. Casleton, A. Wan, N. Erjavic, S. Berman, M.M. Peet, D.M. Aukes, X. He, S.C. Pratt, R.E. Fisher, H. Marvi, New insights on the control and function of octopus suckers. Adv. Intel. Syst. **2**, 1900154 (2020)
4. S. Baik, D.W. Kim, Y. Park, T.-J. Lee, S.H. Bhang, C. Pang, A wet-tolerant adhesive patch inspired by protuberances in suction cups of octopi. Nature **546**(7658), 396–400 (2017)
5. S. Baik, J. Kim, H.J. Lee, T.H. Lee, C. Pang, Highly adaptable and biocompatible octopus-like adhesive patches with meniscus-controlled unfoldable 3d microtips for underwater surface and hairy skin. Adv. Sci. **5**(8), 1800100 (2018)
6. S. Baik, H.J. Lee, D.W. Kim, J.W. Kim, Y. Lee, C. Pang, Bioinspired adhesive architectures: from skin patch to integrated bioelectronics. Adv. Mat. **31**, 1803309 (2019)
7. M.D. Bartlett, N. Kazem, M.J. Powell-Palm, X. Huang, W. Sun, J.A. Malen, C. Majidi, High thermal conductivity in soft elastomers with elongated liquid metal inclusions. Proc. Natl. Acad. Sci. **114**, 201616377 (2017)
8. H. Bing-Shan, W. Li-Wen, F. Zhuang, Z. Yan-zheng, Bio-inspired miniature suction cups actuated by shape memory alloy. Int. J. Adv. Robot. Syst. **6**(3), 29 (2009)
9. Q. Bone, E.R. Brown, M. Usher, The structure and physiology of cephalopodss muscle fibers, in *Cephalopod Neurobiology: Neuroscience Studies in Squid, Octopus and Cuttlefish*, ed. by N.J. Abbot, R. Williamson, L. Maddock (Oxford University Press, Oxford, 1995)
10. P.R. Boyle, Neural control of cephalopod behavior, in *The Mollusca* (Elsevier, Amsterdam, 1986), pp. 1–99
11. B.U. Budelmann, The cephalopod nervous system: What evolution has made of the molluscan design, in *The Nervous Systems of Invertebrates: An Evolutionary and Comparative Approach* (Springer, Berlin, 1995), pp. 115–138
12. R.A. Byrne, M.J. Kuba, D.V. Meisel, Lateralized eye use in *Octopus vulgaris* shows antisymmetrical distribution. Anim. Behav. **68**(5), 1107–1114 (2004)
13. R.A. Byrne, M.J. Kuba, D.V. Meisel, U. Griebel, J.A. Mather, Does *Octopus vulgaris* have preferred arms? J. Comp. Psychol. **120**(3), 198 (2006)
14. M. Calisti, A. Arienti, M.E. Giannaccini, M. Follador, M. Giorelli, M. Cianchetti, B. Mazzolai, C. Laschi, P. Dario, Study and fabrication of bioinspired octopus arm mockups tested on a multipurpose platform, in *3rd IEEE RAS & EMBS International Conference on Biomedical Robotics and Biomechatronics* (2010), pp. 461–466
15. M. Calisti, M. Giorelli, G. Levy, B. Mazzolai, B. Hochner, C. Laschi, P. Dario, An octopus-bioinspired solution to movement and manipulation for soft robots. Bioinspir. Biomim. **6**(3), 036002 (2011)
16. M. Calisti, F. Corucci, A. Arienti, C. Laschi, Dynamics of underwater legged locomotion: modeling and experiments on an octopus-inspired robot. Bioinspir. Biomim. **10**(4), 046012 (2015)
17. M.K. Choi, O.K. Park, C. Choi, S. Qiao, R. Ghaffari, J. Kim, D.J. Lee, M. Kim, W. Hyun, S.J. Kim, H.J. Hwang, S.-H. Kwon, T. Hyeon, N. Lu, D.-H. Kim, Cephalopod-inspired miniaturized suction cups for smart medical skin. Adv. Healthc. Mater. **5**(1), 80–87 (2016)
18. M. Cianchetti, A. Arienti, M. Follador, B. Mazzolai, P. Dario, C. Laschi, Design concept and validation of a robotic arm inspired by the octopus. Mater. Sci. Eng. C **31**(6), 1230–1239 (2011)
19. M. Cianchetti, M. Follador, B. Mazzolai, P. Dario, C. Laschi, Design and development of a soft robotic octopus arm exploiting embodied intelligence, in *IEEE International Conference on Robotics and Automation* (2012), pp. 5271–5276
20. M. Cianchetti, M. Calisti, L. Margheri, M. Kuba, C. Laschi, Bioinspired locomotion and grasping in water: the soft eight-arm octopus robot. Bioinspir. Biomim. **10**(3), 035003 (2015)
21. A. Crespi, A.J. Ijspeert, Online optimization of swimming and crawling in an amphibious snake robot. IEEE Trans. Robot. **24**(1), 75–87 (2008)
22. T. Deng, H. Wang, W. Chen, X. Wang, R. Pfeifer, Development of a new cable-driven soft robot for cardiac ablation, in *IEEE International Conference on Robotics and Biomimetics (ROBIO)* (2013), pp. 728–733

23. T. Doi, S. Wakimoto, K. Suzumori, K. Mori, Proposal of flexible robotic arm with thin McKibben actuators mimicking octopus arm structure, in *IEEE/RSJ International Conference on Intelligent Robots and Systems (IROS)* (2016), pp. 5503–5508
24. J.K. Finn, T. Tregenza, M.D. Norman, Defensive tool use in a coconut-carrying octopus. Curr. Biol. **19**(23), R1069–R1070 (2009)
25. G. Fiorito, C. von Planta, P. Scotto, Problem solving ability of *Octopus vulgaris* Lamarck (Mollusca, Cephalopoda). Behav. Neural Biol. **53**(2), 217–230 (1990)
26. G. Fiorito, G.B. Biederman, V.A. Davey, F. Gherardi, The role of stimulus preexposure in problem solving by *Octopus vulgaris*. Anim. Cogn. **1**(2), 107–112 (1998)
27. M. Follador, M. Cianchetti, C. Laschi, Development of the functional unit of a completely soft octopus-like robotic arm, in *4th IEEE RAS & EMBS International Conference on Biomedical Robotics and Biomechatronics (BioRob)* (2012), pp. 640–645
28. M. Follador, F. Tramacere, B. Mazzolai, Dielectric elastomer actuators for octopus inspired suction cups. Bioinspir. Biomim. **9**(4), 046002 (2014)
29. J. Fras, M. Macias, Y. Noh, K. Althoefer, Fluidical bending actuator designed for soft octopus robot tentacle, in *IEEE International Conference on Soft Robotics (RoboSoft)* (2018), pp. 253–257
30. J. Fras, Y. Noh, M. Macias, H. Wurdemann, K. Althoefer, Bio-inspired octopus robot based on novel soft fluidic actuator, in *IEEE International Conference on Robotics and Automation (ICRA)* (2018), pp. 1583–1588
31. Z. Gong, Z. Xie, X. Yang, T. Wang, L. Wen, Design, fabrication and kinematic modeling of a 3d-motion soft robotic arm, in *IEEE International Conference on Robotics and Biomimetics (ROBIO)* (2016), pp. 509–514
32. P. Graziadei, Muscle receptors in cephalopods. Proc. R. Soc. London, Ser. B Biol. Sci. **161**(984), 392–402 (1965)
33. P.P.C. Graziadei, H.T. Gagne, Sensory innervation in the rim of the octopus sucker. J. Morphol. **150**(3), 639–679 (1976)
34. G.-Y. Gu, J. Zhu, L.-M. Zhu, X. Zhu, A survey on dielectric elastomer actuators for soft robots. Bioinspir. Biomim. **12**(1), 011003 (2017)
35. E. Guglielmino, I. Godage, L. Zullo, D.G. Caldwell, A pragmatic bio-inspired approach to the design of octopus-inspired arms, in *IEEE/RSJ International Conference on Intelligent Robots and Systems* (2013), pp. 4577–4582
36. T. Horvat, K. Karakasiliotis, K. Melo, L. Fleury, R. Thandiackal, A.J. Ijspeert, Inverse kinematics and reflex based controller for body-limb coordination of a salamander-like robot walking on uneven terrain, in *IEEE/RSJ International Conference on Intelligent Robots and Systems (IROS)* (2015), pp. 195–201
37. B. Hu, Z. Dong, Z. Fu, Y. Zhao, A shape memory alloy actuated miniature bio-inspired multi-cavity negative pressure suction cup, in *Earth & Space 2008: Engineering, Science, Construction, and Operations in Challenging Environments* (2008), pp. 1–8
38. C.L. Huffard, Locomotion by *Abdopus aculeatus* (Cephalopoda: Octopodidae): walking the line between primary and secondary defenses. J. Exp. Biol. **209**(19), 3697–3707 (2006)
39. C.L. Huffard, F. Boneka, R.J. Full, Underwater bipedal locomotion by octopuses in disguise. Science **307**(5717), 1927–1927 (2005)
40. L.M. Hvorecny, J.L. Grudowski, C.J. Blakeslee, T.L. Simmons, P.R. Roy, J.A. Brooks, R.M. Hanner, M.E. Beigel, M.A. Karson, R.H. Nichols, J.B. HolmJean, G. Boal, Octopuses *(Octopus bimaculoides)* and cuttlefishes *(Sepia pharaonis, S. officinalis)* can conditionally discriminate. Anim. Cogn. **10**(4), 449–459 (2007)
41. F. Ilievski, A.D. Mazzeo, R.F. Shepherd, X. Chen, G.M. Whitesides, Soft robotics for chemists. Angew. Chem. **123**(8), 1930–1935 (2011)
42. Z. Jing, L. Qiao, H. Pan, Y. Yang, W. Chen, An overview of the configuration and manipulation of soft robotics for on-orbit servicing. Sci. China Inform. Sci. **60**(5), 050201 (2017)
43. S. Johnsen, E.J. Balser, E.A. Widder, Light-emitting suckers in an octopus. Nature **398**(6723), 113 (1999)

44. W.M. Kier, The arrangement and function of molluscan muscle. Mollusca Form Funct. **11**(21), 211–252 (1988)
45. W.M. Kier, The musculature of coleoid cephalopod arms and tentacles. Front. Cell Dev. Biol. **4**, 10 (2016)
46. W.M. Kier, A.M. Smith, The morphology and mechanics of octopus suckers. Biol. Bul. **178**(2), 126–136 (1990)
47. W.M Kier, A.M. Smith, The structure and adhesive mechanism of octopus suckers. Integr. Comp. Biol. **42**(6), 1146–1153 (2002)
48. W.M. Kier, K.K. Smith, Tongues, tentacles and trunks: the biomechanics of movement in muscular-hydrostats. Zool. J. Linnean Soc. **83**(4), 307–324 (1985)
49. S. Kim, C. Laschi, B. Trimmer, Soft robotics: a bioinspired evolution in robotics. Trends Biotechnol. **31**(5), 287–294 (2013)
50. S.Y. Kim, R. Baines, J. Booth, N. Vasios, K. Bertoldi, R. Kramer-Bottiglio, Reconfigurable soft body trajectories using unidirectionally stretchable composite laminae. Nat. Commun. **10**(1), 1–8 (2019)
51. J.-S. Koh, D.-Y. Lee, S.-W. Kim, K.-J. Cho, Deformable soft wheel robot using hybrid actuation, in *IEEE/RSJ International Conference on Intelligent Robots and Systems* (2012), pp. 3869–3870
52. I.M. Koo, K. Jung, J.C. Koo, J.-D. Nam, Y.K. Lee, H.R. Choi, Development of soft-actuator-based wearable tactile display. IEEE Trans. Robot. **24**(3), 549–558 (2008)
53. M. Kuba, D.V. Meisel, R.A. Byrne, U. Griebel, J.A. Mather, Looking at play in *Octopus vulgaris*. Berl. Paläont. Abh. **3**,163–169 (2003)
54. C. Laschi, B. Mazzolai, V. Mattoli, M. Cianchetti, P. Dario, Design of a biomimetic robotic octopus arm. Bioinspir. Biomim. **4**(1), 015006 (2009)
55. C. Laschi, M. Cianchetti, B. Mazzolai, L. Margheri, M. Follador, P. Dario, Soft robot arm inspired by the octopus. Adv. Robot. **26**(7), 709–727 (2012)
56. H. Lee, D.-S. Um, Y. Lee, S. Lim, H.-J. Kim, H. Ko, Octopus-inspired smart adhesive pads for transfer printing of semiconducting nanomembranes. Adv. Mat. **28**(34), 7457–7465 (2016)
57. G. Levy, B. Hochner, Embodied organization of *Octopus vulgaris* morphology, vision, and locomotion. Front. Physiol. **8**, 164 (2017)
58. G. Levy, T. Flash, B. Hochner, Arm coordination in octopus crawling involves unique motor control strategies. Current Biol. **25**(9), 1195–1200 (2015)
59. J. Mann, E.M. Patterson, Tool use by aquatic animals. Philos. Trans. R. Soc. B Biol. Sci. **368**(1630), 20120424 (2013)
60. R.V. Martinez, J.L. Branch, C.R. Fish, L. Jin, R.F. Shepherd, R.M.D. Nunes, Z. Suo, G.M. Whitesides, Robotic tentacles with three-dimensional mobility based on flexible elastomers. Adv. Mat. **25**(2), 205–212 (2013)
61. J.A. Mather, How do octopuses use their arms? J. Com. Psychol. **112**(3), 306 (1998)
62. J.A. Mather, M.J. Kuba, The cephalopod specialties: complex nervous system, learning, and cognition. Can. J. Zool. **91**(6), 431–449 (2013)
63. B. Mazzolai, L. Margheri, M. Cianchetti, P. Dario, C. Laschi, Soft-robotic arm inspired by the octopus: Ii. from artificial requirements to innovative technological solutions. Bioinsp. Biomim. **7**(2), 025005 (2012)
64. B. Mazzolai, A. Mondini, F. Tramacere, G. Riccomi, A. Sadeghi, G. Giordano, E.D. Dottore, M. Scaccia, M. Zampato, S. Carminati, Octopus-inspired soft arm with suction cups for enhanced grasping tasks in confined environments. Adv. Intel. Syst. **1**, 1900041 (2019)
65. W. McMahan, I.D. Walker, Octopus-inspired grasp-synergies for continuum manipulators, in *IEEE International Conference on Robotics and Biomimetics* (2009), pp. 945–950
66. W. McMahan, V. Chitrakaran, M. Csencsits, D. Dawson, I.D. Walker, B.A. Jones, M. Pritts, D. Dienno, M. Grissom, C.D. Rahn, Field trials and testing of the OctArm continuum manipulator, in *Proceedings IEEE International Conference on Robotics and Automation (ICRA)* (2006), pp. 2336–2341
67. T. Moriyama, Y.-P. Gunji, Autonomous learning in maze solution by octopus. Ethology **103**(6), 499–513 (1997)

68. N. Nesher, G. Levy, F.W. Grasso, B. Hochner, Self-recognition mechanism between skin and suckers prevents octopus arms from interfering with each other. Current Biol. **24**(11), 1271–1275 (2014)
69. A. Packard, Sucker display of octopus. Nature **190**(4777), 736 (1961)
70. Q. Pei, M. Rosenthal, S. Stanford, H. Prahlad, R. Pelrine, Multiple-degrees-of-freedom electroelastomer roll actuators. Smart Mat. Struct. **13**(5), N86 (2004)
71. R. Pfeifer, H.G. Marques, F. Iida, Soft robotics: The next generation of intelligent machines, in *Proceedings of the Twenty-Third International Joint Conference on Artificial Intelligence* (AAAI Press, Hilversum, 2013), pp. 5–11
72. P. Polygerinos, N. Correll, S.A. Morin, B. Mosadegh, C.D. Onal, K. Petersen, M. Cianchetti, M.T. Tolley, R.F. Shepherd, Soft robotics: review of fluid-driven intrinsically soft devices; manufacturing, sensing, control, and applications in human-robot interaction. Adv. Eng. Mat. **19**(12), 1700016 (2017)
73. F. Renda, M. Giorelli, M. Calisti, M. Cianchetti, C. Laschi, Dynamic model of a multibending soft robot arm driven by cables. IEEE Trans. Robot. **30**(5), 1109–1122 (2014)
74. J.N. Richter, B. Hochner, M.J. Kuba, Pull or push? octopuses solve a puzzle problem. PloS one **11**(3), e0152048 (2016)
75. H. Rodrigue, W. Wang, D.-R. Kim, S.-H. Ahn, Curved shape memory alloy-based soft actuators and application to soft gripper. Comp. Struct. **176**, 398–406 (2017)
76. C.H.F. Rowell, Activity of interneurones in the arm of octopus in response to tactile stimulation. J. Exp. Biol. **44**(3), 589–605 (1966)
77. D. Rus, M.T. Tolley, Design, fabrication and control of soft robots. Nature **521**(7553), 467 (2015)
78. A. Sadeghi, A. Mondini, E.D. Dottore, A.K. Mishra, B. Mazzolai, Soft legged wheel-based robot with terrestrial locomotion abilities. Front. Robot. AI **3**, 73 (2016)
79. S. Sareh, K. Althoefer, M. Li, Y. Noh, F. Tramacere, P. Sareh, B. Mazzolai, M. Kovac, Anchoring like octopus: biologically inspired soft artificial sucker. J. R. Soc. Interface **14**(135), 20170395 (2017)
80. M. Sfakiotakis, A. Kazakidi, A. Chatzidaki, T. Evdaimon, D.P. Tsakiris, Multi-arm robotic swimming with octopus-inspired compliant web, in *IEEE/RSJ International Conference on Intelligent Robots and Systems* (2014), pp. 302–308
81. M. Sfakiotakis, A. Kazakidi, D.P. Tsakiris, Octopus-inspired multi-arm robotic swimming. Bioinspir. Biomim. **10**(3), 035005 (2015)
82. N. Sholl, A. Moss, W.M. Kier, K. Mohseni, A soft end effector inspired by cephalopod suckers and augmented by a dielectric elastomer actuator. Soft Robot. **6**, 356–367 (2019)
83. P.K. Singh, C.M. Krishna, Continuum arm robotic manipulator: a review. Univers. J. Mech. Eng. **2**(6), 193–198 (2014)
84. G. Sumbre, Y. Gutfreund, G. Fiorito, T. Flash, B. Hochner, Control of octopus arm extension by a peripheral motor program. Science **293**(5536), 1845–1848 (2001)
85. G. Sumbre, G. Fiorito, T. Flash, B. Hochner, Octopuses use a human-like strategy to control precise point-to-point arm movements. Current Biol. **16**(8), 767–772 (2006)
86. T. Takahashi, M. Suzuki, S. Aoyagi, Octopus bioinspired vacuum gripper with micro bumps, in *IEEE 11th Annual International Conference on Nano/Micro Engineered and Molecular Systems (NEMS)* (2016), pp. 508–511
87. T. Tomokazu, S. Kikuchi, M. Suzuki, S. Aoyagi, Vacuum gripper imitated octopus sucker-effect of liquid membrane for absorption, in *IEEE/RSJ International Conference on Intelligent Robots and Systems (IROS)* (2015), pp. 2929–2936
88. F. Tramacere, L. Beccai, F. Mattioli, E. Sinibaldi, B. Mazzolai, Artificial adhesion mechanisms inspired by octopus suckers, in *IEEE International Conference on Robotics and Automation* (2012), pp. 3846–3851
89. F. Tramacere, L. Beccai, M. Kuba, A. Gozzi, A. Difone, B. Mazzolai, The morphology and adhesion mechanism of *Octopus vulgaris* suckers. PLoS One **8**(6), e65074 (2013)

90. F. Tramacere, E. Appel, B. Mazzolai, S.N. Gorb, Hairy suckers: the surface microstructure and its possible functional significance in the octopus vulgaris sucker. Beilst. J. Nanotechnol. **5**(1), 561–565 (2014)
91. F. Tramacere, A. Kovalev, T. Kleinteich, S.N. Gorb, B. Mazzolai, Structure and mechanical properties of *Octopus vulgaris* suckers. J. R. Soc. Interface **11**(91), 20130816 (2014)
92. F. Tramacere, M. Follador, N.M. Pugno, B. Mazzolai, Octopus-like suction cups: from natural to artificial solutions. Bioinspir. Biomim. **10**(3), 035004 (2015)
93. R. Villanueva, A. Guerra, Food and prey detection in two deep-sea cephalopods: *Opisthoteuthis agassizi* and *O. vossi* (Octopoda: Cirrata). Bull. Mar. Sci. **49**(1–2), 288–299 (1991)
94. I.D. Walker, D.M. Dawson, T. Flash, F.W. Grasso, R.T. Hanlon, B. Hochner, W.M. Kier, C.C. Pagano, C.D. Rahn, Q.M. Zhang, Continuum robot arms inspired by cephalopods, in *Unmanned Ground Vehicle Technology VII*, vol. 5804 (2005), pp. 303–314. International Society for Optics and Photonics
95. M.J. Wells, Tactile discrimination of shape by octopus. Q. J. Exp. Psychol. **16**(2), 156–162 (1964)
96. M.J. Wells, Tactile discrimination of surface curvature and shape by the octopus. J. Exp. Biol. **41**(2), 433–445 (1964)
97. J.J. Wilker, Materials science: how to suck like an octopus. Nature **546**(7658), 358 (2017)
98. D.M. Wilson, Nervous control of movement in cephalopods. J. Exp. Biol. **37**(1), 57–72 (1960)
99. J.Z. Young, *The Anatomy of the Nervous System of Octopus Vulgaris* (Clarendon Press, Oxford, 1971)
100. J. Yu, R. Ding, Q. Yang, M. Tan, W. Wang, J. Zhang, On a bio-inspired amphibious robot capable of multimodal motion. IEEE/ASME Trans. Mech. **17**(5), 847–856 (2012)

Chapter 11
Octopus-Inspired Arm Movements

Feng Ling and Eva Kanso

11.1 Introduction

Octopus arms generate a remarkable diversity of complex movements. Their adaptability to a wide range of functions has attracted considerable attention in biomechanics and neuroscience. Octopus arms are examples of muscular hydrostats, where the three-dimensional arrangement of the musculature (see Fig. 11.1a) serves both to generate the force and to provide support for movement [1, 24, 25, 28, 30]. The sensorimotor control of the arms seems to be hierarchically organized, for the octopus nervous system is divided morphologically into three main parts: the central brain, the nerve ring and eight nerve ganglia at the base of the arms, and the nervous system along the arms [46, 52]. The arm nervous system, amounting to two-thirds of the neurons in the octopus nervous system, may function autonomously, and it can generate coordinated stereotypical movements, such as those observed in typical fetch and crawl behaviors [3, 14, 17, 31, 34, 45, 47]. The extreme compliance of the arms and the highly distributed nervous system of the octopus present unique challenges to understanding the neurological control principles that underlie such stereotypical arm movements, let alone more general behavior. Given the infinite-dimensional nature of arm configurations, their neurological control is considered a burden on the nervous system, requiring impractically powerful computations [33]. Current research explores an embodied intelligence approach to octopus control [15]. This concept considers the whole animal as being highly adaptable and self-organizing with potential autonomy and learned motion primitives at each level of organization of the nervous system.

F. Ling · E. Kanso (✉)
Department of Aerospace and Mechanical Engineering, University of Southern California, Los Angeles, CA, USA
e-mail: kanso@usc.edu

D. A. Paley, N. M. Wereley (eds.), *Bioinspired Sensing, Actuation, and Control in Underwater Soft Robotic Systems*, https://doi.org/10.1007/978-3-030-50476-2_11

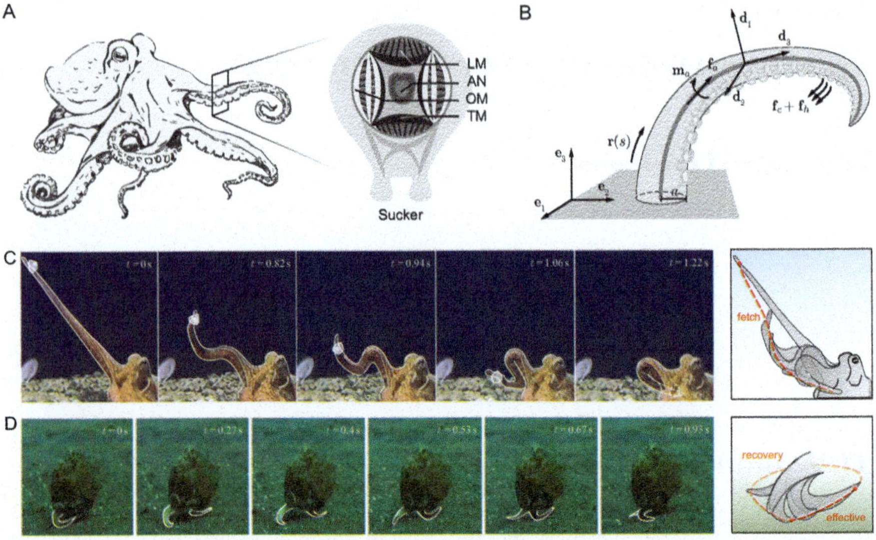

Fig. 11.1 Fetching and crawling maneuver of an octopus arm. (**a**) A typical octopus and composition of its muscular hydrostat. Here AN is the axial nerve cord, LM represents longitudinal muscle fiber, TM transverse muscle fiber, and OM oblique muscle fiber. Combined action of these muscle fibers results in active stretching, bending, twisting, and changes to its local rigidity along the octopus arm. (**b**) Rod model of an octopus arm. Specifically, we treat the octopus arm as an elastic filament with centerline coordinate $\mathbf{r}(s, t)$ and variable radius a, clamped at its base, subjected to an active tension $\mathbf{f}_a$, active bending/twisting moment $\mathbf{m}_a$, contact forces $\mathbf{f}_c$, and hydrodynamic drag forces and $\mathbf{f}_h$. The local director frame $\{\mathbf{d}_i\}$ $(i = 1, 2, 3)$ encodes the angular velocity and curvature information of the filament. (**c**) The octopus forms multiple joints along its arm when fetching objects from a distance [47]. Experimental images reported in [47] are shown with the fetching arm outlined in white. The traced arm shapes are stacked on the rightmost panel with the dashed red line indicating the trajectory traveled by the point of contact on the arm. (**d**) Certain octopus species crawl bipedally along the seafloor [17]. Snapshots of such crawling motion reported in [17] are reproduced with the visible portion of limbs outlined in white. The rightmost panel depicts the motion of one of these limbs and the trajectory described by its tip point

Once understood, the control mechanisms of soft-bodied octopus could be relevant for the design and development of soft robotic systems that operate autonomously in aquatic, potentially cluttered, environments. Indeed, there has been a growing effort among roboticists and engineers to study the octopus biomechanics and control as a source of inspiration for the design and construction of soft robotic arms [10, 11, 18–20, 37, 38, 43, 49, 50].

This work studies the mechanical control of octopus arms during typical fetch and crawl maneuvers; see Fig. 11.1c, d. When fetching an object from a distance, an octopus arm develops multiple pointed joints as shown in Fig. 11.1c; the joints are roughly fixed in location and their movements remain essentially planar [47]. In the rightmost panel of Fig. 11.1c, we superimpose these images on the same frame. The superimposed images show that the distal portion of the arm translates towards the target (octopus mouth) in a nearly straight fashion in the early phase of the fetch maneuver. Although this motion is internally actuated by the arm's musculature, the arm trajectory carries a striking similarity to Euler buckling of an elastic beam or filament. In Euler buckling, one end of the filament is compressed by a force pointing in a fixed direction; when the force exceeds a certain threshold, the filament buckles and bends in sinusoidal waves with wave numbers depending on the degree of the dominating mode excited by the compressive force.

When moving on the seafloor, several octopus species have been shown to crawl in a bipedal fashion [17]. This crawling gait is realized by two arms that form a single bending joint in an alternating lock-step rhythm. Compared to the fetch maneuver, the location of the bent joints translates along the arm based on its phase in the crawling cycle, which is crucial to effectively transfer forces to the sea floor through the distal portion of arm and to help maintain a clearance distance away from the seafloor for the octopus body and/or its payload. In Fig. 11.1d, we show snapshots from a crawling motion taken from a video recording of [17]. In the rightmost panel of Fig. 11.1d, we superimpose the images of one of the arms. The superimposed images suggest that the arm tip traces out a periodic trajectory consisting of a nearly straight portion while in contact with the seafloor, which seems to serve as an effective stroke for force generation, and a fast curved portion that quickly brings the arm back to its initial configuration. The arm movement in other examples of octopus crawling exhibits similar features [8, 13].

To study these fetch and crawl motions, we model the octopus arm as an elastic filament free to undergo three-dimensional deformations. We then formulate a systematic approach aimed at computing the internal actuation along the filament required for producing filament motions akin to the fetch and crawl motions observed experimentally. Our approach is reminiscent of recent ideas presented in the context of embodied intelligence and machine learning, where the control of tendon-driven humanoids is learned from training experiments prescribing the motion of the end-effector; see, for example, [32, 42]. Specifically, our approach consists of the following four steps, as depicted in Fig. 11.2,

1. *Experimental observation and video/image processing:* extract key features of octopus arm movement from fetch and crawl experiments as described above.
2. *Design tractable forward problems:* prescribe arm tip motions based on experimental observations and solve the filament equations of motion to obtain full arm deformations
3. *Compute inverse dynamics:* find internal actuations that are compatible with the full arm deformations obtained in step 2.
4. *Validate and test:* modify the forward problem to produce full arm deformations based on the internal actuation obtained in step 3, and repeat under nominal and modified environmental conditions to test for robustness.

The organization of this chapter is as follows. Section 11.2 formulates the general equations governing the three-dimensional deformation of an elastic filament submerged in an otherwise quiescent fluid. We present the results of the forward and inverse problems for the fetch and crawl motions in Sect. 11.3. Section 11.4

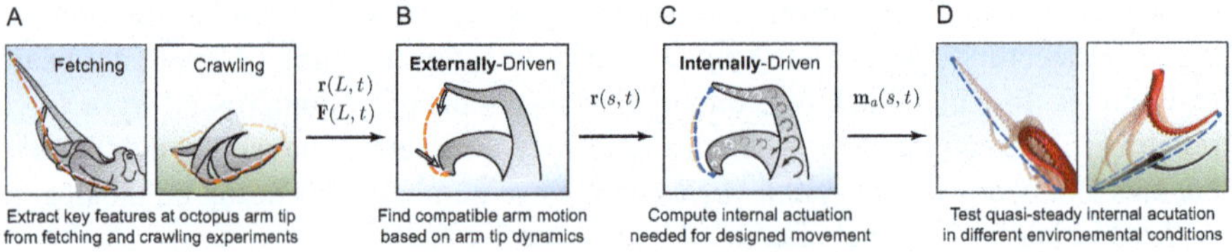

Fig. 11.2 Bioinspired framework for actuation strategy of an octopus arm. To obtain an internal actuation profile for the octopus arm, we first design arm-tip dynamics based on key features summarized from the experimental observations. For example, we can fix the trajectory of the arm tip so that the crawling motion is reproduced. The designed tip movements are then put into our *externally driven* elastic filament model to obtain the trajectory of the entire octopus arm. Next, we compute the internal active bending moment $\mathbf{m}_a$ necessary to produce the arm movement based on an inverse solver. The solution is then analyzed in detail in an octopus arm simulator. For example, one can change the height of the sea bed for the crawling example and see how well the actuation strategy performs. (**a**) Observe. (**b**) Design. (**c**) Invert. (**d**) Validate and test

discusses the relevance of our approach and our main findings in the context of octopus biomechanics and soft robotic systems.

11.2 Mathematical Model of an Octopus Arm

For simplicity, we model the octopus arm as an inextensible and nonshearable elastic filament of length L and radius a. The radius a may vary with arclength $s \in [0, L]$, measured from the base point $s = 0$ representing the point of attachment of the arm to the octopus body to the arm's free tip $s = L$; see Fig. 11.1b. The centerline of the filament is described by the position vector $\mathbf{r}(s, t)$, which varies with arclength s and time t, relative to a fixed inertial frame $(\mathbf{e}_1, \mathbf{e}_2, \mathbf{e}_3)$. The instantaneous distribution of velocities along the centerline is given by $\mathbf{v}(s, t) = \partial \mathbf{r}/\partial t$. The local orientation of the filament is described by an orthonormal material frame $(\mathbf{d}_1, \mathbf{d}_2, \mathbf{d}_3)$, with $\mathbf{d}_3 = \partial \mathbf{r}/\partial s = \mathbf{t}$ the tangential unit vector. The rate of change of $\mathbf{d}_i$ $(i = 1, 2, 3)$ in time is given by $\dot{\mathbf{d}}_i = \boldsymbol{\omega} \times \mathbf{d}_i$, where the dot denotes differentiation with respect to t, and $\boldsymbol{\omega}$ is the angular velocity vector. The change of $\mathbf{d}_i$ along the filament is given by $\mathbf{d}_i' = \boldsymbol{\kappa} \times \mathbf{d}_i$, where the prime denotes differentiation with respect to s and $\boldsymbol{\kappa}$ is a generalized curvature vector recording the infinitesimal change in orientation along the filament. Specifically, given the rotation matrix $\mathbf{Q}(s, t)$ that maps vectors expressed in the inertial frame to their counterparts in the filament material frame, $\boldsymbol{\omega}$ is the vector representation of the skew-symmetric matrix $\boldsymbol{\omega}^{\times} = \mathbf{Q}(\dot{\mathbf{Q}})^{\mathsf{T}}$ in the material frame, whereas $\boldsymbol{\kappa}$ is that of the skew-symmetric matrix $\boldsymbol{\kappa}^{\times} = \mathbf{Q}(\mathbf{Q}')^{\mathsf{T}}$, with $()^{\mathsf{T}}$ denoting the transpose operator.

For general three-dimensional deformations, the balance of forces and moments on a cross section of the filament is given by the Cosserat equations, expressed in the current deformed configuration [4, 5, 7, 9, 29, 36],

$$\frac{\partial \mathbf{N}}{\partial s} + \mathbf{f}_h + \mathbf{f}_c + \mathbf{f}_a = \frac{\partial(\rho A \mathbf{v})}{\partial t},$$
$$\frac{\partial \mathbf{M}}{\partial s} + \mathbf{t} \times \mathbf{N} + \mathbf{m}_a = \frac{\partial(\rho \mathbf{J} \boldsymbol{\omega})}{\partial t}. \quad (11.1)$$

Here, ρ, A, and $\mathbf{J}$ are the density, cross-sectional area, and moment of area of the filament, $\mathbf{f}_h$ and $\mathbf{f}_c$ are the hydrodynamic and contact (when applicable) forces per unit length, $\mathbf{f}_a$ and $\mathbf{m}_a$ are the actuation forces and moments per unit length due to muscle activity, and $\mathbf{N}$ and $\mathbf{M}$ are the internal elastic forces and moments.

To close the system of equations in (11.1), constitutive relations are needed. For linearly elastic materials, $\mathbf{N} = \mathbf{K}\mathbf{s}$ and $\mathbf{M} = \mathbf{B}\boldsymbol{\kappa}$ relate the internal force $\mathbf{N}$ and moment $\mathbf{M}$ to the strain vector $\mathbf{s} = e\mathbf{t} - \mathbf{d}_3$ and curvature vector $\boldsymbol{\kappa}$, respectively, with $\mathbf{K}$ and $\mathbf{B}$ being the axial/shear and bending/twist stiffness tensors, and $e = \mathrm{d}s/\mathrm{d}\hat{s}$ the extension of local arclength s with respect to a strain-free reference configuration with arclength $\hat{s}$. For filaments with homogeneous circular cross-sections, $\mathbf{K}$ and $\mathbf{B}$ are constant matrices of the form $\mathbf{K} = K\mathtt{diag}([4/3, 4/3, (1+\nu)/2])$ and $\mathbf{B} = B\mathtt{diag}([1, 1, 1/(1+\nu)])$, where K is the shear stiffness, B is the bending rigidity, and ν the Poisson's ratio. For a more detailed discussion, see [6, 9].

While cephalopod arms are likely to exhibit inhomogeneous and potentially nonlinear material properties due to their complex muscle composition and variation in activity along the arm, establishing an experimentally founded baseline for this dependence remains a challenge for current and future research. Therefore, we limit our discussion to inextensible and nonshearable filaments with linear material properties; specifically, we implement the inextensibility condition in weak form, by setting $K \gg B$; for implementation details, see the Appendix at the end of this chapter. For further discussion on the inextensible limit, see [40].

In order to model the fluid forces acting along the octopus arm, we first estimate the Reynolds number—a dimensionless number indicating the relative importance of inertial versus viscous fluid forces. Considering a typical arm of length $L = 1$–10 cm moving at a speed of $V = \|\mathbf{v}\| \approx 0.1$–$10$ cm/s in an otherwise still fluid, the Reynolds number is given by $\mathrm{Re} = \rho V L/\mu$ with μ taken to be the viscosity of water, and is of the order $O(10)$ to $O(10^4)$. For this range, the flow is unsteady, but not necessarily turbulent. Thus it is difficult to judge *a priori* the relationship between the fluid forces and the octopus arm velocities without detailed experiments or direct numerical simulations that couple the unsteady fluid motion to the movement of the octopus arms themselves [2]. In analytical models of fluid–structure interactions, the fluid forces are considered to depend linearly on velocity at lower Re, and they are typically represented by a quadratic dependence on velocity at higher Re [48]. Here, we assume linear dependence and take advantage of the small aspect ratio of the filament ($a/L \ll 1$) to write $\mathbf{f}_h = -\mu C_d \mathbf{A}\mathbf{v}$, where C_d is a drag coefficient and $\mathbf{A}$ is a matrix operator that adjusts the drag force to be larger in the transverse direction to the filament, consistent with the slenderness assumption. Results of our work were also validated and tested for a quadratic drag of the form $\mathbf{f}_h = -2a\rho_f C_d \mathbf{A}\mathbf{v}|\mathbf{v}|$, where ρ_f represents the fluid density and $2a$ arises from dividing the total projected area $2aL$ by the length of the arm L. Both models ignore added-mass effects assuming that for a slender object, drag forces are dominant [12].

We model the contact force $\mathbf{f}_c = -\nabla V$ as the negative gradient of a repulsion potential V that prevents self-penetration or penetration of the arm with surrounding boundaries when present. Specifically, V is a smooth bump function such that the repulsion force is of the form

$$V(\delta\mathbf{r}) = \begin{cases} -\alpha \exp[1/(\|\delta\mathbf{r}\| - \beta)] & \|\delta\mathbf{r}\| \leq \beta, \\ 0 & \text{otherwise}, \end{cases} \tag{11.2}$$

where $\delta\mathbf{r}$ is the relative position vector from the centerline of the filament to the point of obstruction (e.g., centerline position of another point on the filament), α is a scaling factor, and β indicates the distance where the contact potential is zero. In the following simulations, the scaling constant is set to $\alpha = 10^{15}$. For wall contact, $\beta = 5a$, whereas for self-contact $\beta = 3.5\bar{a}$, with $\bar{a}$ denoting the mean arm radius between points of contact. Other contact models, including directional friction and sticking action of octopus suckers [26, 27], could also be implemented in this arm simulation framework [9].

In the following study, we focus on the quasi-steady version of the Cosserat rod model, akin to those proposed by [43]. Namely, we ignore second-order inertial effects by setting the right-hand sides of Eq. (11.1) to zero. We also ignore active stretching/contraction forces $\mathbf{f}_a$. We restrict the octopus arm to have free boundary conditions at its tip at $s = L$ and to be clamped at its base at $s = 0$, i.e.,

$$\mathbf{N}|_{s=L} = \mathbf{M}|_{s=L} = \mathbf{0}, \quad \mathbf{v}|_{s=0} = \mathbf{0}, \quad \mathbf{t}|_{s=0} = \pm\mathbf{e}_3, \tag{11.3}$$

with the sign in the last equality above depending on the initial orientation of the arm: $\mathbf{t}|_{s=0} = \mathbf{e}_3$ for the fetch maneuver, and $\mathbf{t}|_{s=0} = -\mathbf{e}_3$ for the crawl maneuver. This leads to well-posed forward and inverse problems: (1) the forward problem consists of setting the active moment density $\mathbf{m}_a(s,t)$ due to muscle activity to be identically zero and solving for the arm movement $\mathbf{r}(s,t)$ given properly chosen conditions at the arm tip; and (2) the inverse problem consists of computing the active moment density $\mathbf{m}_a(s,t)$ needed to produce such arm movements given no constraints on the arm tip.

We solve the forward and inverse problems in non-dimensional form using the arm length L as the characteristic length scale and B/L^2 as the characteristic force scale. An intrinsic time scale $\tau = \mu C_d L^4/B$ is obtained from balancing the elastic and linear drag forces. The dimensionless radius $\tilde{a} = a/L$ is the slenderness ratio, which we consider to be $a/L = (1.5 - s)^2/50$ to reflect that the radius of the arm decreases with s. Dimensionless time is given by $\tilde{t} = t/\tau$, and the dimensionless active moment density is given by $\tilde{\mathbf{m}}_a = \mathbf{m}_a L^2/B$. In the following, we drop the tilde notation and consider all quantities to be in their dimensionless form. It is worth noting that the time scale τ is reminiscent of the elasto-hydrodynamic time scale used in the dimensionless Sperm number, typically defined as the ratio of this time scale and another intrinsic time scale such as that associated with an applied oscillatory motion or force, see, e.g., [51].

11.3 Fetch and Crawl with Internal Active Moments

This section details our protocol used to obtain the full arm movement $\mathbf{r}(s,t)$ and the required internal actuation $\mathbf{m}_a(s,t)$ inspired by the fetch and crawl behaviors shown in Fig. 11.1c, d.

To set-up the associated forward problem for the fetch maneuver depicted in Fig. 11.1c, we abstracted the octopus arm movement as an elastic filament undergoing Euler buckling under a compressive load $-300F_o\mathbf{e}_3$ on the filament, with both ends clamped, i.e., $\mathbf{t}|_{s=0} = \mathbf{t}|_{s=L} = \mathbf{e}_3$, and with the arm tip allowed to translate along the z-axis, so that $\mathbf{v}|_{s=0} = \mathbf{0}$ and $\mathbf{v}|_{s=L} = v_L\mathbf{e}_3$. Note that the compressive force $-300F_o$ was chosen large enough to excite the fourth modal shape of the buckled arm.

We solved for the passive kinematics $\mathbf{r}(s,t)$ of the entire filament by first inverting the quasi-steady moment balance equation to obtain internal elastic force

$$\mathbf{N} = \mathbf{t} \times \frac{\partial \mathbf{M}}{\partial s} + \Lambda \mathbf{t}. \tag{11.4}$$

Here, both $\mathbf{M} = \mathbf{B}\boldsymbol{\kappa}$ and $\Lambda = (\mathbf{K}\mathbf{s}) \cdot \mathbf{t}$ can be readily calculated knowing $\mathbf{r}$ at a given instant in time. Next, we calculated the centerline velocity $\mathbf{v}$ from the passive quasi-steady force balance equation

$$\frac{\partial \mathbf{N}}{\partial s} - \mu C_d \mathbf{A}\mathbf{v} + \mathbf{f}_c = 0, \tag{11.5}$$

and, given the velocity $\mathbf{v}$, we updated the filament configuration $\mathbf{r}$. We used this procedure to propagate the initial data $\mathbf{r}(s,0)$ forward until the desired final integration time. For further implementation details in the strain-free reference configuration, see the Appendix.

The configuration $\mathbf{r}(s,t)$ of the filament is shown in Fig. 11.3a at five representative instants in time. The time units of these snapshots is shown normalized with respect to the total integration time $T = 0.01$ in the forward Euler buckling problem. The elastic filament spontaneously buckles into three bent sections, similar to the joints observed in Fig. 11.1c based on [47].

Given the desired arm kinematics $\mathbf{r}(s,t)$, we setup the inverse problem to solve for the active moment $\mathbf{m}_a(s,t)$ based on the desired kinematics $\mathbf{r}(s,t)$. Given $\mathbf{r}$ and $\mathbf{v}$ along the filament at a specific time, we computed the instantaneous $\mathbf{N}$ based on Eq. (11.5) by integrating from the distal tip. We then used the calculated $\mathbf{N}$ and the constitutive relation $\mathbf{M} = \mathbf{B}\boldsymbol{\kappa}$ to compute the active moment

$$\mathbf{m}_a = -\frac{\partial \mathbf{M}}{\partial s} - \mathbf{t} \times \mathbf{N}. \tag{11.6}$$

The components m_1 and m_2 of the active moment density $\mathbf{m}_a = m_1\mathbf{d}_1 + m_2\mathbf{d}_2$ are shown in Fig. 11.3b at five representative instants in time. Here, we restricted

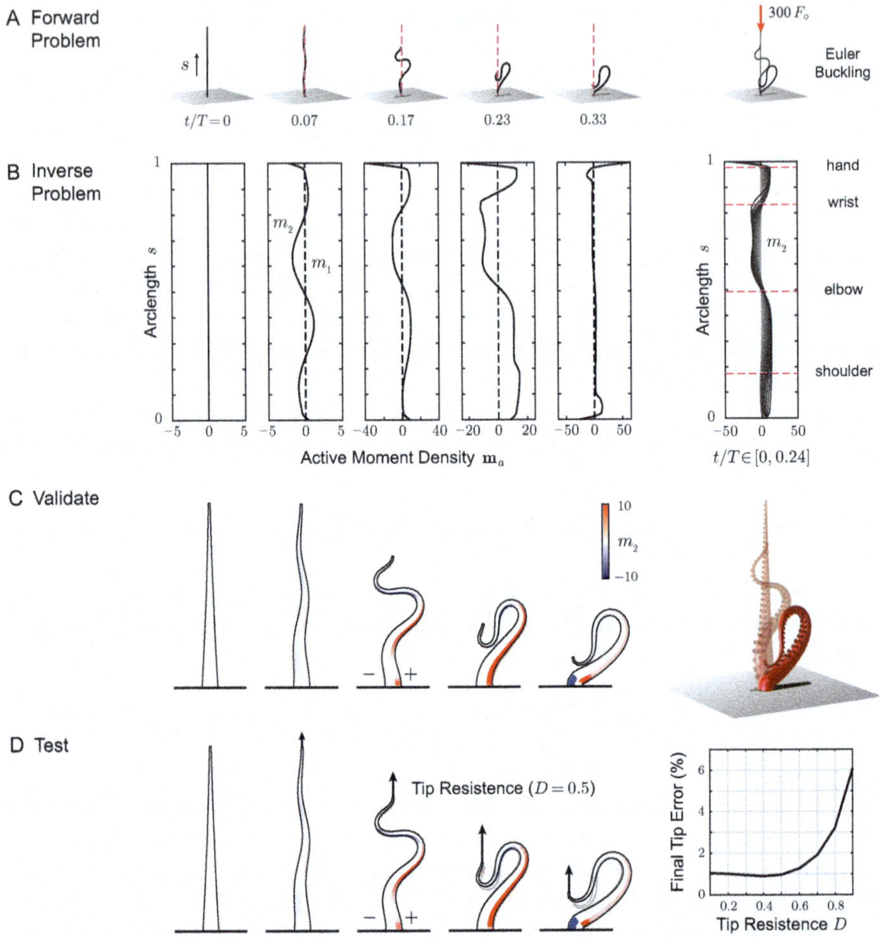

Fig. 11.3 **Reproduction of the fetch maneuver.** (**a**) Five representative snapshots of the planar Euler buckling of an initially straight rod. (**b**) Active moment density $\mathbf{m}_a$ solved from the inverse problem at the corresponding time instances. Evolution of the active component m_2 before self-contact occurs is shown in the rightmost panel. Formation of still joints mimic those observed in Fig. 11.1c. (**c**) The simulated octopus arm movement based on the internal moment profiles is shown at the same time snapshots. The colormap indicates positive or negative active moment density m_2 in red or blue, respectively. (**d**) The robustness of this solution is tested under different tip boundary conditions corresponding to different values of viscous drag due to the fetched object. The final tip error is computed by evaluating $(\mathbf{r}_{\text{modified}} - \mathbf{r}_{\text{euler}})/(\mathbf{r} - \mathbf{r}_{\text{euler}})$ at $s = L$ and $t = T$, where $\mathbf{r}_{\text{modified}}$ is the solution under modified testing condition, $\mathbf{r}$ is the solution using active moments based on the inverse problem, and $\mathbf{r}_{\text{euler}}$ is the Euler buckling solution based on the original forward problem.

the buckling motion to be planar to match experimental observation of octopus fetch [47], thus m_1 is identically zero. The active moment m_2 initially grows at a wave number 4, reflecting that in the forward problem, we excited the fourth modal shape of the buckled arm. In order to maintain the orientation of the tip and base, m_2 becomes sharper near $s = 1$. After self-contact around $t/T = 0.24$, the moment resets almost everywhere except at the base and the tip to maintain their orientation. The locations along the arm at which both active moments m_1 and m_2 are zero can be viewed as joints along the arm, since they coincide with points where local curvature changes sign. These joint locations essentially remained constant for the fetching maneuver for the bulk of the motion (rightmost panel of Fig. 11.3b), akin to observations of [47].

Next, using the obtained active moment $\mathbf{m}_a(s, t)$, we solved the forward problem again (the quasi-steady version of (11.1)) to ensure that the internal actuation profile obtained in the backward problem indeed produces an arm movement consistent with the desired motion designed in the first forward problem. This step is in essence similar to the forward problem described above but for non-zero active moment density. In the first forward problem, $\mathbf{t}$ and $\mathbf{v}$ were prescribed at the tip while in this step, the boundary conditions (11.3) are applied. It is worth noting that we used $\lambda\mathbf{m}_a(L, t)$ at the arm tip instead of $\mathbf{m}_a(L, t)$, where the constant λ was solved via a binary search. This step was necessary because the constraint on the arm tip in the forward problem essentially imposed a different boundary condition at $s = L$ than that of (11.3).

The filament configuration $\mathbf{r}(s, t)$ under the action of the internal active moment density $\mathbf{m}_a(s, t)$ is shown in Fig. 11.3c at the same five representative instances in time. It is evident that the fetch movement is successfully recovered when the environmental conditions are the same as those used during the first forward problem. In addition, the shape of the active moment density could inform us about the necessary muscle activation to achieve such movement. In Fig. 11.3c, we plot the positive active moment m_2 on the right side of the arm in red, and negative on the left in blue. Assuming that these internal moments arise from differential contraction of internal muscle fibers, one could interpret the colormap in Fig. 11.3c as an indication of the location and magnitude of contraction of the longitudinal muscle fibers necessary for reproducing the corresponding arm movement.

Lastly, we tested the robustness of this actuation profile under modified tip boundary conditions that represent differing amounts of drag force induced by objects grabbed at the tip. Specifically, we impose an additional drag force $F_{\text{tip}} = -D\mathbf{v}$ at the arm tip $s = 1$, while using the same actuation moments. We found that the arm performed the fetching maneuver with different effectiveness depending on the value of D; see right panel of Fig. 11.3d. An example solution for $D = 0.5$ is shown in Fig. 11.3d, where the final tip location terminates slightly above the original free-tip solution. In all these tests, the control parameter λ is kept constant at 2. The deviation between solutions under these modified conditions is computed by finding the relative tip error of these simulations compared to the case without any tip resistance ($D = 0$ and tip is force-free), using the original Euler buckling solution from the forward problem as a baseline.

Turning our attention to the crawl maneuver in Fig. 11.1d, we emulated the arm movement during crawling by fixing the base point $s = 0$ of the octopus arm at a distance $H/L = 0.7$ above a flat seabed and prescribing the motion of the tip point $(x(L,t), y(L,t), z(L,t))$ for a period of time $T = 0.1$. Specifically, we considered

$$y(L,t) = \frac{2}{3}\left[\cos(2\pi f(t)) - 1\right] + \frac{2}{\pi}, \quad z(L,t) = x(L,t) + \frac{2}{\pi},$$

where

$$x(L,t) = \begin{cases} 0 & 0 < f \leq 0.5 \\ \frac{1}{3}\sin(2\pi f) & 0.5 < f \leq 1 \end{cases},$$

and

$$f(t) = \begin{cases} \frac{2}{3}t & 0 < t \leq 0.75T \\ 2t - T & 0.75T < t \leq T \end{cases}.$$

These constraints were chosen such that the tip of the crawling arm moves in a straight line backwards during its effective stroke, then quickly swings forward along an elliptical arc during the recovery stroke; see the dashed red trajectory in Fig. 11.4a. The initial configuration of the filament was set to $x(s,0) = 0$, $y(s,0) = (2L/\pi)(1 - \cos(\pi s/2))$, and $z(s,0) = (2L/\pi)\sin(\pi s/2)$.

We solved the forward problem given this initial filament configuration and tip motion. We made sure that the arm returned to its original configuration by linearly interpolating the shape at the final time step with the initial condition. The filament movement $\mathbf{r}(s,t)$ is shown in Fig. 11.4a at five snapshots in time. We then used this motion for the inverse problem to compute the active moment density $\mathbf{m}_a(s,t)$ necessary to reproduce the desired crawling motion; see Fig. 11.4b. In Fig. 11.4c, the active moment density $\mathbf{m}_a$ was used to recover the crawling movement, first for the same conditions as the initial forward problem with a sea floor located at $H/L = 0.7$ below the base point of the octopus arm. Then we raised the seabed such that the arm's base point was located at $H/L = 0.6$ above the new sea floor. Interestingly, upon conforming the initial configuration to be compatible with a new sea floor height ($\mathbf{r}_{\text{modified}}(s,0) = \mathbf{r}(s,0) + s\Delta H/L\, \mathbf{e}_3$), the arm performed very similar crawling motion given the *exact* same active bending moment $\mathbf{m}_a$, modulo the multiplier λ at the arm tip discussed above. The values of λ we used were in the range between 1.5 and 2 and were roughly linear with respect to $\Delta H/L$. These values were selected to make sure the resulting arm movement is exactly periodic. The robustness to variations in seabed height is attributed to the compliance built into the elastic nature of the arm. Average tip tracking error under different amount of raised seabed height $\Delta H/L$ is shown in the rightmost panel of Fig. 11.4d.

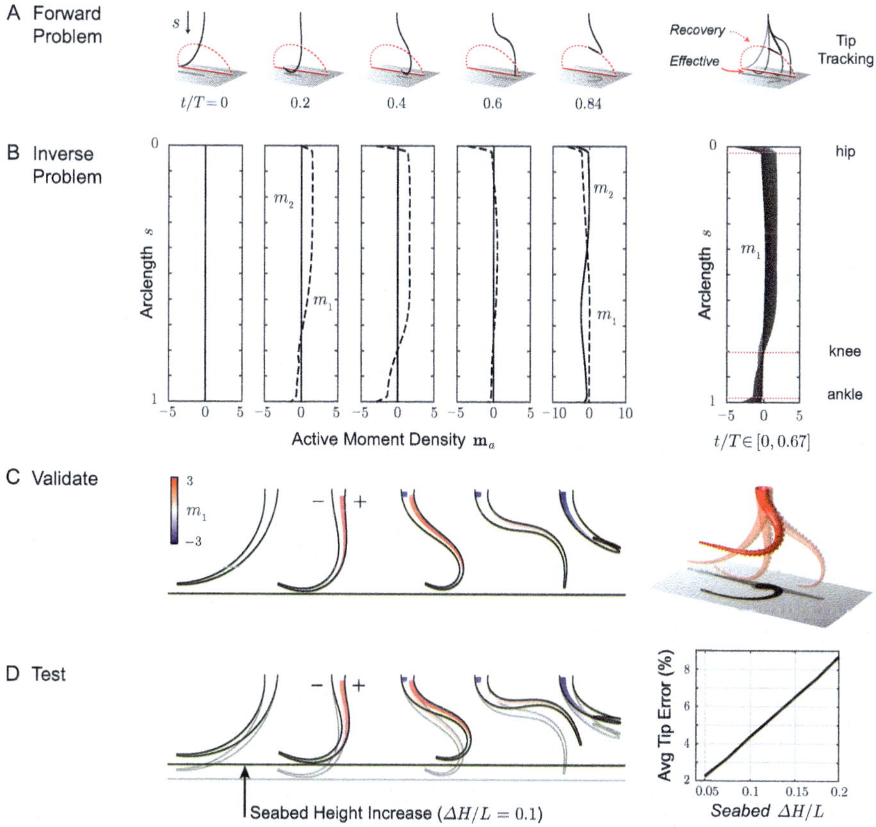

Fig. 11.4 Reproduction of the crawling maneuver. (**a**) Five representative snapshots of the crawling cycle with constrained tip. The effective stroke is when the tip is constrained to move along the bottom surface, and the recovery stroke is constructed from a tilted semi-circle. (**b**) Active moment density $\mathbf{m}_a$ solved from the inverse problem at the corresponding time instances. Evolution of the active component m_1 before the initiation of recovery stroke is shown in the rightmost panel. Formation of a slightly moving knee joint mimic those observed in Fig. 11.1d. (**c**) The simulated octopus arm movement based on the internal moment profiles is shown at the same time snapshots. The colormap indicates positive or negative active moment density m_2 in red or blue, respectively. (**d**) The robustness of this actuation is tested with respect to different seabed height. Note that the active moment at the arm tip and the initial arm configuration are adjusted accordingly to be compatible with the higher sea bed ($\mathbf{r}_{\text{modified}}(s, 0) = \mathbf{r}(s, 0) + s\Delta H/L\, \mathbf{e}_3$, where $\mathbf{r}_{\text{modified}}$ is the solution under the modified seabed condition, and $\mathbf{r}$ is the solution using active moments based on the inverse problem). The average tip error is computed by evaluating $(\mathbf{r}_{\text{modified}} - s\Delta H/L\, \mathbf{e}_3 - \mathbf{r}_{\text{tracking}})/(\mathbf{r} - \mathbf{r}_{\text{tracking}})$ at $s = L$ and averaging over the entire period. Here $\mathbf{r}_{\text{tracking}}$ is the tip tracking solution from the original forward problem

11.4 Concluding Remarks

Starting from experimental observations, we extracted minimal features of an octopus arm undergoing fetch and crawl maneuvers, and we used these features to reproduce the arm movements in an internally actuated continuum model. Specifically, we found that when fetching, the arm trajectory carries similarities to Euler buckling, and during bipedal crawling, the arm tip traces a periodic trajectory. We used this knowledge to construct forward problems to compute full arm deformations and we calculated the internal actuation required for such deformations by solving an inverse dynamics problem. These actuation strategies can be viewed as an abstraction of the motor control instructed by the octopus nervous system. Lastly, we tested the robustness of these results and we found that flexibility enables the arm to adapt to changes in the environment with minimal or no modification to the internal actuation.

A few comments on the advantages and limitations of our approach are in order. Our continuum model of the octopus arm allows for general three-dimensional deformations, including stretching and bending, and can be readily extended to account for shear and twist. To control these deformations, we developed a novel physics-based approach that is informed by experimental data; see Fig. 11.2. This approach fits nicely with current trends in the field of embodied intelligence, where the control laws are intimately coupled to the morphology and biomechanics of the system [15, 33, 42]. Further, this approach can be used in conjunction with machine-learning techniques [23, 32], where we envision the actuation strategies obtained here could form initial baseline policies for accelerated reinforcement learning of robust fetching and crawling policies under more complex and changing environmental conditions.

Future extensions of this work would account for unsteady arm motions [16] and more advanced fluid–structure interactions [21, 22, 39, 41, 44]. It would also account for higher fidelity models of the internal musculature, including time-dependent material properties [53]. Such models could use intrinsic curvatures, twists, and stretches of an elastic filament to encode the internal actuation of octopus arm muscles. Alternatively, future efforts could also focus on incorporating the detailed actuation mechanisms used in soft robotic systems such as tendon-driven [43] or pneumatic actuation [49]. Lastly, more accurate models of the self-interaction and interaction with the environment, including anisotropic friction and modeling of octopus suckers, should be incorporated.

Acknowledgments This work is partially supported via a Basic Research Challenge Grant from the Office of Naval Research, ONR Award Number N00014-17-1-2062, and a research grant from the Army Research Office, ARO Award Number W911NF-16-1-0074.

Appendix: Numerical Implementation

To solve for the filament dynamics in three dimensions, we numerically integrate the governing equations (11.1), together with the corresponding boundary conditions. To this end, we discretize the filament's centerline into $n+1$ vertices $\mathbf{r}_1, \ldots, \mathbf{r}_{n+1}$ and n straight edges $\boldsymbol{\ell}_i = \mathbf{r}_{i+1} - \mathbf{r}_i$, where $i = 1, \ldots, n$, in the spirit of [7, 9]. The unit tangent to edge i is defined as $\mathbf{t}_i = \boldsymbol{\ell}_i/\ell_i$, where $\ell_i = \|\boldsymbol{\ell}_i\|$. The translational velocities $\mathbf{v}_1, \ldots \mathbf{v}_{n+1}$ are assigned to vertices. The body frames $\{\mathbf{d}_{1,i}, \mathbf{d}_{2,i}, \mathbf{d}_{3,i} = \mathbf{t}_i\}$ and the rotation matrices $\mathbf{Q}_i$ are naturally assigned to edges.

To obtain a discrete representation of the generalized curvature vector $\boldsymbol{\kappa}$, we start from the definition of the associated skew-symmetric matrix $\boldsymbol{\kappa}^{\times} = \mathbf{Q}(\mathbf{Q}')^{\mathsf{T}}$. The solution to this first-order differential equation is of the form $\mathbf{Q}(s + \Delta s) = \exp(-\boldsymbol{\kappa}^{\times}\Delta s)\mathbf{Q}(s)$, where Δs is a segment of constant curvature. Thus, writing $\mathbf{Q}_i = \exp(-\boldsymbol{\kappa}_i^{\times}\bar{\ell}_i)\mathbf{Q}_{i-1}$, we define the discrete curvature matrix as $\boldsymbol{\kappa}_i^{\times} = -\log\left(\mathbf{Q}_i\mathbf{Q}_{i-1}^{\mathsf{T}}\right)/\bar{\ell}_i$, where $\bar{\ell}_i = \frac{1}{2}(\ell_{i-1} + \ell_i)$ is a Voronoi integration domain from the midpoint of the previous edge to that of the next edge. Given $\boldsymbol{\kappa}_i^{\times}$, we can readily evaluate $\boldsymbol{\kappa}_i$ and the discrete elastic moments $\mathbf{M}_i$ in the filament material frame.

When computing derivatives of quantities expressed in the body frame (i.e., on edges), we need to consider appropriate frame transport. For example, the derivative of the internal moment is computed as follows:

$$\left(\frac{\partial \mathbf{M}}{\partial s}\right)_i = \mathcal{D}\big[\mathbf{B}_i\boldsymbol{\kappa}_i\big] + \mathcal{A}\big[\underbrace{(\boldsymbol{\kappa}_i\bar{\ell}_i) \times (\mathbf{B}_i\boldsymbol{\kappa}_i)}_{\text{frame transport}}\big], \tag{11.7}$$

where $\mathcal{D}$ is a forward difference operator and $\mathcal{A}$ an averaging operator that maps quantities from adjacent vertices to the in-between edge. Specifically, when expressed in matrix form, the operators $\mathcal{D}$ and $\mathcal{A}$ have entries $\mathcal{D}_{ij} = \delta_{i,j+1} - \delta_{ij}$ and $\mathcal{A}_{ij} = \left(\delta_{i,j+1} + \delta_{ij}\right)/2$.

We solve the force and moment quantities in the strain-free reference configuration. Therefore, we need to adjust all quantities expressed in the current geometric configuration to the strain-free reference configuration denoted by $\hat{(\cdot)}$. Provided that cross-sections retain their circular shapes at all times, we need only to account for a local dilation scalar $e_i = \ell_i/\hat{\ell}_i$ at appropriate places in the discretized equations to ensure consistency. Specifically, due to the conservation of volume of an infinitesimal volume element, a dilation in edge length translates to a shrinkage of cross-sectional area; namely, we have $\ell_i A_i = \hat{\ell}_i\hat{A}_i$ which implies that $A_i = \hat{A}_i/e_i$. Consequently, the discrete bending/twisting rigidity tensor $\hat{\mathbf{B}}_i$ and tensile stiffness $\hat{K}_i$ in the stain-free reference configuration are related to $\mathbf{B}_i$ and K_i in the current configuration via $\mathbf{B}_i = \hat{\mathbf{B}}_i/e_i^2$ and $K_i = \hat{K}_i/e_i$. All length quantities need to scaled by a factor of e_i when converted to the strain-free reference configuration. In addition, because we use the integration domain $\bar{\ell}_i$, all quantities that vary with arc-length over the Voronoi region need to be properly averaged. For example, the dilation factor e is averaged over $\overline{\ell}_i$ using $\langle e_i \rangle = \left(e_i \hat{\ell}_i + e_{i-1} \hat{\ell}_{i-1} \right) / 2\hat{\overline{\ell}}_i$. To adjust the generalized curvature with respect to the dilation factor, we write $\boldsymbol{\kappa}_i = \hat{\boldsymbol{\kappa}}_i / \langle e_i \rangle$.

The inextensibility condition is enforced weakly by setting $K \approx 10^5 B$, requiring the use of standard time integrators for stiff equations (MATLAB's `ode15s`) to propagate the filament position $\mathbf{r}_i$ forward in time. We enforce boundary conditions strongly by fixing $\mathbf{r}$ and $\mathbf{t}$ at the appropriate end vertices. Results of this numerical method have been validated against standard analytical results (e.g., Euler buckling and Mitchell's instability thresholds, deflection curve of nonshearable cantilever beam). For more details, see the supplemental material accompanying [35].

References

1. D.S. Adams, R. Keller, M. Koehl, The mechanics of notochord elongation, straightening and stiffening in the embryo of Xenopus laevis. Development **110**(1), 115–130 (1990)
2. N. Agre, S. Childress, J. Zhang, L. Ristroph, Linear drag law for high-Reynolds-number flow past an oscillating body. Phys. Rev. Fluids **1**(3), 033202 (2016)
3. J. Altman, Control of accept and reject reflexes in the octopus. Nature **229**(5281), 204 (1971)
4. S.S. Antman, Kirchhoff's problem for nonlinearly elastic rods. Quart. Appl. Math. **32**(3), 221–240 (1974)
5. S.S. Antman, *Nonlinear Problems of Elasticity* (Springer, Berlin, 1995)
6. B. Audoly, Y. Pomeau, Elasticity and geometry, in *Peyresq Lectures on Nonlinear Phenomena* (World Scientific, Singapore, 2000), pp. 1–35
7. M. Bergou, M. Wardetzky, S. Robinson, B. Audoly, E. Grinspun, Discrete elastic rods. ACM Trans. Graph. **27**(3), 63 (2008)
8. G. Duane, T. Evans, C. Poe, D. Royle, A coconut octopus uses tools to snatch a crab. YouTube (2017). https://youtu.be/BFda1MZ54G4
9. M. Gazzola, L. Dudte, A. McCormick, L. Mahadevan, Forward and inverse problems in the mechanics of soft filaments. R. Soc. Open Sci. **5**(6), 171628 (2018)
10. F. Giorgio-Serchi, G.D. Weymouth, Drag cancellation by added-mass pumping. J. Fluid Mech. **798**, R3 (2016)
11. N. Giri, I. Walker, Continuum robots and underactuated grasping. Mech. Sci. **2**(1):51–58 (2011)
12. R. Godoy-Diana, B. Thiria, On the diverse roles of fluid dynamic drag in animal swimming and flying. J. R. Soc. Interface. **15**(139), 20170715 (2018)
13. L. Groskin, C. Baker, C. Huffard, J. Grubich, Run, octopus, run! YouTube (2015). https://youtu.be/23qzi88k3aM
14. Y. Gutfreund, T. Flash, G. Fiorito, B. Hochner, Patterns of arm muscle activation involved in octopus reaching movements. J. Neurosci. **18**(15), 5976–5987 (1998)
15. B. Hochner, An embodied view of octopus neurobiology. Curr. Biol. **22**(20):R887–R892 (2012)
16. S. Hoshyari, H. Xu, E. Knoop, S. Coros, M. Bächer, Vibration-minimizing motion retargeting for robotic characters. ACM Trans. Graph. **38**(4), 102 (2019)
17. C.L. Huffard, F. Boneka, R.J. Full, Underwater bipedal locomotion by octopuses in disguise. Science **307**(5717), 1927–1927 (2005)
18. B.A. Jones, I.D. Walker, Practical kinematics for real-time implementation of continuum robots. IEEE Trans. Robot. **22**(6), 1087–1099 (2006)
19. B.A. Jones, I.D. Walker, Kinematics for multisection continuum robots. IEEE Trans. Robot. **22**(1), 43–55 (2006)

20. R. Kang, D.T. Branson, E. Guglielmino, D.G. Caldwell, Dynamic modeling and control of an octopus inspired multiple continuum arm robot. Comput. Math. Appl. **64**(5), 1004–1016 (2012)
21. E. Kanso, Swimming due to transverse shape deformations. J. Fluid Mech. **631**, 127–148 (2009)
22. E. Kanso, J.E. Marsden, C.W. Rowley, J.B. Melli-Huber, Locomotion of articulated bodies in a perfect fluid. J. Nonlinear Sci. **15**(4), 255–289 (2005)
23. Ł. Kidziński, S.P. Mohanty, C.F. Ong, Z. Huang, S. Zhou, A. Pechenko, et al. Learning to run challenge solutions: adapting reinforcement learning methods for neuromusculoskeletal environments, in The NIPS'17 Competition: Building Intelligent Systems (Springer, Berlin, 2018), pp. 121–153
24. W.M. Kier, The diversity of hydrostatic skeletons. J. Exp. Biol. **215**(8), 1247–1257 (2012)
25. W.M. Kier, The musculature of coleoid cephalopod arms and tentacles. Front. Cell Dev. Biol. **4**, 10 (2016)
26. W.M. Kier, A.M. Smith, The morphology and mechanics of octopus suckers. Biol. Bull. **178**(2), 126–136 (1990)
27. W.M. Kier, A.M. Smith, The structure and adhesive mechanism of octopus suckers. Integr. Comp. Biol. **42**(6), 1146–1153 (2002)
28. W.M. Kier, J.T. Thompson, Muscle arrangement, function and specialization in recent coleoids. Berliner Paläobiologische Abhandlungen **3**, 141–162 (2003)
29. G. Kirchhoff, Ueber das Gleichgewicht und die Bewegung eines unendlich dünnen elastischen Stabes. J. für die reine und angewandte Mathematik. **56**, 285–313 (1859)
30. M. Koehl, K.J. Quillin, C.A. Pell, Mechanical design of fiber-wound hydraulic skeletons: the stiffening and straightening of embryonic notochords. Am. Zool. **40**(1), 28–041 (2000)
31. M.J. Kuba, R.A. Byrne, D.V. Meisel, J.A. Mather, Exploration and habituation in intact free moving octopus vulgaris. Int. J. Comp. Psychol. **19**(4), 426–438 (2006)
32. S. Lee, M. Park, K. Lee, J. Lee, Scalable muscle-actuated human simulation and control. ACM Trans. Graph. **38**(4), 73 (2019)
33. G. Levy, B. Hochner, Embodied organization of Octopus vulgaris morphology, vision, and locomotion. Frontiers in Physiology **8**, 164 (2017)
34. G. Levy, T. Flash, B. Hochner, Arm coordination in octopus crawling involves unique motor control strategies. Curr. Biol. **25**(9), 1195–1200 (2015)
35. F. Ling, H. Guo, E. Kanso, Instability-driven oscillations of elastic microfilaments. J. R. Soc. Interface **15**(149), 20180594 (2018)
36. A.E.H. Love, A treatise on the mathematical theory of elasticity (2013), Cambridge university press
37. W. McMahan, I.D. Walker, Octopus-inspired grasp-synergies for continuum manipulators, in *2008 IEEE International Conference on Robotics and Biomimetics* (IEEE, Piscataway, 2009), pp. 945–950
38. W. McMahan, V. Chitrakaran, M. Csencsits, D. Dawson, I.D. Walker, B.A. Jones, et al., Field trials and testing of the OctArm continuum manipulator, in *Proceedings 2006 IEEE International Conference on Robotics and Automation, ICRA 2006* (IEEE, Piscataway, 2006), pp. 2336–2341
39. D. Mokbel, H. Abels, S. Aland, A phase-field model for fluid–structure interaction. J. Comput. Phys. **372**, 823–840 (2018)
40. Neukirch S, et al., Vibrations of post-buckled rods: the singular inextensible limit. J. Sound Vib. **331**(3), 704–720 (2012)
41. C.S. Peskin, The immersed boundary method. Acta Numer. **11**, 479–517 (2002)
42. R. Pfeifer, J. Bongard, *How the Body Shapes the Way We Think: A New View of Intelligence* (MIT Press, Cambridge, 2006)

43. F. Renda, M. Cianchetti, M. Giorelli, A. Arienti, C. Laschi, A 3D steady-state model of a tendon-driven continuum soft manipulator inspired by the octopus arm. Bioinspir. Biomim. **7**(2), 025006 (2012)
44. C.H. Rycroft, Y. Sui, E. Bouchbinder, An Eulerian projection method for quasi-static elastoplasticity. J. Comput. Phys. **300**, 136–166 (2015)
45. G. Sumbre, Y. Gutfreund, G. Fiorito, T. Flash, B. Hochner, Control of octopus arm extension by a peripheral motor program. Science **293**(5536), 1845–1848 (2001)
46. G. Sumbre, G. Fiorito, T. Flash, B. Hochner, Neurobiology: motor control of flexible octopus arms. Nature **433**(7026), 595 (2005)
47. G. Sumbre, G. Fiorito, T. Flash, B. Hochner, Octopuses use a human-like strategy to control precise point-to-point arm movements. Curr. Biol. **16**(8), 767–772 (2006)
48. S. Vogel, *Life in Moving Fluids: The Physical Biology of Flow* (Princeton University Press, Princeton, 1994)
49. I.D. Walker, D.M. Dawson, T. Flash, F.W. Grasso, R.T. Hanlon, B. Hochner, et al., Continuum robot arms inspired by cephalopods, in *Unmanned Ground Vehicle Technology VII*, vol. 5804 (International Society for Optics and Photonics, 2005), pp. 303–315
50. G. Weymouth, V. Subramaniam, M. Triantafyllou, Ultra-fast escape maneuver of an octopus-inspired robot. Bioinspir. Biomim. **10**(1), 016016 (2015)
51. C.H. Wiggins, R.E. Goldstein, Flexive and propulsive dynamics of elastica at Low Reynolds number. Phys. Rev. Lett. **80**, 3879–3882 (1998)
52. J.Z. Young, *Anatomy of the nervous system of octopus vulgaris* (Clarendon Press,Oxford, 1971)
53. X. Zhang, F.K. Chan, T. Parthasarathy, M. Gazzola, Modeling and simulation of complex dynamic musculoskeletal architectures. Nat. Commun. **10**(1), 1–12 (2019)

Chapter 12
Decentralized Estimation and Control of a Soft Robotic Arm

Sachin Shivakumar, Daniel M. Aukes, Spring Berman, Ximin He, Rebecca E. Fisher, Hamidreza Marvi, and Matthew M. Peet

12.1 Introduction

Consider a rigid robotic manipulator with a single segment of fixed length r allowed to rotate about the origin in a 2D plane. The tip of the segment can reach the points on the circle centered at the origin with radius r as shown in Fig. 12.1a. Note that there does not exist any configuration of the segment in which the tip can be inside or outside this circle. As the segment is rigid, the reachable set of the tip of the segment is limited. However, if the robot was made of two segments of length $\frac{1}{2}r$ each connected by a pin joint, as shown in Fig. 12.1b, with a constraint that the joint angle be greater than 90°, then the tip of the robot can reach every point whose

S. Shivakumar · M. M. Peet (✉)
Department of Mechanical and Aerospace Engineering, Arizona State University, Tempe, AZ, USA
e-mail: sshivak8@asu.edu; mpeet@asu.edu

D. M. Aukes
The Polytechnic School, Ira A. Fulton Schools of Engineering, Arizona State University, Mesa, AZ, USA

S. Berman · H. Marvi
School for Engineering of Matter, Transport and Energy, Arizona State University, Tempe, AZ, USA

X. He
Department of Material Science and Engineering, University of California, Los Angeles, Los Angeles, CA, USA

R. E. Fisher
School of Life Science, Arizona State University, Tempe, AZ, USA

Department of Basic Medical Sciences, University of Arizona College of Medicine—Phoenix, Phoenix, AZ, USA

D. A. Paley, N. M. Wereley (eds.), *Bioinspired Sensing, Actuation, and Control in Underwater Soft Robotic Systems*, https://doi.org/10.1007/978-3-030-50476-2_12

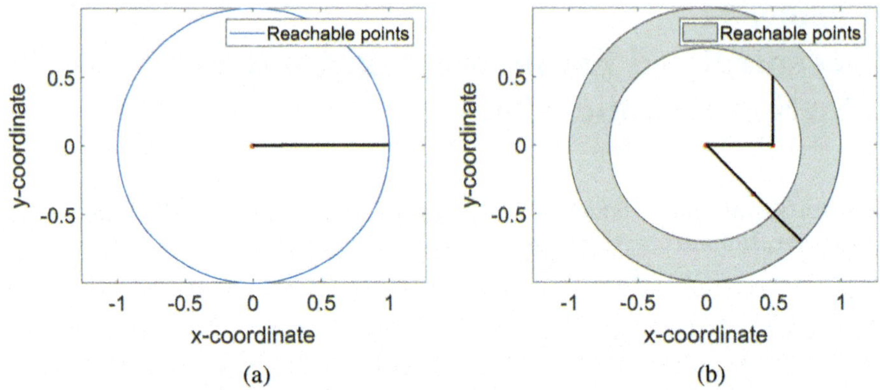

Fig. 12.1 Figure on the left shows the locations reachable by the tip of a robot arm with one segment allowed to rotate about the origin. On the right, the robot arm is made of two segments and can reach all of the points in the shaded region. (**a**) Robot with one segment. (**b**) Robot with two segments

distance is between $\frac{1}{\sqrt{2}}r$ and r. Increasing the number of segments further leads to an increase in the configuration space and the points that the tip can reach. A continuum robotic arm is an arm with an infinite number of infinitesimally small segments and the tip of the robot can reach every point in the circle of radius r. In addition, the added degrees of freedom in a continuum robot arm allow the robot to take more difficult geometric configurations. Conventionally, robots built with highly flexible components are called soft robots. Soft robots can be used to perform tasks that require complex motions and high precision, for instance, soft robots can be used in medical surgeries [3], autonomous underwater exploration [6], and manufacturing [17]. To control or observe a soft robot arm, we will obtain a linear model for the dynamics. Next, we will use a newly developed computational framework to design H_∞-optimal controllers and observers. Finally, we will show how the controller or observer can be implemented in a decentralized manner.

Modeling a Soft Robot Several control-oriented models have been developed for soft robots that approximate the soft robot as a series of segments governed by the kinematics of rigid or deformable objects [5, 8]. However, approximating a soft robot as a series of rigid arm segments requires us to use a large number of segments to accurately represent large deformations. Furthermore, control of these models in a decentralized manner is computationally intractable. On the other hand, modeling the soft robot as a single continuously deformable body with large deformations leads to nonlinear stress–strain relationships that depend on spatial derivatives of the displacement of the arm—implying that the dynamics of the system are governed by a nonlinear function of the spatial derivatives of the distributed state, i.e., a nonlinear PDE. Unfortunately, the control of a nonlinear PDE is often suboptimal and ad hoc. For this reason, we propose to linearize the nonlinear dynamics about a collection of setpoints. A setpoint, in the case of a PDE, can be any pre-specified state that satisfies the PDE and the boundary conditions.

There is no general method to linearize PDEs with nonlinear terms. However, in [1], a method was given for the special case when the nonlinear terms are polynomials in the states and their spatial derivatives. In this case, the original nonlinear PDE system is approximated by a system of linear PDEs about a collection of setpoints. This is the approach taken in Sect. 12.2.2 of this manuscript.

Observation and Control of a Linear Soft Robot Model Optimal control of linear PDEs with discretization is challenging. For this problem, we turn to the recently developed Partial Integral Equation (PIE) framework [12]. Using the PIE framework described in [12] and [15], our PDE can be represented as

$$\mathcal{T}\dot{\mathbf{x}}(t) = \mathcal{A}\mathbf{x}(t) + \mathcal{B}_1 w(t) + \mathcal{B}_2 u(t),$$

where $\mathcal{T}$, $\mathcal{A}$, $\mathcal{B}_i$ are Partial Integral (PI) operators of the form

$$\left(\mathcal{P}_{\{N_i\}}\mathbf{y}\right)(s) := N_0(s)\mathbf{y}(s) + \int_a^s N_1(s,\theta)\mathbf{y}(\theta)d\theta + \int_s^b N_2(s,\theta)\mathbf{y}(\theta)d\theta$$

because the PDE has a single spatial dimension. Optimal control and estimation of PIEs can be performed using a generalization of Linear Matrix Inequalities (LMIs). We refer to this framework as the Linear PI Inequality (LPI) framework. Once our linearized PDEs have been converted to PIEs, in Sect. 12.3.2, the PIETOOLS toolbox [11] is used to find optimal continuum estimators and feedback controllers.

Decentralized Implementation of the Controller Continuum observers and feedback controllers are naturally distributed along the robot arm. Implementing such distributed controllers and observers using large numbers of embedded sensors and actuators may be challenging unless some mechanism can be found to decentralize the calculation of feedback gains and error corrections. A central processing hub that collects sensor information and decides actuator inputs is undesirable because using a central hub requires long wires connecting each segment to the hub. The wiring alone required for such an implementation has the potential to exceed the mass of the robot arm itself. Furthermore, communication over long wires can introduce potentially destabilizing delay into the system. In absence of a central processing hub, unless there is some decentralization of the controller (or observer), the coordination of n actuators (or sensors) would require $n(n-1)$ interconnections and hence has issues similar to an implementation with a central processing hub. Therefore, a decentralized observer/controller implementation is preferred. Unfortunately, the design of a decentralized H_∞-optimal controller or estimator is known to be NP-hard, except in the case of Quadratically Invariant systems [7].

Although the design of a finite-dimensional H_∞-optimal decentralized controller or observer is an NP-hard non-convex optimization problem, Sect. 12.4 shows that controllers and observers of the PI form admit a naturally decentralized implementation. Specifically, the number of interconnections in this implementation scales as $2n$ (as opposed to $n(n-1)$).

Summary of Main Results The main contribution of this work is to apply the techniques developed to design optimal observers and controllers for a linear PDE to a soft robot manipulator. We model the dynamics of the robot arm as a nonlinear PDE and discuss one of the strategies to linearize the nonlinear PDE. Then the linear PDE is converted to a linear PIE. The design of a distributed H_∞-optimal observer or controller for the PIE is posed as a convex-optimization problem with LPI constraints. The optimization problem is then solved using PIETOOLS. Furthermore, a decentralized implementation of these observers and controllers is proposed.

Note on Continuum Controllers and Observers The methods discussed in this chapter aid in the design of the optimal controllers and observers without approximating the PDE by an ODE. The approximation is done at the implementation stage after the design of the optimal controller or observer. We will use a finite number of equispaced actuators and sensors. However, other non-equispaced strategies, e.g., [4], can be implemented in a decentralized manner due to the form of the PI operators. The key point is that the design process needs to be performed only once and different discretization strategies for the observer can be used—which is not the case if the PDE is discretized prior to the design stage. Furthermore, models that approximate a PDE by a finite-dimensional system before finding a controller or observer, e.g., a hyper-redundant continuum robotic arm [19], often result in a controller that requires significant computational effort in motion planning, thus making it impractical in real-time applications.

12.2 Modeling a Soft Robot Arm

This section introduces a nonlinear PDE model that describes the dynamics of a soft robot arm, followed by linearization of the system to obtain a linear PDE model.

12.2.1 Nonlinear Euler Beam Model

The nonlinear model of the soft robot arm is derived using the following assumptions. The robot is assumed to move in a 2D plane. The deformations are defined by the axial and transverse coordinates. The derivation of this model can be found in [9]. The dynamics of this system are

$$\begin{aligned}
&\partial_x \left(EAe(x,t)\cos(\theta(x,t))\right) + \partial_x \left(\frac{\partial_x(EI\partial_x\theta(x,t))\sin(\theta(x,t))}{1+e(x,t)}\right) \\
&\quad = \rho\ddot{u}(x,t) + \partial_x \left(\frac{J\ddot{\theta}(x,t)\sin(\theta(x,t))}{1+e(x,t)}\right),
\end{aligned}$$

$$\partial_x(EAe(x,t)\sin(\theta(x,t))) - \partial_x\left(\frac{\partial_x(EI\partial_x\theta(x,t))\cos(\theta(x,t))}{1+e(x,t)}\right) + f(x,t)$$
$$= \rho\ddot{w}(x,t) + \partial_x\left(\frac{J\ddot{\theta}(x,t)\cos(\theta(x,t))}{1+e(x,t)}\right),$$

where u is the axial deformation, w is the transverse deformation, E is the Young's modulus, A is the area of cross section, I is the moment of inertia about the axis perpendicular to the plane of motion, J is the rotational moment of inertia about the axial direction, ρ is the mass per unit length, e is the axial stretch, and θ is the rotation of the cross-section.

For this study, assume rotary inertia J is small and the arm is isotropic with uniform cross section. These assumptions are not required however, and can be relaxed if necessary. Next, simplify the equations of motion to obtain

$$EA\partial_x(e(x,t)\cos(\theta(x,t))) + EI\partial_x\left(\frac{\partial_x^2\theta(x,t)\sin(\theta(x,t))}{1+e(x,t)}\right) = \rho\ddot{u}(x,t)$$
$$EA\partial_x(e(x,t)\sin(\theta(x,t))) - EI\partial_x\left(\frac{\partial_x^2\theta(x,t)\cos(\theta(x,t))}{1+e(x,t)}\right) + f(x,t) = \rho\ddot{w}(x,t). \tag{12.1}$$

Now, replace the geometric relations $\tan(\theta) = \frac{\partial_x w}{1+\partial_x u}$ and $e = (((1+\partial_x u)^2 + (\partial_x w^2))^{\frac{1}{2}} - 1)$ in Eq. (12.1). Taylors' series expansion is then used to rewrite the nonlinear terms as polynomials and terms with degree higher than 2 are truncated. This yields a nonlinear PDE with second-order derivatives in time t and derivatives up to fourth-order in the spatial variable, x, i.e.,

$$\begin{aligned}\rho\ddot{u}(x,t) - EA\partial_x^2 u(x,t) &= EA\partial_x w(x,t)\partial_x^2 w(x,t)\\ &\quad + EI(\partial_x^2 w(x,t)\partial_x^3 w(x,t) + \partial_x w(x,t)\partial_x^4 w(x,t)),\\ \rho\ddot{w}(x,t) + EI\partial_x^4 w(x,t) &= EA(\partial_x^2 u(x,t)\partial_x w(x,t) + \partial_x u(x,t)\partial_x^2 w(x,t))\\ &\quad + EI(\partial_x^2 u(x,t)\partial_x^2 w(x,t) + \partial_x u(x,t)\partial_x^3 w(x,t))\\ &\quad + EI(\partial_x^3(\partial_x u(x,t)\partial_x w(x,t))) + f(x,t).\end{aligned} \tag{12.2}$$

The solutions to Eq. (12.2), $u, w : [0,\infty) \to W^{2,4}[0,L]$, are twice Fréchet differentiable in time for all $t > 0$ where

$$W^{2,4}[0,L] := \left\{x \in L_2[0,L] : \frac{\partial^k}{\partial s^k}x \in L_2[0,L],\ k \leq 4\right\}$$

and $L_2[X]$ is the space of Lesbegue integrable functions on X. Furthermore, the solutions satisfy the following boundary conditions:

$$u(0,t)=0,\quad \partial_x u(0,t)=0,\quad \partial_x^2 u(L,t)=0,\quad \partial_x^3 u(L,t)=0,$$
$$w(0,t)=0,\quad \partial_x w(0,t)=0,\quad \partial_x^2 w(L,t)=0,\quad \partial_x^3 w(L,t)=0.$$

These boundary conditions imply the arm is fixed at one end and free at the other. Equation (12.2) has polynomial nonlinearity, i.e., nonlinear terms that are products and derivatives of functions.

12.2.2 Setpoint Linearization of the Soft Robot Arm Dynamics

Typically, for nonlinear ODE systems, Jacobian linearization is used to linearize the system. Unfortunately, Jacobian linearization cannot be directly used for nonlinear PDEs. We can, however, use a similar approach [1] by considering deviations from a pre-specified solution (setpoint) to obtain a linear PDE that approximates the behavior of the system in the neighborhood of the chosen setpoint. Consider a nonlinear PDE,

$$\dot{\mathbf{x}}(t)=\mathcal{A}(\mathbf{x}(t),u(t)),$$

where $\mathcal{A}: X\times U\to X$ is a nonlinear function that is locally Lipschitz. A solution $\mathbf{x}$ near the setpoint $\mathbf{x}_e$ can be written as $\mathbf{x}=\mathbf{x}_e+\epsilon\mathbf{w}$. Since the setpoint is a solution to the PDE, $\dot{\mathbf{x}}_e(t)=\mathcal{A}(\mathbf{x}_e(t),u(t))$. Then, we can find

$$\epsilon\dot{\mathbf{w}}(t)=\mathcal{A}(\mathbf{x}_e(t)+\epsilon\mathbf{w}(t),u(t))-\mathcal{A}(\mathbf{x_e}(t),u(t)).$$

We expand $\mathcal{A}(\mathbf{x}_e(t)+\epsilon\mathbf{w}(t),u(t))$ using a Taylors' series expansion. The linear approximation is obtained by truncating the terms with degree of ϵ greater than 1.

Applying this technique to the nonlinear PDE in Eq. (12.2) with setpoint u_e and w_e, we obtain the following linear PDE in which the independent variables are omitted for brevity:

$$\begin{aligned}\rho\ddot{\tilde{u}}&=EA\partial_x^2\tilde{u}+(EA\partial_x^2 w_e+EI\partial_x^4 w_e)\partial_x\tilde{w}+(EA\partial_x w_e+EI\partial_x^3 w_e)\partial_x^2\tilde{w}\\&\quad+EI\partial_x^2 w_e\partial_x^3\tilde{w}+EI\partial_x w_e\partial_x^4\tilde{w}\\ \rho\ddot{\tilde{w}}&=(EA\partial_x^2 u_e+EI\partial_x^4 u_e)\partial_x\tilde{w}+(EA\partial_x u_e+EI\partial_x^2 u_e+3EI\partial_x^3 u_e)\partial_x^2\tilde{w}\\&\quad+(EI\partial_x u_e+3EI\partial_x^3 u_e)\partial_x^3\tilde{w}+(-EI+EI\partial_x u_e)\partial_x^4\tilde{w}\\&\quad+(EA\partial_x^2 w_e+EI\partial_x^3 w_e+EI\partial_x^4 w_e)\partial_x\tilde{u}+(EA\partial_x w_e\\&\quad+EI\partial_x^2 w_e+3EI\partial_x^3 w_e)\partial_x^2\tilde{u}\\&\quad+3EI\partial_x^2 w_e\partial_x^3\tilde{u}+EI\partial_x w_e\partial_x^4\tilde{u}+f.\end{aligned}\tag{12.3}$$

Finally, define new states for the system Eq. (12.3) as $\mathbf{x} = \left[\tilde{u}\ \dot{\tilde{u}}\ \tilde{w}\ \dot{\tilde{w}}\right]^T$. The PDE can now be written as

$$\dot{\mathbf{x}}(t) = \sum_{j=0}^{4} A_j(s)\partial_s^j \mathbf{x}(t) + B_1(s) f(t), \tag{12.4}$$

where

$$A_0(x) = \begin{bmatrix} 0\,1\,0\,0 \\ 0\,0\,0\,0 \\ 0\,0\,0\,1 \\ 0\,0\,0\,0 \end{bmatrix},$$

$$A_1(x) = \frac{1}{\rho}\begin{bmatrix} 0 & 0 & 0 & 0 \\ 0 & 0 & (EA\partial_x^2 w_e + EI\partial_x^4 w_e) & 0 \\ 0 & 0 & 0 & 0 \\ (EA\partial_x^2 w_e + EI\partial_x^3 w_e + EI\partial_x^4 w_e) & 0 & (EA\partial_x^2 u_e + EI\partial_x^4 u_e) & 0 \end{bmatrix},$$

$$A_2(x) = \frac{1}{\rho}\begin{bmatrix} 0 & 0 & 0 & 0 \\ EA & 0 & (EA\partial_x w_e + EI\partial_x^3 w_e) & 0 \\ 0 & 0 & 0 & 0 \\ (EA\partial_x w_e + EI\partial_x^2 w_e + 3EI\partial_x^3 w_e) & 0 & (EA\partial_x u_e + EI\partial_x^2 u_e + 3EI\partial_x^3 u_e) & 0 \end{bmatrix},$$

$$A_3(x) = \frac{1}{\rho}\begin{bmatrix} 0 & 0 & 0 & 0 \\ 0 & 0 & EI\partial_x^2 w_e & 0 \\ 0 & 0 & 0 & 0 \\ 3EI\partial_x^2 w_e & 0 & (EI\partial_x u_e + 3EI\partial_x^2 u_e) & 0 \end{bmatrix},$$

$$A_4(x) = \frac{1}{\rho}\begin{bmatrix} 0 & 0 & 0 & 0 \\ 0 & 0 & EI\partial_x w_e & 0 \\ 0 & 0 & 0 & 0 \\ EI\partial_x w_e & 0 & (-EI + EI\partial_x u_e) & 0 \end{bmatrix}, \quad B_1(x) = \frac{1}{\rho}\begin{bmatrix} 0 \\ 0 \\ 0 \\ 1 \end{bmatrix}. \tag{12.5}$$

The next section presents a way to convert a linear PDE of this format to a PIE representation using the A_i's and B_i's defined above. The A_i's and B_i's for our model are obtained by substituting the setpoint functions u_e and w_e illustrated in Fig. 12.2.

12.3 Estimation and Control of Linear PDEs

This section presents the PIE representation of a PDE following [15]. PIE representation has the advantage that it is defined by bounded linear operators (no spatial derivatives). Furthermore, H_∞-optimal observers and feedback controllers can be

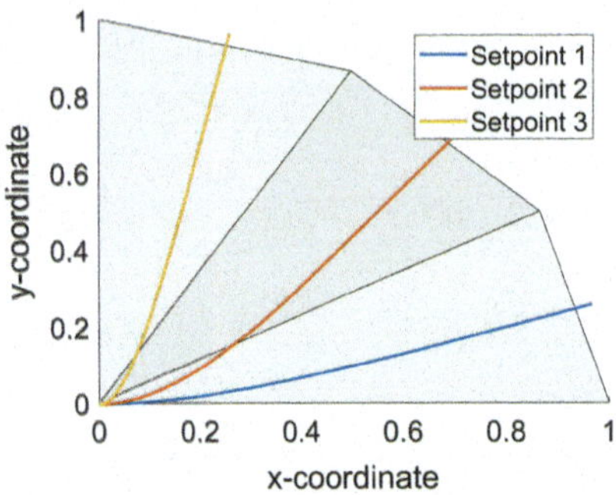

Setpoint 1: $u_e(x) = 0,$
$w_e(x) = 0.55x^2 - 0.38x^3 + 0.09x^4$
Setpoint 2: $u_e(x) = 0,$
$w_e(x) = 2.8x^2 - 2.6x^3 + 0.94x^4$
Setpoint 3: $u_e(x) = 0,$
$w_e(x) = 28.8x^2 - 74.2x^3 + 71.7x^4$

Fig. 12.2 Figure on the left shows three functions chosen as setpoints. The scheduling variable is the y-coordinate of the tip of the arm. The conic sections in gray show the operation zone for which the corresponding setpoints are provided on the right

obtained for systems in the PIE framework using newly developed computational tools such as PIETOOLS [11].

12.3.1 Partial Integral Operators

As introduced in [12], a PI operator is a parameterized linear mapping between infinite-dimensional spaces $\mathbb{R}^m \times L_2^n \to \mathbb{R}^p \times L_2^q$. We define two classes of PI-operators, 3-PI and 4-PI. As the nomenclature suggests, 3-PI operators, denoted as $\mathcal{P}_{\{N_i\}} : L_2^m[a,b] \to L_2^n[a,b]$, are parameterized by 3 matrix-valued functions $N_0 : [a,b] \to \mathbb{R}^{n\times m}$ and $N_1, N_2 : [a,b] \times [a,b] \to \mathbb{R}^{n\times m}$. 3-PI operators are bounded linear maps between the normed spaces $L_2^m[a,b]$ and $L_2^n[a,b]$, endowed with the standard L_2 inner product:

$$\left(\mathcal{P}_{\{N_i\}}\mathbf{y}\right)(s) := N_0(s)\mathbf{y}(s) + \int_a^s N_1(s,\theta)\mathbf{y}(\theta)d\theta + \int_s^b N_2(s,\theta)\mathbf{y}(\theta)d\theta. \tag{12.6}$$

Similarly, 4-PI operators are bounded linear operators between $\mathbb{R}^m \times L_2^n[a,b]$ and $\mathbb{R}^p \times L_2^q[a,b]$ and are parameterized by the matrix P, the matrix-valued functions Q_1, Q_1 and the 3-PI operator $\mathcal{P}_{\{N_i\}}$ where $P : \mathbb{R}^m \to \mathbb{R}^p$, $Q_1 : [a,b] \to \mathbb{R}^{p\times n}$, $Q_2 : [a,b] \to \mathbb{R}^{q\times m}$ and $\mathcal{P}_{\{R_i\}} : L_2^n[a,b] \to L_2^q[a,b]$:

$$\mathcal{P}\begin{bmatrix} P, & Q_1 \\ Q_2, & \{R\} \end{bmatrix}\begin{bmatrix} x \\ \mathbf{y} \end{bmatrix}(s) := \begin{bmatrix} Px + \int_a^b Q_1(s)\mathbf{y}(s)ds \\ Q_2(s)x + \mathcal{P}_{\{R_i\}}\mathbf{y}(s) \end{bmatrix}. \tag{12.7}$$

12.3.2 Partial Integral Equations

Partial Integral Equations (PIEs) take the form

$$\begin{aligned}\mathcal{T}\dot{\mathbf{x}}(t) &= \mathcal{A}\mathbf{x}(t) + \mathcal{B}_1 w(t) + \mathcal{B}_2 u(t),\\ z(t) &= \mathcal{C}_1\mathbf{x}(t) + \mathcal{D}_{11} w(t) + \mathcal{D}_{12} u(t),\\ y(t) &= \mathcal{C}_2\mathbf{x}(t) + \mathcal{D}_{21} w(t) + \mathcal{D}_{22} u(t),\end{aligned} \tag{12.8}$$

where $\mathcal{T}, \mathcal{A}, \mathcal{B}_i, \mathcal{C}_i$, and $\mathcal{D}_{ij}$ are 4-PI operators. As shown in [12], any linear PDE with a single spatial variable can be represented as a PIE. PIEs are useful in analysis and control of linear PDEs because they do not have boundary conditions or continuity constraints. In the following example, we provide the PI operators that reformulate a fourth-order linear PDE (of the form Eq. (12.4)) as a PIE.

Consider a linear PDE of the form

$$\begin{aligned}\dot{x}(s,t) &= \sum_{j=0}^{4} A_j(s)\partial_s^j x(s,t) + B_1(s)w(t) + B_2(s)u(t),\\ Bx_b(t) &= 0,\end{aligned} \tag{12.9}$$

where $A_i : \mathbb{R} \to \mathbb{R}^{n\times n}$, $B_i : \mathbb{R} \to \mathbb{R}^n$, $\mathrm{rank}(B) = 4n$ and

$$\begin{aligned}x_b(t) = \mathrm{col}(&x(a,t), \partial_s x(a,t), \partial_s^2 x(a,t), \partial_s^3 x(a,t), x(b,t), \partial_s x(b,t),\\ &\partial_s^2 x(b,t), \partial_s^3 x(b,t)).\end{aligned}$$

Define the 3-PI operators

$$\mathcal{T} = \mathcal{P}_{\{G_i\}}, \mathcal{A} = \mathcal{P}_{\{H_i\}}, \mathcal{B}_i = \mathcal{P}_{\{B_i,0,0\}}, \mathcal{C}_i = 0, \mathcal{D}_{ij} = 0, \tag{12.10}$$

where

$$G_0(s) = 0, \quad G_1(s,\theta) = -K_0(s)(BT(b))^{-1}BQ(b,\theta) + L_0(s,\theta),$$

$$G_2(s,\theta) = -K_0(s)(BT(b))^{-1}BQ(b,\theta),$$

$$H_0(s) = A_4(s), \quad H_1(s,\theta) = -\sum_{j=0}^{3}(A_j(s)K_j(s)(BT(b))^{-1}BQ(b,\theta) + L_j(s,\theta)),$$

$$H_2(s,\theta) = -\sum_{j=0}^{3} A_j(s)K_j(s)(BT(b))^{-1}BQ(b,\theta),$$

$$T(s) = \begin{bmatrix} I & 0 & 0 & 0 \\ 0 & I & 0 & 0 \\ 0 & 0 & I & 0 \\ 0 & 0 & 0 & I \\ I & s-a & (s-a)^2/2 & (s-a)^3/3! \\ 0 & I & s-a & (s-a)^2/2! \\ 0 & 0 & I & s-a \\ 0 & 0 & 0 & I \end{bmatrix}, \quad Q(s,\theta) = \begin{bmatrix} 0 \\ 0 \\ 0 \\ 0 \\ (s-\theta)^3/3! \\ (s-\theta)^2/2! \\ s-\theta \\ I \end{bmatrix},$$

$$K_j(s) = T_{\{j+5\}}(s),\ L_j(s,\theta) = Q_{\{j+5\}}(s,\theta). \tag{12.11}$$

The row index j for T and Q stands for the jth subpartition row shown above.

Then for any x that satisfies Eq. (12.9), $\mathbf{x} = \partial_s^4 x$ satisfies Eq. (12.8) for the 3-PI operators as defined in Eq. (12.10). Using A_i and B_i as defined in Eq. (12.5), we define the 3-PI operators Eq. (12.10) to find the PIE form of the PDE Eq. (12.3).

12.3.3 Linear Partial Integral Inequalities

Optimization problems with PI operator variables and Linear PI Inequality constraints can be solved using the software package PIETOOLS [11]. These optimization problems are referred to as Linear PI Inequalities (LPIs) and take the form

$$\mathcal{P}\begin{bmatrix} P_0, & Q_0 \\ Q_0^T, & \{R_{0i}\} \end{bmatrix} + \sum_{k=1}^{N} x_j \mathcal{P}\begin{bmatrix} P_k, & Q_k \\ Q_k^T, & \{R_{ki}\} \end{bmatrix} \succcurlyeq 0, \tag{12.12}$$

where the decision variable is $x \in \mathbb{R}^N$ and $\mathcal{P}\begin{bmatrix} P_k, & Q_k \\ Q_k^T, & \{R_{ki}\} \end{bmatrix} : \mathbb{R}^m \times L_2^n[a,b] \to \mathbb{R}^m \times L_2^n[a,b]$ is a given self-adjoint 4-PI operator for $0 \leq k \leq N$ and $k \in \mathbb{Z}$.

12.3.4 H_∞-Optimal Observer Design

This subsection extends the LMI result used for design of H_∞-optimal observers of ODEs to an LPI for observation of systems defined by PIEs.

12.3.4.1 H_∞-Optimal Observer Design for Linear ODEs

Consider the problem of designing the H_∞ optimal observer for a linear ODE whose dynamics are governed by the equations

$$\begin{aligned}\dot{x}(t) &= Ax(t) + Bw(t), \qquad x(0) = 0,\\ z(t) &= C_1x(t) + D_{11}w(t),\\ y(t) &= C_2x(t) + D_{21}w(t), \end{aligned} \tag{12.13}$$

where $A : \mathbb{R}^n \to \mathbb{R}^n$, $B : \mathbb{R} \to \mathbb{R}^n$, $C_i : \mathbb{R}^n \to \mathbb{R}$, $D_{i1} : \mathbb{R} \to \mathbb{R}$, $w \in L_2([0, \infty))$ is a disturbance, y is the measured output, and z is the regulated output. Let the observer dynamics be

$$\begin{aligned}\dot{\hat{x}}(t) &= A\hat{x}(t) + L(C_2\hat{x}(t) - y(t)),\\ \hat{z}(t) &= C_1\hat{x}(t), \end{aligned} \tag{12.14}$$

where $L : \mathbb{R} \to \mathbb{R}^n$, $\hat{x}$ is the estimated state, and $z_e(t) = \hat{z}(t) - z(t)$ is the error in the estimate of regulated output. The closed-loop dynamics of the error system with state $e(t) := \hat{x}(t) - x(t)$ as determined by Eqs. (12.13) and (12.14) are

$$\dot{e}(t) = (A + LC_2)e(t) - (B + LD_{21})w(t), \qquad z_e(t) = C_1e(t) - D_{11}w(t). \tag{12.15}$$

Using the KYP Lemma [2], we formulate the following LMI that provides necessary and sufficient conditions for the existence of an H_∞-optimal observer with $\frac{\|z_e\|_{L_2}}{\|w\|_{L_2}} \leq \gamma$. The LMI used to find H_∞-optimal observer for this system is

$$P > 0$$

$$\begin{bmatrix} -\gamma I & -D_{11}^T & -(PB + ZD_{21})^T \\ *^T & -\gamma I & C_1 \\ *^T & *^T & (PA + ZC_2)^T + PA + ZC_2 \end{bmatrix} \leq 0, \tag{12.16}$$

where the observer gains are $L = P^{-1}Z$.

12.3.4.2 H_∞-Optimal Observer Design for Linear PIEs

We generalize the LMI in Sect. 12.3.4.1 to an LPI yielding continuum observers for systems modeled using PIEs. Specifically, suppose the system to be observed is defined by Eq. (12.8). We propose an observer structure similar to Eq. (12.14). The closed-loop error dynamics are

$$\begin{aligned}\mathcal{T}\dot{\mathbf{e}}(t) &= (\mathcal{A} + \mathcal{L}\mathcal{C}_2)\mathbf{e}(t) - (\mathcal{B} + \mathcal{L}\mathcal{D}_{21})w(t), \qquad \mathbf{e}(0) = 0\\ z_e(t) &= \mathcal{C}_1\mathbf{e}(t) - \mathcal{D}_{11}w(t), \end{aligned} \tag{12.17}$$

where $\mathcal{T} : L_2^n[a,b] \rightarrow L_2^n[a,b]$, $\mathcal{A} : L_2^n[a,b] \rightarrow L_2^n[a,b]$, $\mathcal{B} : \mathbb{R} \rightarrow L_2^n[a,b]$, $C_i : L_2^n[a,b] \rightarrow \mathbb{R}$, $\mathcal{D}_{i1} : \mathbb{R} \rightarrow \mathbb{R}$ and $\mathcal{L} : \mathbb{R} \rightarrow L_2^n[a,b]$ are PI operators. The LPI constraints for the PIE are given below.

Theorem *Suppose there exists bounded linear operators $\mathcal{P} : L_2^n[a,b] \rightarrow L_2^n[a,b]$ and $\mathcal{Z} : \mathbb{R} \rightarrow L_2^n[a,b]$, such that $\mathcal{P}$ is coercive and*

$$\begin{bmatrix} -\gamma I & -\mathcal{D}_{11}^* & -(\mathcal{P}\mathcal{B}+\mathcal{Z}\mathcal{D}_{21})^*\mathcal{T} \\ (\cdot)^* & -\gamma I & C_1 \\ (\cdot)^* & (\cdot)^* & (\mathcal{P}\mathcal{A}+\mathcal{Z}C_2)^*\mathcal{T}+\mathcal{T}^*(\mathcal{P}\mathcal{A}+\mathcal{Z}C_2) \end{bmatrix} \preccurlyeq -\epsilon I \tag{12.18}$$

where $\epsilon > 0$ and $I : L_2 \rightarrow L_2$ is an identity operator. Then $\mathcal{P}^{-1}$ exists and is a bounded linear operator and, for $\mathcal{L} = \mathcal{P}^{-1}\mathcal{Z}$ and any $w \in L_2([0,\infty))$, any solution for the system Eq. (12.17) *satisfies* $\|z_e\|_{L_2} < \gamma\|w\|_{L_2}$. □

For proof, please refer to [18].

12.3.5 H_∞-Optimal Controller Synthesis

Similar to Sect. 12.3.4, we state the LMI for finding an H_∞ -optimal controller for a linear ODE system and then present the extension to an LPI.

12.3.5.1 H_∞-Optimal Controller Synthesis for Linear ODEs

Consider an ODE

$$\begin{aligned} \dot{x}(t) &= Ax(t) + B_1 w(t) + B_2 u(t), \\ z(t) &= C_1 x(t) + D_{11} w(t) + D_{12} u(t), \end{aligned} \tag{12.19}$$

where $A : \mathbb{R}^n \rightarrow \mathbb{R}^n$, $B_i : \mathbb{R} \rightarrow \mathbb{R}^n$, $C_1 : \mathbb{R}^n \rightarrow \mathbb{R}$, $D_{1i} : \mathbb{R} \rightarrow \mathbb{R}$, $w \in L_2([0,\infty))$ is a disturbance, u is the input, and z is the regulated output. If there exists a $P > 0$, where $P \in \mathbb{S}^n$, and $Z : \mathbb{R}^n \rightarrow \mathbb{R}$ such that

$$\begin{bmatrix} PA^T + AP + Z^T B_2^T + B_2 Z & B_1 & PC_1^T + Z^T D_{12}^T \\ *^T & -\gamma I & D_{11}^T \\ *^T & *^T & -\gamma I \end{bmatrix} \leq 0,$$

then for $u(t) = ZP^{-1}x(t)$, the solutions of Eq. (12.19) satisfy $\|z\|_{L_2} \leq \gamma\|w\|_{L_2}$ for $w \in L_2([0,\infty))$.

12.3.5.2 H_∞-Optimal Controller Synthesis for Linear PIEs

Similar to the ODE shown earlier, consider a PIE system

$$\begin{aligned} \mathcal{T}\dot{\mathbf{x}}(t) &= \mathcal{A}\mathbf{x}(t) + \mathcal{B}_1 w(t) + \mathcal{B}_2 u(t), \qquad \mathbf{x}(0) = 0, \\ z(t) &= \mathcal{C}_1\mathbf{x}(t) + \mathcal{D}_{11} w(t) + \mathcal{D}_{12} u(t), \end{aligned} \tag{12.20}$$

where $\mathcal{T} : L_2^n[a,b] \to L_2^n[a,b]$, $\mathcal{A} : L_2^n[a,b] \to L_2^n[a,b]$, $\mathcal{B}_i : \mathbb{R} \to L_2^n[a,b]$, $\mathcal{C}_1 : L_2^n[a,b] \to \mathbb{R}$ and $\mathcal{D}_{1i} : \mathbb{R} \to \mathbb{R}$ are PI operators. Then the following LPI can be used to find H_∞-optimal controller gains for a linear PIE.

Theorem *Suppose there exists bounded linear operators* $\mathcal{P} : L_2^n[a,b] \to L_2^n[a,b]$ *and* $\mathcal{Z} : L_2^n[a,b] \to \mathbb{R}$, *such that* $\mathcal{P}$ *is coercive and*

$$\begin{bmatrix} -\gamma I & \mathcal{D}_{11}^* & (\mathcal{P}\mathcal{C}_1^* + \mathcal{Z}^*\mathcal{D}_{12}^*) \\ (\cdot)^* & -\gamma I & \mathcal{B}_1 \\ (\cdot)^* & (\cdot)^* & (\cdot)^* + \mathcal{T}(\mathcal{A}\mathcal{P} + \mathcal{B}_2\mathcal{Z})^* \end{bmatrix} \preccurlyeq -\epsilon I \tag{12.21}$$

where $\epsilon > 0$ *and* $I : L_2 \to L_2$ *is an identity operator. Then there exists a bounded and coercive linear operator* $\mathcal{P}^{-1}$. *Further, for* $u = \mathcal{K}\mathbf{x}$, *where* $\mathcal{K} = \mathcal{Z}\mathcal{P}^{-1}$, *and any* $w \in L_2$, *any solution for Eq.* (12.20) *satisfies* $\|z\|_{L_2} \le \gamma \|w\|_{L_2}$. □

For proof, please refer to [16].

12.3.6 *PIETOOLS: A MATLAB Toolbox for Handling PI Operators*

The PI-operators form a *-subalgebra, i.e., operations such as addition, concatenation, composition, and adjoint operations on PI operators are closed and result in another linear PI operator. If the operators $\mathcal{P}$ and $\mathcal{Z}$ are parameterized by PI operators, then the constraints in Eqs. (12.18) and (12.21) are LPIs. Furthermore, positivity constraints on LPIs can be enforced using LMIs [14]; the numerical implementation is done in MATLAB by using PIETOOLS [11]. PIETOOLS borrows some functions for handling polynomials from SOSTOOLS [13] and can use a variety of SDP solvers such as SeDuMi, Mosek, etc.

12.4 Decentralized Implementation of the Controller and Observer

This section presents a method for decentralized implementation of controllers and observers in PI form. To calculate the gains $\mathcal{K}$ and $\mathcal{L}$, the inverse operator $\mathcal{P}^{-1}$ must be computed. A self-adjoint, bounded and coercive operator in 4-PI form has an inverse that is necessarily a 4-PI operator. Specifically, for a self-adjoint, bounded and coercive 4-PI operator $\mathcal{P}\begin{bmatrix} P, & Q \\ Q^T, & \{R_i\} \end{bmatrix}$, if $R_1 = R_2$, then we have an exact formula for the inverse $\mathcal{P}^{-1} = \mathcal{P}\begin{bmatrix} \hat{P}, & \hat{Q}_1 \\ \hat{Q}_2, & \{\hat{R}_i\} \end{bmatrix}$ as presented in Theorem 8 of [10]. However, to find $\mathcal{P}^{-1}$, we need to find $R_0(s)^{-1}$, which is a rational function. Instead of calculating $R_0(s)^{-1}$ exactly, we approximate it numerically by a polynomial and calculate the inverse $\mathcal{P}^{-1}$. The controller ($\mathcal{K} = \mathcal{Z}\mathcal{P}^{-1}$) or observer ($\mathcal{L} = \mathcal{P}^{-1}\mathcal{Z}$) gains are in 3-PI form

$$(\mathcal{K}\mathbf{x})(s) = K_0(s)\mathbf{x}(s) + \int_a^s K_1(s,\theta)\mathbf{x}(\theta)d\theta + \int_s^b K_2(s,\theta)\mathbf{x}(\theta)d\theta,$$

$$(\mathcal{L}\mathbf{x})(s) = L_0(s)\mathbf{x}(s) + \int_a^s L_1(s,\theta)\mathbf{x}(\theta)d\theta + \int_s^b L_2(s,\theta)\mathbf{x}(\theta)d\theta,$$

where the K_i's and L_i's are matrix-valued polynomials of appropriate dimensions.

For the implementation, we discretize the domain $[0, L]$ by an equispaced grid with n grid points. The distributed state $x(\cdot, t)$ is approximated by the vector of states $x_i(t) = x(s_i, t)$, where $s_i = i\Delta s$, $i \in \{0, \dots, n\}$ and $\Delta s = L/(n+1)$. Then

$$K_n x_i(t) := (\mathcal{K}x)(s_i, t) \approxeq K_0(s_i)x_i(t) + K_{1a}(s_i)\sum_{j=0}^{i} K_{1b}(s_j)x_j(t)\Delta s$$
$$+ K_{2a}(s_i)\sum_{j=i+1}^{n} K_{2b}(s_j)x_j(t)\Delta s,$$

where K_n is the finite-dimensional approximation of $\mathcal{K}$ on the grid with n segments. Note that the K_i's are polynomials and hence can be factored as $K_i(s,\theta) = K_{ia}(s)K_{ib}(\theta)$.

Typically, a full-state feedback requires n^2 information transfers to find the inputs $u_i(t) = K_n x_i(t)$ for all i. A system is said to be decentralized if it operates based on local information. More specifically, in this work, the implementation is said to be decentralized if the input for any node is calculated using information available at a node or its immediate neighbors. As discussed below, an observer/controller of 3-PI form can indeed be implemented such that inputs are calculated using information from immediately adjacent nodes (local information) (Fig. 12.3).

Fig. 12.3 Information transfer needed to find input $u_i(t)$

Each node i has access to the information $K_0(s_i)x_i(t)$, $K_{1a}(s_i)$, and $K_{2a}(s_i)$. The information needed from other segments are the cumulative values $c_i(t) = \sum_{j=0}^{i-1} K_{1b}(s_j)x_j(t)$ and $d_{i+1}(t) = \sum_{j=i+1}^{n} K_{2b}(s_j)x_j(t)$. Suppose node i receives the sum $c_i(t)$ from the node $i-1$. Then we find $c_{i+1}(t) = c_i(t) + K_{1b}(s_i)x_i(t)$ and send the information to node $i+1$. In the opposite direction, we find $d_i(t) = d_{i+1}(t) + K_{2b}(s_i)x_i(t)$ and transmit the information to node $i-1$. Then every node uses only local information to determine the input

$$u_i(t) = K_0(s_i)x_i(t) + K_{1a}(s_i)c_i(t)\Delta s + K_{2a}(s_i)d_{i+1}(t)\Delta s$$

and the exchange of information can be achieved with just $2n$ information transfers. Thus, we do not require communication between non-adjacent nodes. Further, delay arising from communication between distal nodes is avoided and communication overhead is dramatically reduced.

12.4.1 Observer Implementation in MATLAB

Using PIETOOLS, we design the H_∞-optimal observer for each of the setpoints shown in Fig. 12.2, which results in 3 different observer gains corresponding to the different linearized PDEs. The linearized observers are implemented as a single gain-scheduled observer where the active observer is determined by the sector in which the tip of the arm is located. The domain [0, 1] is divided into a equispaced grid with 10 grid points. The even-order derivatives in the PDE equation (12.2) are approximated by a second-order central difference expression, whereas the odd-order derivatives are approximated by using first-order upwind-biased difference expressions. The discretization methods are not discussed in detail because it is not the focus of this work and other discretization methods can be used to approximate the solution of PDE. A MATLAB ode solver was used to solve the finite-difference approximation of the PDE using the initial conditions

$$\hat{x}_i(s,0) = x_1(s,0) = x_2(s,0) = x_4(s,0) = 0, x_3(s,0) = (2s^2 - (8/3)s^3 + s^4)$$

and input disturbance $w(t) = \frac{\sin(t)}{10t}$, where x is the system state, and $\hat{x}$ is the observer state.

The two observed outputs y_1 and y_2 are the errors in transverse displacement of the tip and average error in the estimate of the PIE state. The regulated output is the average error in the estimate of the PIE states:

$$y(t) = \begin{bmatrix} \hat{x}(L,t) - x(L,t) \\ \int_0^L \mathbf{e}(\eta,t)d\eta \end{bmatrix}, \quad z(t) = \int_0^L \mathbf{e}(s,t)ds, \quad \mathbf{e} = \partial_s^4 \hat{x} - \partial_s^4 x.$$

The outputs can be expressed in PIE form, Eq. (12.8), by defining C_i and $\mathcal{D}_{ij}$ as

$$\mathcal{D}_{ij} = 0,\ C_1 = \mathcal{P}_{\{0,I,I\}},\ C_2 = \mathcal{P}_{\{0,C_2,C_2\}},\ C_2(s,\theta) = \begin{bmatrix} G_1(L,\theta) \\ I \end{bmatrix}, \tag{12.22}$$

where G_1 is as defined in Eq. (12.11). The material properties and dimensions for the soft robot arm are set as $E = 5 \times 10^3\,\mathrm{N/m^2}$, $r = 0.025\,\mathrm{m}$, $L = 1\,\mathrm{m}$, and $\rho = 1.1 \times 10^3\,\mathrm{kg/m^3}$.

The following is a stepwise procedure to implement the observer:

1. Define the PI operators $\mathcal{T}$, $\mathcal{A}$, $\mathcal{B}_i$ as in Eq. (12.5) and C_i, $\mathcal{D}_{ij}$ as in Eq. (12.22).
2. Solve the LPI (12.18) using PIETOOLS to find $\mathcal{P}$ and $\mathcal{Z}$.
3. Find the observer $\mathcal{L} = \mathcal{P}^{-1}\mathcal{Z}$ for each setpoint.
4. Find the finite-dimensional approximation L of the observer $\mathcal{L}$.
5. Use the finite-difference approximation of the PDE equation (12.4) to find A, B, and C_i.
6. Implement the closed-loop error dynamics Eq. (12.15) in MATLAB.

As seen in Fig. 12.4a, the error asymptotically converges to zero and the approximation of the actual shape of the arm is shown in Fig. 12.4b. Figure 12.5 shows that the gain-scheduled observer converges for different initial conditions. If the LPI (12.18) is solved with a large positive value for ϵ, then, for the obtained observer, the observer state converges to the system state faster. However, using large values for ϵ results in high observer gain values and higher H_∞-norm.

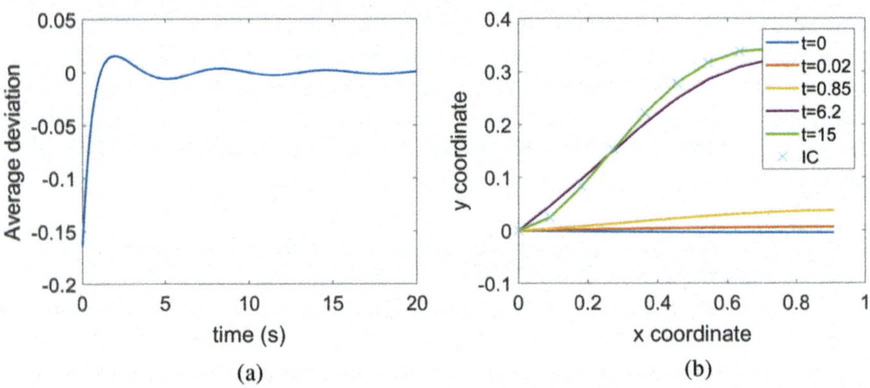

Fig. 12.4 The figure on the left shows the time-evolution of the regulated output, which in this case, is the average error in the estimate of the state. On the right, is the estimate of the transverse displacement at various times. (**a**) $z_3(t)$ vs t. (**b**) $\hat{x}_3(s,t)$ vs s at different t values

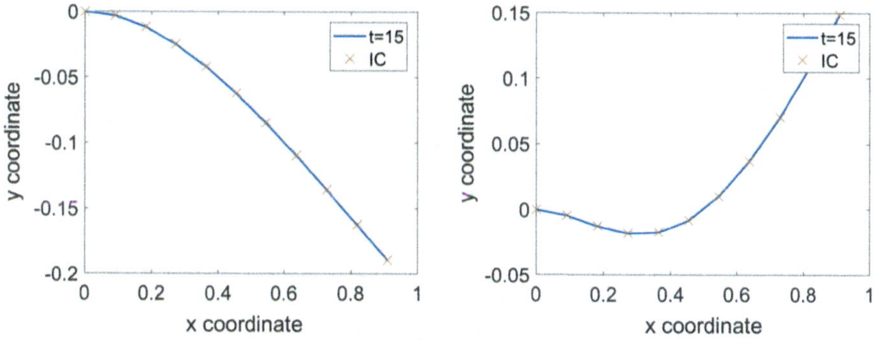

(a) (b)

Fig. 12.5 Observer state convergence for various initial conditions. (**a**) Fourth-order polynomial initial condition. (**b**) Fifth-order polynomial initial condition

12.5 Conclusion

This chapter discusses a method for linearizing a nonlinear PDE about setpoints. We introduce the PIE representation of a linear PDE and pose the design of H_∞-optimal observer and controller gains as an optimization problem with PI variables and LPI constraints. A computational tool, PIETOOLS, is used to solve the resulting LPI optimization problems. Further, we present a method to implement the proposed continuum controllers or observers in a decentralized manner. Finally, we apply the methods developed for linear coupled PDE systems to a model of a soft robot arm. A MATLAB implementation shows convergence and stability of the observer.

Acknowledgments This work was supported by Office of Naval Research Award N00014-17-1-2117 and National Science Foundation under grant No. 1739990.

References

1. A.S. Banach, W.T. Baumann, Gain-scheduled control of nonlinear partial differential equations, in *1990 IEEE Conference on Decision and Control* (1990), pp. 387–392
2. S. Boyd, L. El Ghaoui, E. Feron, V. Balakrishnan, *Linear Matrix Inequalities in System and Control Theory*, vol. 15 (SIAM, 1994)
3. M. Cianchetti, T. Ranzani, G. Gerboni, T. Nanayakkara, K. Althoefer, P. Dasgupta, A. Menciassi, Soft robotics technologies to address shortcomings in today's minimally invasive surgery: the STIFF-FLOP approach. Soft Robot. **1**(2), 122–131 (2014)
4. N. Darivandi, K. Morris, A. Khajepour, An algorithm for LQ optimal actuator location. Smart Mater. Struct. **22**(3), 035001 (2013)
5. I.S. Godage, G.A. Medrano-Cerda, D.T. Branson, E. Guglielmino, D.G. Caldwell, Modal kinematics for multisection continuum arms. Bioinspiration Biomimetics **10**(3), 035002 (2015)

6. R. Kang, A. Kazakidi, E. Guglielmino, D.T. Branson, D.P. Tsakiris, J.A. Ekaterinaris, D.G. Caldwell, Dynamic model of a hyper-redundant, octopus-like manipulator for underwater applications, in *2011 IEEE/RSJ International Conference on Intelligent Robots and Systems* (2011), pp. 4054–4059
7. L. Lessard, S. Lall, Quadratic invariance is necessary and sufficient for convexity, in *2011 American Control Conference* (2011), pp. 5360–5362
8. Matsuno, F., K. Suenaga, Control of redundant 3D snake robot based on kinematic model, in *2003 IEEE International Conference on Robotics and Automation (Cat. No. 03CH37422)*, vol. 2 (2003), pp. 2061–2066
9. A.H. Nayfeh, P.F. Pai, *Linear and Nonlinear Structural Mechanics* (Wiley, London, 2008)
10. M.M. Peet, K. Gu, SOS for systems with multiple delays: Part 1. H_∞-optimal control, in *2019 American Control Conference* (2019), pp. 3849–3856
11. M.M. Peet, S. Shivakumar, PIETOOLS DDF for networks with delay (2020). https://codeocean.com/capsule/0447940/. Accessed 20 April 2020
12. M.M. Peet, S. Shivakumar, A. Das, S. Weiland, Discussion paper: A new mathematical framework for representation and analysis of coupled PDEs. IFAC-PapersOnLine **52**(2), 132–137 (2019)
13. S. Prajna, A. Papachristodoulou, P. Seiler, P.A. Parrilo, SOSTOOLS and its control applications, in *Positive Polynomials in Control* (Springer, 2005), pp. 273–292
14. S. Shivakumar, A. Das, M.M. Peet, PIETOOLS: A MATLAB toolbox for manipulation and optimization of partial integral operators. arXiv:1910.01338 (2019)
15. S. Shivakumar, A. Das, S. Weiland, M.M. Peet, A generalized LMI formulation for input-output analysis of linear systems of ODEs coupled with PDEs, in *2019 IEEE Conference on Decision and Control* (2019), pp. 280–285
16. S. Shivakumar, A. Das, Weiland, S., M.M. Peet, Duality and H_∞-optimal control of coupled ODE-PDE systems. arXiv:2004.03638 (2020)
17. V. Vikas, F. Cohen, R. Grassi, C. Sözer, B. Trimmer, Design and locomotion control of a soft robot using friction manipulation and motor–tendon actuation. IEEE Trans. Robot. **32**(4), 949–959 (2016)
18. S. Wu, S. Shivakumar, M.M. Peet, C. Hua, H_∞-optimal observer design for linear systems with delays in states, outputs and disturbances. arXiv:2004.04482 (2020)
19. T. Zheng, D.T. Branson, R. Kang, M. Cianchetti, E. Guglielmino, M. Follador, G.A. Medrano-Cerda, I.S. Godage, D.G. Caldwell, Dynamic continuum arm model for use with underwater robotic manipulators inspired by octopus vulgaris, in *2012 IEEE International Conference on Robotics and Automation* (2012), pp. 5289–5294

Chapter 13
Modeling Soft Swimming Robots using Discrete Elastic Rod Method

Weicheng Huang, Zachary Patterson, Carmel Majidi, and M. Khalid Jawed

13.1 Introduction

Recent progress in the field of soft robotics [1] has led to new classes of bio-inspired swimming robots that are largely composed of soft and elastically deformable materials. These include robots powered with hydraulics [2], ionic polymer–metal composites (IPMCs) [3], dielectric elastomer actuators (DEAs) [4], and shape memory alloys (SMAs) [5]. SMA-powered actuators are especially promising for swimming robots since the heat used to activate the SMA can be quickly dissipated by convection through the surrounding water. An example of a soft swimming robot inspired by the seastar is presented in Fig. 13.1. The limbs of the robot are composed of soft silicone elastomer that are embedded with SMA wires. When electrical current is delivered to the wires, they cause the limbs to bend and generate a forward thrust.

There are a variety of computational tools that can be used to model soft robots like the seastar inspired robot described above. These include reduced-order finite element analysis (FEA) formulations [6–9] as well as 3D representations using voxel-based discretization [10, 11]. Recently, there has been interest in modeling soft robots as a combination of slender soft appendages using elastic rod theory [12–14]. When combined with discrete differential geometry (DDG) numerical tools used within the computer graphics community, these rod theories can be used to achieve rapid simulation runtimes.

W. Huang · M. K. Jawed
Mechanical and Aerospace Engineering, University of California, Los Angeles, CA, USA
e-mail: khalidjm@seas.ucla.edu

Z. Patterson · C. Majidi (✉)
Mechanical Engineering, Carnegie Mellon University, Pittsburgh, PA, USA
e-mail: cmajidi@andrew.cmu.edu

D. A. Paley, N. M. Wereley (eds.), *Bioinspired Sensing, Actuation, and Control in Underwater Soft Robotic Systems*, https://doi.org/10.1007/978-3-030-50476-2_13

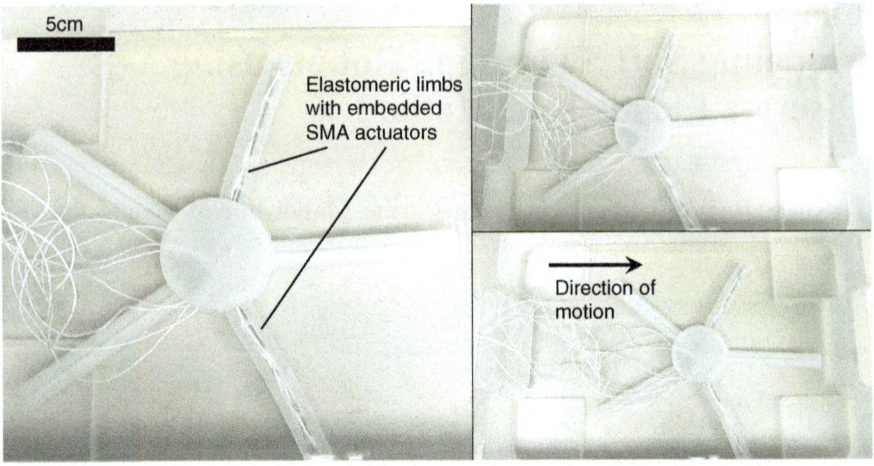

Fig. 13.1 SMA-powered seastar-inspired soft robot

The DDG-based approach to simulation has been used to model elastic slender structures, e.g., rods [15–17], ribbons [18], plates [19], shells [20], X-shells [21], viscous threads [17, 22], and viscous sheets [23]. The method of discrete elastic rods (DER) is emerging as an especially popular tool for rapid simulation of engineered systems composed of slender filamentary structures. Goldberg et al. used a 2D version of DER to simulate a crawling, caterpillar-inspired soft robot in which each robot segment was modeled as an elastic rod element [24].

This chapter shows how DER can be used to simulate seastar-inspired soft swimming robots composed of radiating elastic limbs that flap in response to actuator stimulation. Figure 13.2 presents several snapshots of a seastar-inspired soft robot moving through a prescribed regular pentagon trajectory from simulations. Here, the actuated limbs are highlighted as blue, and the corresponding actuating sequence of the presented trajectory is in Fig. 13.7b. Our analysis is limited to two dimensions, although it can be generalized into 3D as detailed in [15], which presents a comprehensive description of the DER formulation and its application to robotics modeling. We begin with a brief background on DER in Sect. 13.2, followed by an overview of the mathematical formulation used to describe the kinematics, elastic energies, and time-marching scheme of the rod system in Sect. 13.3. Section 13.4 applies this model to simulate the dynamics of a robot composed of five radiating limbs moving in a fluid. Finally, concluding remarks and avenues for future research are presented in Sect. 13.5.

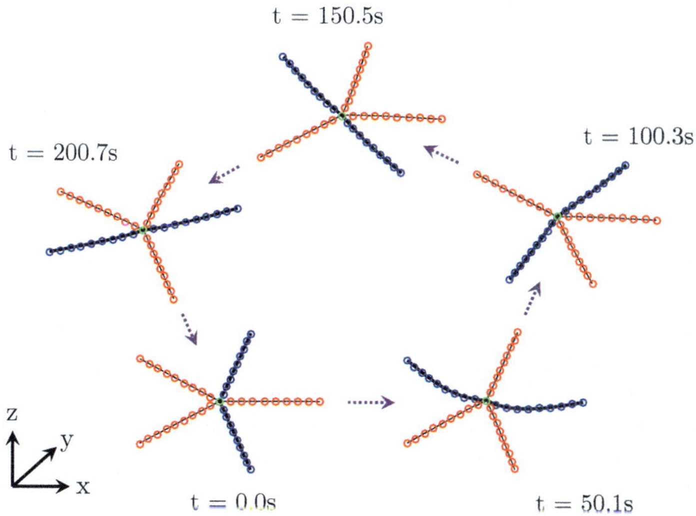

Fig. 13.2 DER-based simulation of a seastar-inspired soft robot following a pentagonal trajectory. The blue bold lines are the actuated limbs

13.2 Background

Kirchhoff introduced a celebrated model for elastic rods in 1859 [25]. A recent and related development in the mechanics of elastic rods came in 2008 with the introduction of discrete elastic rods (DER) [16, 17]—a computationally efficient algorithm that can seamlessly capture the geometrically nonlinear dynamics of elastic rods. Although the method was originally introduced by the computer graphics community, it has been successfully employed in the context of mechanical engineering problems since 2014 [26]. The original DER papers [16, 17] are fairly short in length, yet packed with advanced concepts from discrete differential geometry. A detailed exposition of this method is presented in [15] and generalizations of this modeling approach have been developed for plates [19], shells [20], ribbons [18], and gridshells [21].

For many soft robotic systems, DER is capable of much faster computational speeds than more conventional modeling techniques. This improvement is because the subcomponents of the robot are treated as 1D elements (capable of deformation in two- or three-dimensions) rather than volumetric 3D bodies. The reduction in dimensionality allows for the system of partial differential equations typically used to model 3D elastic bodies to be replaced with a much simpler system of ordinary differential equations. With the analytical formulation of energy variation (related to elastic force) and second variation (known as the Hessian matrix), the equations of motion of the dynamic system can be implicitly updated with a larger time-marching step. For some soft robotic systems, like the seastar inspired robot studied here, locomotion can be accurately simulated using a reasonable temporal–spatial discretization in faster than real time. That is, the time required on a single thread of a standard CPU to perform the simulation is faster than the wall-clock time it would take for the robot to perform the physical motion, which is a first step towards a major challenge, i.e., developing online robot controllers, in soft robotic engineering.

13.3 Discrete Elastic Rod Method

Each limb of the soft robot is treated as a planar elastic rod (cf. elastic beam) that bends in the x–y plane. In the discrete setting of DER shown schematically in Fig. 13.3, the rod centerline is discretized into N nodes: $\mathbf{x}_0, \ldots, \mathbf{x}_{N-1}$, where $\mathbf{x}_i \equiv [x_i,\ y_i]^T$, superscript T denotes the transposition operation, and $0 \leq i \leq N-1$, such that the total degree of freedom (DOF) vector is size of $2N$,

$$\mathbf{q} = [x_0, y_0, x_1, y_1, \ldots, x_{N-1}, y_{N-1}]^T. \tag{13.1}$$

This space discretization results into $N-1$ edge vectors: $\mathbf{e}^0, \ldots, \mathbf{e}^{N-2}$, where $\mathbf{e}^i = \mathbf{x}_{i+1} - \mathbf{x}_i$, and $0 \leq i \leq N-2$. Hereafter, we use subscripts to denote quantities associated with nodes, e.g., $\mathbf{x}_i$, and superscripts when associated with edges, e.g., $\mathbf{e}^i$. Based on this kinematic representation, the remainder of this section discusses the formulation of elastic energies, elastic forces, and the time-stepping procedure of the rod system.

An elastic rod is modeled as a mass–spring system, with a lumped mass at each node. Associated with each node and edge is a discrete bending and stretching energy, respectively. In the case of a 3D rod, we would also have to account for the elastic twisting energy and related rotational inertia. For a rod with a constant Young's modulus E, cross-section area A, and second moment of inertia I, the elastic stretching energy associated with the ith edge is

$$E_s^i = \frac{1}{2} EA \left(\epsilon^i\right)^2 \left\| \bar{\mathbf{e}}^i \right\|, \tag{13.2}$$

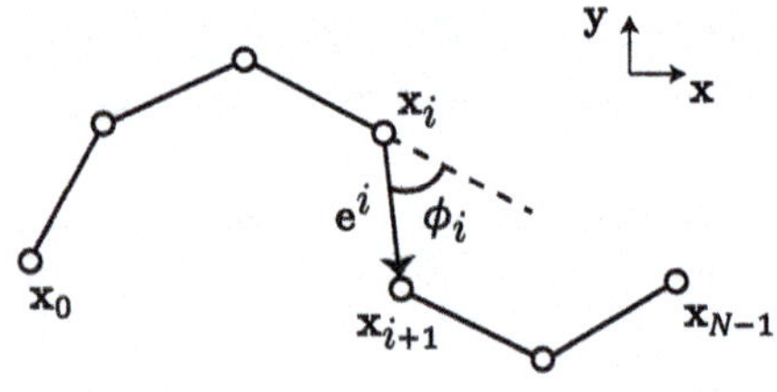

Fig. 13.3 Schematic of a discrete rod in 2D

where $\epsilon^i = \|\mathbf{e}^i\|/\|\bar{\mathbf{e}}^i\| - 1$ is the axial strain of the ith edge, and notation with the bar on top indicates the quantity is evaluated in the undeformed configuration, e.g., $\|\bar{\mathbf{e}}^i\|$ is the undeformed length of the ith edge.

The bending energy of the ith node is measured by the misalignment between two consecutive edges $\{\mathbf{e}^{i-1}, \mathbf{e}^i\}$, i.e.,

$$E_{b,i} = \frac{1}{2} EI(\kappa_i - \bar{\kappa}_i)^2 \Delta l_i, \tag{13.3}$$

where $\kappa_i = 2\tan(\phi_i/2)/\Delta l_i$ is the discrete curvature at the ith node and $\Delta l_i = (\|\bar{\mathbf{e}}^{i-1}\| + \|\bar{\mathbf{e}}^i\|)/2$ is its Voronoi length. The turning angle ϕ_i is

$$\phi_i = \tan^{-1} \frac{\left(\mathbf{e}^{i-1} \times \mathbf{e}^i\right) \cdot \hat{\mathbf{n}}_z}{\mathbf{e}^{i-1} \cdot \mathbf{e}^i}, \tag{13.4}$$

where $\hat{\mathbf{n}}_z$ is the unit vector along the z-axis, i.e., the numerator on the right-hand side is the z-component of the cross-product between $\mathbf{e}^{i-1}$ and $\mathbf{e}^i$. Note that ϵ^i, κ_i, and ϕ_i represent the finite difference approximations of stretch, curvature, and turning angle.

The total potential energy of the rod system is obtained by summing the elastic stretching energies over all the edges and bending energies over all the nodes (except the first and the last nodes where the bending energies are always zero), i.e.,

$$E = \sum_{i=0}^{N-2} E_s^i + \sum_{i=1}^{N-2} E_{b,i}. \tag{13.5}$$

Next, the internal elastic force vector $\mathbf{F}^{\text{int}}$ is the negative gradient of the total potential energy:

$$\mathbf{F}^{\text{int}} = -\frac{\partial E}{\partial \mathbf{q}}. \tag{13.6}$$

Specifically, the non-zero force component of the ith node, $[F_{2i}^{\text{int}}, F_{2i+1}^{\text{int}}]^T$, is only related to five neighboring energy terms: E_s^{i-1}, E_s^i, $E_{b,i-1}$, $E_{b,i}$, and $E_{b,i+1}$. We have

$$F_{2i}^{\text{int}} = -\frac{\partial}{\partial x_i}\left(E_s^{i-1} + E_s^i + E_{b,i-1} + E_{b,i} + E_{b,i+1}\right) \quad \text{and} \tag{13.7a}$$

$$F_{2i+1}^{\text{int}} = -\frac{\partial}{\partial y_i}\left(E_s^{i-1} + E_s^i + E_{b,i-1} + E_{b,i} + E_{b,i+1}\right) \tag{13.7b}$$

Next, we turn to the formulation of the external force vector. The external force is comprised of the force from the external environment, e.g., gravity, damping force from a viscous medium [27], or the actuation force from an external electromagnetic field [28]. Here, we consider the hydrodynamic force from fluid experienced by soft robotic limbs as the external force in our numerical framework. The discretized version of a fluid drag force applied on the ith node is [29, 30]

$$\mathbf{F}_i^{\text{ext}} = -\frac{1}{2}\rho_f C_d D \|\mathbf{v}_i\| \mathbf{v}_i \Delta l_i, \tag{13.8}$$

where ρ_f is the density of fluid medium, $D = 2r_0$ is the diameter of the rod, C_d is the drag coefficient, and $\mathbf{v}_i \equiv [\dot{x}_i, \dot{y}_i]^T$ is the relative velocity of the ith node and fluid. The external force vector (size $2N$) on the entire rod is $\mathbf{F}^{\text{ext}} = \left[\mathbf{F}_0^{\text{ext}}, \mathbf{F}_1^{\text{ext}}, \ldots, \mathbf{F}_{N-1}^{\text{ext}}\right]^T$.

Finally, we discuss the time-marching scheme. We update the equations of motion of a discrete rod system by a first-order, implicit Euler method, because of its unconditional convergence and numerical stability [31]. We solve the following $2N$ equation of motions and update the DOF vector $\mathbf{q}$ and its velocity (time derivative of DOF) $\mathbf{v} = \dot{\mathbf{q}}$ from time step t_k to $t_{k+1} = t_k + h$ (h is the time step size) based on the following statement of force balance:

$$\mathbf{f} \equiv \mathbb{M}\frac{[\mathbf{q}(t_{k+1}) - \mathbf{q}(t_k) - h\mathbf{v}(t_k)]}{h^2} - \mathbf{F}^{\text{int}}(t_{k+1}) - \mathbf{F}^{\text{ext}}(t_{k+1}) = \mathbf{0}, \tag{13.9}$$

where $\mathbb{M}$ is the diagonal mass matrix comprised of the lumped masses, $\mathbf{F}^{\text{int}}$ is the internal elastic force given by Eqs. (13.6) and (13.14), $\mathbf{F}^{\text{ext}}$ is the external force formulated in Eq. (13.8), and t_k denotes evaluation of the quantity at time t_k.

We use the Newton–Raphson method to solve this set of nonlinear equations of motion. At each time step t_{k+1}, we first guess a new solution on the basis of the previous state, i.e.,

$$\mathbf{q}^{(1)}(t_{k+1}) = \mathbf{q}(t_k) + h\mathbf{v}(t_k). \tag{13.10}$$

Then, we optimize the solutions by gradient decent, such that the new solution at the $(k+1)$th step is

$$\mathbf{q}^{(n+1)}(t_{k+1}) = \mathbf{q}^{(n)}(t_{k+1}) - \mathbb{J}^{(n)} \backslash \mathbf{f}^{(n)}, \tag{13.11}$$

where $\mathbb{J}$ is the Jacobian matrix associated with Eq. (13.9),

$$\mathbb{J} \equiv \frac{\partial \mathbf{f}}{\partial \mathbf{q}} = \frac{1}{h^2}\mathbb{M} + \frac{\partial^2 E}{\partial \mathbf{q} \partial \mathbf{q}} - \frac{\partial \mathbf{F}^{\text{ext}}}{\partial \mathbf{q}}. \tag{13.12}$$

A detailed formulation of the Hessian matrix associated with the elastic energy, i.e., the second term $\partial^2 E/\partial \mathbf{q}^2$ in Eq. (13.12), is in [17]. The Jacobian associated with the external force, $\frac{\partial \mathbf{F}^{\text{ext}}}{\partial \mathbf{q}}$, can be re-written as $\frac{1}{h}\frac{\partial \mathbf{F}^{\text{ext}}}{\partial \mathbf{v}}$ and trivially computed from Eq. (13.8). Importantly, the Jacobian $\mathbb{J}$ is a banded matrix and the time complexity of this algorithm is $O(N)$ for a single rod system (the computational time linearly scales with the number of node). This computational efficiency has motivated its application in the animation industry (e.g., hair simulation for movies) as well as its adoption in mechanical engineering.

13.4 Simulation

Here, the main steps of the algorithm for modeling the dynamics of a soft swimming robot with the DER method are outlined below. We first introduce the geometric discretization of the seastar-inspired soft robot composed of a number of SMA-based actuators. Next, the modeling of the actuators is discussed, followed by a discussion of the locomotion of the seastar-inspired soft robot in two special cases: a straight path and a regular pentagonal trajectory. Finally, the computational efficiency of the DER-based numerical framework is detailed.

In Fig. 13.4a, we model the seastar-inspired soft robot with five planar elastic rods connected at the center point, $\mathbf{x}_0 \equiv [x_0, \ y_0]^T$. Each limb, with arc length $L = 0.1$ m, is discretized into N nodes and $N-1$ edges. In the current study, we choose $N = 10$, resulting in an edge length of $\Delta L \approx 1.11$ cm. As all five limbs share the center node, $\mathbf{x}_0$, the total number of nodes in the soft robot system is $n = 5 \times N - 4$. The ith node on jth limb is denoted as $\mathbf{x}_{\{j,i\}}$. The material properties used in this numerical study are as follows: Young's modulus $E = 1$MPa, material density $\rho_m = 1000\,\text{kg/m}^3$, fluid density $\rho_f = \rho_m$ (neutrally buoyant), rod radius $r_0 = 1.6$ mm (and, therefore, second moment of inertia $I = \pi r_0^4/4$ and diameter $D = 2r_0$), and drag coefficient $C_d = 1.0$.

Besides the bending and stretching energies formulated in the DER method for each limb, the bending energy at the center node requires special treatment. As shown in Fig. 13.4b, the bending energy between the first limb and second limb at the center node is

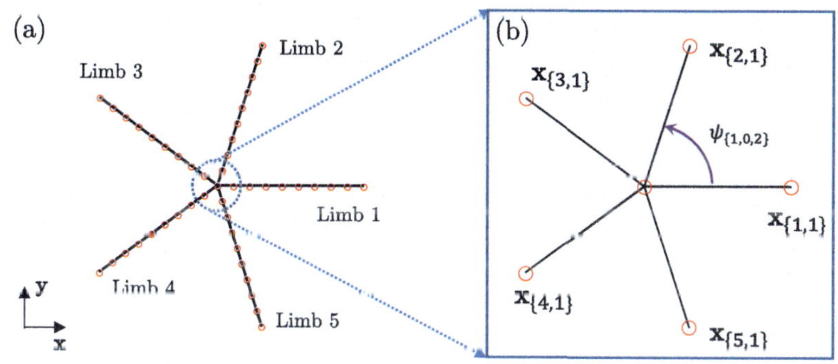

Fig. 13.4 (a) Geometric discretization of seastar-inspired soft robot. (b) Zoomed-in description of the center point

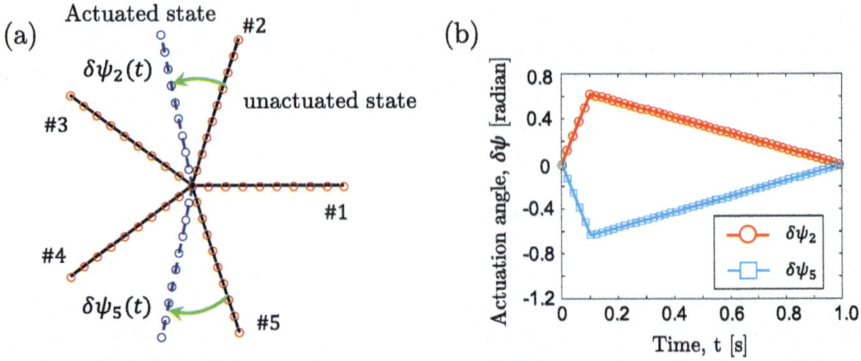

Fig. 13.5 (**a**) Actuated and unactuated state of seastar-inspired robot. (**b**) Actuation signal of #2 and #5 limbs: rotating angle, $\delta\psi$, as a function of time during one actuation cycle

$$E_{b,\{1,0,2\}} = \frac{1}{2} EI \left(\kappa_{\{1,0,2\}} - \bar{\kappa}_{\{1,0,2\}}\right)^2, \tag{13.13}$$

where $\kappa_{\{1,0,2\}}$ is measured by the angle between the first limb and second limb, $\psi_{\{1,0,2\}}$. Similarly, there are four more components to the bending energy at the center node associated with the angles $\psi_{\{2,0,3\}}$, $\psi_{\{3,0,4\}}$, $\psi_{\{4,0,5\}}$, and $\psi_{\{5,0,1\}}$. Overall, the total elastic energy of seastar-inspired soft robot system is

$$E_{\text{total}} = \sum_{j=1}^{5} \left[\left(\sum_{i=0}^{N-2} E_s^i \right)_j + \left(\sum_{i=1}^{N-2} E_{b,i} \right)_j \right] \tag{13.14a}$$

$$+ E_{b,\{1,0,2\}} + E_{b,\{2,0,3\}} + E_{b,\{3,0,4\}} + E_{b,\{4,0,5\}} + E_{b,\{5,0,1\}}. \tag{13.14b}$$

Its first gradient (negative internal force) and second gradient (Hessian matrix) can be analytically formulated in a manner similar to common applications of the DER methods.

Actuation is incorporated into the simulation by varying the natural curvature between two neighboring limbs, e.g., $\bar{\kappa}_{\{1,0,2\}}$ in Eq. (13.13), with time. This variation is measured through characterization of a single SMA-powered actuator, as described next. The electrically activated SMA wire enables rapid transition between the unactuated state and the actuated state [32, 33]. Assume the limbs can rotate by a certain angle as a function of time during the actuating–cooling process, e.g., #2 rotates by an angle $\delta\psi_2(t)$ relative to the rest of the system, as shown in Fig. 13.5a. We define anti-clockwise rotation as positive and clockwise rotation as negative.

Usually, experimental characterization of an actuator can be used to get the actuation function, e.g., $\delta\psi_2(t)$. Here, for simplicity, we assume a piecewise linear function, shown in Fig. 13.5b, to describe the actuated and cooling state. We actuated limbs #2 and #5 simultaneously for the straight locomotion along X-

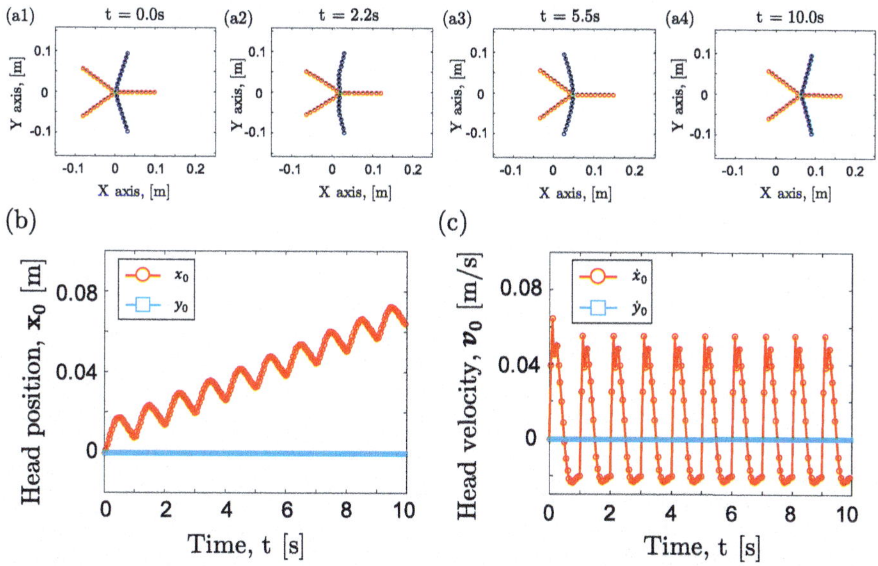

Fig. 13.6 **(a1)–(a4)** Snapshots of swimming seastar-inspired robot at different time steps, t = 0.0 s, 2.2 s, 5.5 s, and, 10.0 s. (**b**) Head node position as a function of time. (**c**) Head node velocity as a function of time

axis, and the maximum actuation angle is $\pi/5$, which is half of the angle between two neighboring limbs in the undeformed configuration. The actuation period in Fig. 13.5b is $\Delta t_0 = 1.0$ s, with actuating time $\Delta t_1 = 0.1$ s and cooling time $\Delta t_2 = 0.9$ s.

By symmetrically actuating limbs #2 and #5 with the input shown in Fig. 13.5b, this bio-inspired swimming robot can move forward along the X-axis. In Fig. 13.6a1–a4, we show multiple snapshots of the seastar-inspired soft robot swimming in water at various time steps, $t \in \{0.0, 2.2, 5.5, 10.0\}$ s. We observe that the robot swims forward during the actuating phase, but moves slightly backward in the cooling state. Overall, the net motion of the robot is forward over an entire cycle. Figs. 13.6b,c show the center node position and velocity, respectively, as functions of time over 10 actuation periods. Because of the symmetry of geometric and actuating conditions, this bio-inspired soft robot has no displacement or velocity components along the Y-axis.

Next, we show a piecewise-linear trajectory of the robot swimming in a fluid medium. The soft robot can follow a straight trajectory along one limb by actuating the two neighboring limbs next to it, e.g., the soft robot can swim along X-axis (the direction of limb #1) by actuating limbs #2 and #5, as discussed in the previous case. A complex nonlinear trajectory, on the other hand, can be achieved by combining multiple straight trajectories that are not parallel. In Fig. 13.2, we provide some snapshots of a seastar-inspired soft robot moving along a prescribed regular pentagonal trajectory, i.e., a closed path that combines five straight trajectories.

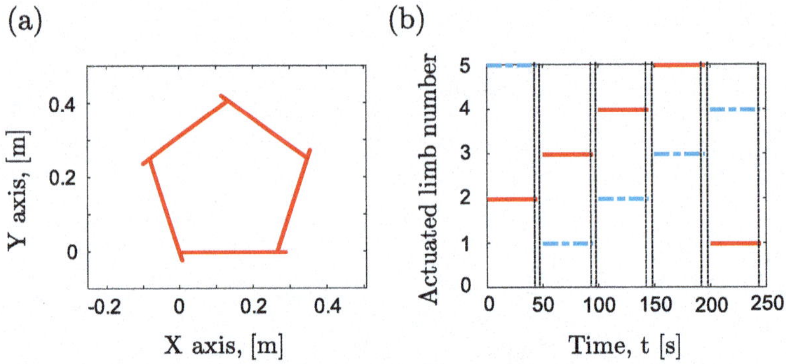

Fig. 13.7 (**a**) Regular pentagon trajectory of seastar-inspired soft robot. (**b**) Actuation input signal for regular pentagon trajectory. Blue dashed line: negative rotating actuation; red solid line: positive rotating actuation

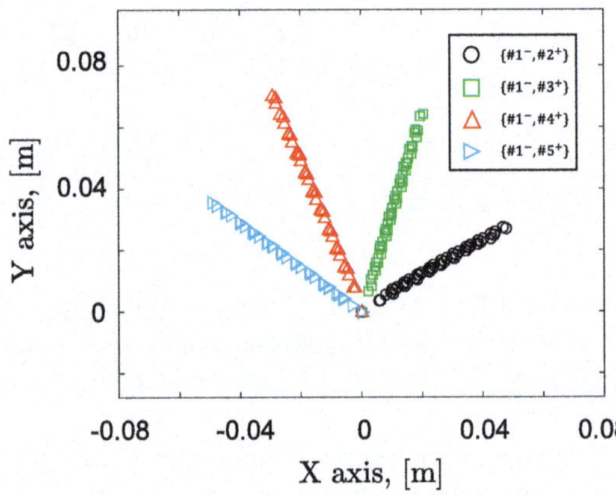

Fig. 13.8 Trajectories of seastar-inspired soft robot when actuating through different combinations of limbs over 10 actuation periods

Figure 13.7a shows the trajectory traced by the center node. Here, we actuated two limbs at a time for each straight line segment. For the presented pentagonal trajectory in the current study, the actuation sequence of the limbs is

$$\{2, 5\};\ \{1, 3\};\ \{2, 4\};\ \{3, 5\};\ \{1, 4\}, \tag{13.15}$$

as shown in Fig. 13.7b, and the actuating function for each limb is in Fig. 13.5b. During each straight path, we actuated two specific limbs in Eq. (13.15) for 45 s, then left the limbs unactuated for 5 s in order to bring the robot to a complete stop. This way, we maintained the same initial conditions for each straight component of the pentagonal trajectory.

Previous investigations mainly focus on a specific actuating condition. Here, exploiting the efficiency and robustness of our simulator, we perform different actuating combinations of soft limbs. In Fig. 13.8, we plot the trajectories of the seastar-inspired soft robot with different actuated limbs, i.e.,

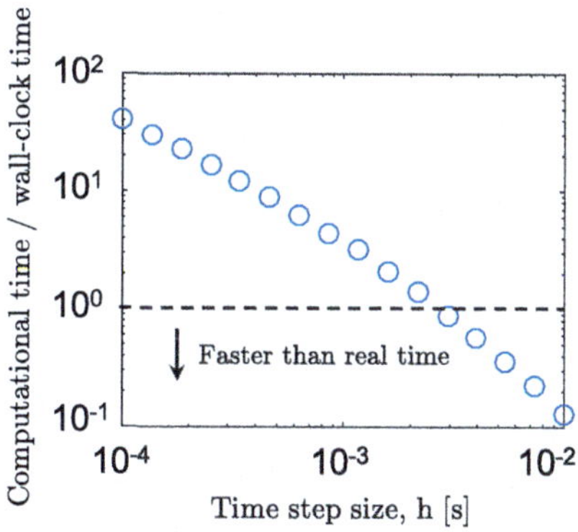

Fig. 13.9 The ratio between computational time and wall-clock time as a function of time step size, h, with a fixed number of vertices, $n = 46$

$\{1^-, 2^+\}$, $\{1^-, 3^+\}$, $\{1^-, 4^+\}$, $\{1^-, 5^+\}$, over 10 actuation periods. Here, positive indicates anti-clockwise rotation and negative indicates clockwise rotation, similar to the previous study. The seastar-inspired soft robot always moves along the bisector of the two actuated limbs, while the translation distance is different. The maximum translation is obtained when actuating limbs #1 and #4 simultaneously, whereas the actuating combination of limbs #1 and #2 has a lower efficiency. Overall, any desired nonlinear trajectory can be fitted by a zig-zag path comprised of some simple straight lines as described above.

Finally, we highlight the computational efficiency of the DER-based soft robotic simulation. For the case of the seastar-inspired swimmer, this numerical framework can achieve real-time simulation speed in which the time required to perform the computation is equal to or faster than the duration of the physical motion being modeled. In Fig. 13.9, with a fixed number of nodes, $n = 46$, the computational time linearly scales with time step size, h. The simulations ran on a single thread of AMD Ryzen 1950X CPU @ 3.4 GHz. The simulator can run faster than real time when the time step size $h \gtrsim 3.0$ ms. Numerical issues associated with large step size appear at $h \gtrsim 20$ ms.

13.5 Conclusion

We have introduced a computational framework to study the dynamics of a seastar-inspired soft robot that is adapted from methods popular in the computer graphics community. This numerical framework integrates elasticity of slender structures, actuation of shape memory alloy, and hydrodynamic loading for fluid–structure interaction, to achieve fast prediction of the motion of seastar-inspired soft robot swimming in a fluid medium. In particular, the DDG-based method presented here can quantitatively predict the robot's motion while running faster than real time on one thread of a contemporary desktop processor. The computational efficiency of this method makes it ideally suited for algorithms that iterate over a wide variety of parameters in order to select a robot with optimized design. Moving forward, it would be interesting to apply this method to simulate limbed locomotion of soft robots in 3D. This includes the untethered quadruped in [33] as well as the soft SMA-powered robots presented in [34]. Future efforts could also explore the development of a data-driven trajectory planning algorithm for the motion control of bio-inspired soft robots. Such an algorithm would combine rapid, DDG-based simulation of robots with training data for mechanical and fluid–structure interactions obtained from empirical measurements and sensor feedback.

Acknowledgments The authors gratefully acknowledge support from the Office of Naval Research (award #N00014-17-1-2063; Program Manager: Dr. Tom McKenna).

References

1. S.I. Rich, R.J. Wood, C. Majidi, Untethered soft robotics. Nature Electron. **1**(2), 102 (2018)
2. R.K. Katzschmann, A.D. Marchese, D. Rus, Hydraulic autonomous soft robotic fish for 3d swimming, in *Experimental Robotics* (Springer, New York, 2016), pp. 405–420
3. X. Tan, D. Kim, N. Usher, D. Laboy, J. Jackson, A. Kapetanovic, J. Rapai, B. Sabadus, X. Zhou, An autonomous robotic fish for mobile sensing, in *Proceedings of the 2006 IEEE/RSJ International Conference on Intelligent Robots and Systems* (IEEE, New York, 2006), pp. 5424–5429
4. T. Li, G. Li, Y. Liang, T. Cheng, J. Dai, X. Yang, B. Liu, Z. Zeng, Z. Huang, Y. Luo, et al., Fast-moving soft electronic fish. Sci. Adv. **3**(4), e1602045 (2017)
5. M.D. Bartlett, N. Kazem, M.J. Powell-Palm, X. Huang, W. Sun, J.A. Malen, C. Majidi, High thermal conductivity in soft elastomers with elongated liquid metal inclusions. Proc. Natl. Acad. Sci. **114**(9), 2143–2148 (2017)
6. C. Duriez, Control of elastic soft robots based on real-time finite element method, in *Proceedings of the 2013 IEEE International Conference on Robotics and Automation* (IEEE, New York, 2013), pp. 3982–3987
7. G. Runge, A. Raatz, A framework for the automated design and modelling of soft robotic systems. CIRP Annals **66**(1), 9–12 (2017)
8. O. Goury, C. Duriez, Fast, generic, and reliable control and simulation of soft robots using model order reduction. IEEE Trans. Robot. **34**(6), 1565–1576 (2018)
9. J. Chenevier, D. González, J.V. Aguado, F. Chinesta, E. Cueto, Reduced-order modeling of soft robots. PloS One **13**(2), e0192052 (2018)
10. J. Hiller, H. Lipson, Dynamic simulation of soft multimaterial 3d-printed objects. Soft Robot. **1**(1), 88–101 (2014)
11. N. Cheney, J. Bongard, H. Lipson, Evolving soft robots in tight spaces, in *Proceedings of the 2015 annual conference on Genetic and Evolutionary Computation* (ACM, New York, 2015), pp. 935–942
12. X. Zhou, C. Majidi, O.M. O'Reilly, Soft hands: an analysis of some gripping mechanisms in soft robot design. Int. J. Solids Struct. **64**, 155–165 (2015)
13. S. Grazioso, G. Di Gironimo, B. Siciliano, A geometrically exact model for soft continuum robots: the finite element deformation space formulation. Soft Robot. **6**(6), 790–811 (2018)
14. F. Renda, F. Giorgio-Serchi, F. Boyer, C. Laschi, J. Dias, L. Seneviratne, A unified multi-soft-body dynamic model for underwater soft robots. Int. J. Robot. Res. **37**(6), 648–666 (2018)
15. M.K. Jawed, A. Novelia, O.M. O'Reilly, *A primer on the Kinematics of Discrete Elastic Rods*. Springer Briefs in Applied Sciences and Technology (Springer, Berlin, 2018)

16. M. Bergou, M. Wardetzky, S. Robinson, B. Audoly, E. Grinspun, Discrete elastic rods. ACM Trans. Graph. **27**(3), 63 (2008)
17. M. Bergou, B. Audoly, E. Vouga, M. Wardetzky, E. Grinspun, Discrete viscous threads. ACM Trans. Graph. **29**(4), 116 (2010)
18. Z. Shen, J. Huang, W. Chen, H. Bao, Geometrically exact simulation of inextensible ribbon, in *Computer Graphics Forum*, vol. 34 (Wiley, New York, 2015), pp. 145–154
19. D. Baraff, A. Witkin, Large steps in cloth simulation, in *Proceedings of the 25th Annual Conference on Computer Graphics and Interactive Techniques* (ACM, New York, 1998), pp. 43–54
20. E. Grinspun, A.N. Hirani, M. Desbrun, P. Schröder, Discrete shells, in *Proceedings of the 2003 ACM SIGGRAPH/Eurographics Symposium on Computer Animation* (Eurographics Association, Aire-la-Ville, 2003), pp. 62–67
21. J. Panetta, M. Konaković-Luković, F. Isvoranu, E. Bouleau, M. Pauly, X-shells: a new class of deployable beam structures. ACM Trans. Graphics (TOG) **38**(4), 83 (2019)
22. B. Audoly, N. Clauvelin, P.-T. Brun, M. Bergou, E. Grinspun, M. Wardetzky, A discrete geometric approach for simulating the dynamics of thin viscous threads. J. Comput. Phys. **253**, 18–49 (2013)
23. C. Batty, A. Uribe, B. Audoly, E. Grinspun, Discrete viscous sheets, in *ACM Transactions on Graphics*, vol. 31 (ACM, New York, 2012), p. 113
24. N.N. Goldberg, X. Huang, C. Majidi, A. Novelia, O.M. O'Reilly, D.A. Paley, W.L. Scott, On planar discrete elastic rod models for the locomotion of soft robots. Soft Robot. **6**(5), 595–610 (2019)
25. G. Kirchhoff, Ueber das gleichgewicht und die bewegung eines unendlich dünnen elastischen stabes. J. Reine. Angew. Math. **56**, 285–313 (1859)
26. M.K. Jawed, F. Da, J. Joo, E. Grinspun, P.M. Reis, Coiling of elastic rods on rigid substrates. Proc. Natl. Acad. Sci. **111**(41), 14663–14668 (2014)
27. M.K. Jawed, N.K. Khouri, F. Da, E. Grinspun, P.M. Reis, Propulsion and instability of a flexible helical rod rotating in a viscous fluid. Phys. Rev. Lett. **115**(16), 168101 (2015)
28. L. Zhang, J.J. Abbott, L. Dong, K.E. Peyer, B.E. Kratochvil, H. Zhang, C. Bergeles, B.J. Nelson, Characterizing the swimming properties of artificial bacterial flagella. Nano Lett. **9**(10), 3663–3667 (2009)
29. Y. Modarres-Sadeghi, M. Païdoussis, C. Semler, A nonlinear model for an extensible slender flexible cylinder subjected to axial flow. J. Fluids Struct. **21**(5–7), 609–627 (2005)
30. M. Mallick, A. Kumar, N. Tamboli, A. Kulkarni, P. Sati, V. Devi, S. Chandar, Study on drag coefficient for the flow past a cylinder. Int. J. Civil Eng. Res. **5**(4), 301–306 (2014)
31. W. Huang, M.K. Jawed, Newmark-beta method in discrete elastic rods algorithm to avoid energy dissipation. J. Appl. Mech. **86**(8), 084501 (2019)
32. X. Huang, K. Kumar, M.K. Jawed, A. Mohammadi Nasab, Z. Ye, W. Shan, C. Majidi, Highly dynamic shape memory alloy actuator for fast moving soft robots. Adv. Mater. Technol. **4**(4), 1800540 (2019)
33. X. Huang, K. Kumar, M. Jawed, Z. Ye, C. Majidi, Soft electrically actuated quadruped (SEAQ)-integrating a flex circuit board and elastomeric limbs for versatile mobility. IEEE Robot. Autom. Lett. **4**(3), 2415-2422 (2019)
34. X. Huang, K. Kumar, M.K. Jawed, A.M. Nasab, Z. Ye, W. Shan, C. Majidi, Chasing biomimetic locomotion speeds: Creating untethered soft robots with shape memory alloy actuators. Sci. Robot. **3**(25), 7557 (2018)

Chapter 14
Distributed Control of a Planar Discrete Elastic Rod for Eel-Inspired Underwater Locomotion

William L. Scott, Prateek Jaya Prakash, and Derek A. Paley

14.1 Introduction

Underwater robotics is a well-explored research domain. With potential applications in underwater rescue missions, equipment maintenance, and surveillance, the last two decades have seen a significant rise in research in this domain, including in eel-like robots. The long and slender structure of such robots provides superior capabilities for access through narrow openings and within confined areas.

Early work in experimental biology [6, 8] focused on quantifying and modeling the traveling-wave swimming gaits of anguilliform (eel-like) swimmers. These observations have since served as a basis for control design and are still widely used. The serpenoid-curve formulation of Hirose [9] states that the angle between neighboring segments in a modular snake-like (terrestrial) or eel-like (underwater) robot vary sinusoidally in time using the same frequency and amplitude at each segment, with a constant phase difference between neighboring segments. Serpenoid trajectories have been applied to a variety of segmented robot morphologies, including wheeled and sliding designs on land [18] and eel-inspired underwater robots with rigid rotating joints [13, 15]. Though most models of eel-like robots view the structure as a discrete series of rigid links, a geometrically exact continuous model was investigated in [2]. A few examples of soft robotic eel-like robots exist, including a super-thin transparent device based on dielectric elastomer actuators [3], and a fiber-reinforced fluidic actuator design [4]. Passive bending dynamics

W. L. Scott · D. A. Paley (✉)
Department of Aerospace Engineering and Institute for Systems Research, University of Maryland, College Park, MD, USA
e-mail: wlscott@umd.edu; dpaley@umd.edu

P. J. Prakash
Institute for Systems Research, University of Maryland, College Park, MD, USA
e-mail: pjayapra@umd.edu

D. A. Paley, N. M. Wereley (eds.), *Bioinspired Sensing, Actuation, and Control in Underwater Soft Robotic Systems*, https://doi.org/10.1007/978-3-030-50476-2_14

are found to aid locomotion of snakes over obstacle-filled terrain [19]. Segmented robots can be robust to mechanical failure because they are modular and highly redundant. On the other hand, one of the main drawbacks is their poor power efficiency for locomotion [12].

While soft actuators themselves are lightweight and flexible, they often require rigid or heavy electronic or mechanical components to provide power, including batteries, circuit boards, and pumps for fluidic actuators. For example, the soft eel-like robot in [4] must include a large rigid head to house the fluid pressure source. Recent examples of advances in remote power delivery in soft robotics include wireless electric charging for a shape-memory alloy (SMA) caterpillar robot [14] and light-based energy delivery in a tensegrity rolling robot with liquid-crystal elastomer actuators [20]. In addition, chemical reactions have been utilized as a source for fluidic pressure in order to build an octopus robot with entirely flexible materials [21].

This chapter presents an abstract model of an eel-like underwater robot, where the body of the robot is represented as a flexible elastic rod with the capability to control the rate of change of the intrinsic curvature (resting shape under no external forces) along the length of the rod. We take as inspiration a class of fluidic bending soft actuators known as PneuNet (pneumatic network) [17] or bellows-style actuators. In a two-channel bellows actuator, the intrinsic curvature varies as fluid is pumped from one chamber to the other, enabling a controlled bending motion. For an eel-inspired robot design, we envision a number of these actuators connected in series, with individual local control and curvature sensing. The input of our control system is the rate of change of intrinsic curvature, which is analogous to the pumped fluid flow rate in a two-chamber bellows actuator. Fluidic soft actuators with water as the working fluid are a good choice for underwater robotics since they can operate in a high-pressure environment over a range of depths [5].

Although many mobile soft robotics systems have relied on open-loop control to achieve locomotion, a model-based approach utilizes closed-loop control to more closely track a desired reference trajectory and increase performance. The main benefits of feedback control are to mitigate the effects of unknown disturbances, such as from external forces, sensor noise, or model inaccuracies. By monitoring the error between the current and desired state, feedback control dynamically regulates the input signal to track a desired reference trajectory, which in this case is a particular time-varying curvature profile.

This chapter presents several control architectures for eel-inspired traveling-wave locomotion. We consider two communication topologies for the series of sensor-actuator modules, or nodes, that make up the robot: a centralized hub-and-spoke topology where a central pattern generator supplies a clock signal to each node, and a distributed circulant topology where each node communicates with only one neighbor. A decentralized framework requires the nodes to interact only with their neighbors. We also consider two types of local curvature control: tracking the intrinsic curvature to a desired value and using PID feedback control to bring the actual curvature at the node to a desired value.

The contributions of this work are (1) an abstract mathematical model for eel-like locomotion in fluid based on the combination of an existing planar discrete elastic rod (PDER) model and a fluid force model; (2) the addition of novel internal damping force expressions to the PDER model; (3) local curvature controllers for trajectory tracking; and (4) a decentralized controller for traveling-wave motion. Simulations of swimming gaits over a range of traveling-wave amplitudes and frequencies show that curvature feedback increases locomotion speed with only slightly increased energy consumption. Furthermore the decentralized feedback approach achieves similar performance as the centralized approach, able it with reduced communication requirements.

The outline of the chapter is as follows. Section 14.2 presents a PDER mathematical model for eel-like robot locomotion, including conservative forces due to bending and stretching motion, internal damping forces, and external fluid forces. Section 14.3 introduces the serpenoid-curve traveling-wave reference trajectory, and describes the centralized approach to node synchronization. Section 14.4 presents local curvature control laws and illustrates performance with numerical simulations. Section 14.5 introduces the decentralized feedback control for undulatory locomotion without a central pattern generator. Section 14.6 summarizes the results and ongoing and future work in this area.

14.2 Dynamic Model for an Eel-Like Soft Robot

This section presents a mathematical model for locomotion in a flexible underwater eel-like robot. We use the Planar Discrete Elastic Rod (PDER) formulation to describe the robot's state as a discrete series of nodes and edges, with internal forces arising from stretching and bending deformation, including a new derivation of bending damping. Our external fluid force model is based on an existing model for a rigid underwater eel-like robot [13], which includes drag and added mass terms.

14.2.1 Planar Discrete Elastic Rod Dynamics

An ideal elastic rod is a continuum structure with an infinite number of degrees of freedom. In order to characterize the behavior of a rod in a manner that is computationally tractable, we utilize a discrete model. The behavior of the elastic rod is suitably approximated using a planar discrete elastic rod formulation [7], which is a specialization of three-dimensional discrete elastic rod theory [1]. The choice of the number of nodes presents a trade-off between model accuracy and computational requirements [10]. For our purposes, we choose a low number of nodes to enable real-time computation. The effects of model error are mitigated by the use of feedback control.

The rod is discretized into a series of N nodes connected by $N-1$ edges. The position of the kth node is

$$\boldsymbol{x}_k = x_k \boldsymbol{E}_1 + y_k \boldsymbol{E}_2, \tag{14.1}$$

where x_k and y_k are Cartesian coordinates, $\boldsymbol{E}_1$ is a unit vector in the horizontal direction, and $\boldsymbol{E}_2$ is a unit vector in the vertical direction. It follows that $\boldsymbol{E}_3 = \boldsymbol{E}_1 \times \boldsymbol{E}_2$ is directed out of the plane. Edge $\boldsymbol{e}^k$ connects nodes $\boldsymbol{x}_k$ and $\boldsymbol{x}_{k+1}$. (Note, edges are denoted with superscript indexing and nodes are denoted by subscript indexing.) We have [7]

$$\boldsymbol{e}^k = \boldsymbol{x}_{k+1} - \boldsymbol{x}_k, \quad \boldsymbol{t}^k = \frac{\boldsymbol{e}^k}{\left\|\boldsymbol{e}^k\right\|}, \quad \text{and } \boldsymbol{n}^k = \boldsymbol{E}_3 \times \boldsymbol{t}^k. \tag{14.2}$$

The vectors $\boldsymbol{t}^k$ and $\boldsymbol{n}^k$ are the unit tangent and unit normal of the kth edge, respectively. The discrete curvature at a node κ_i is defined as a function of the turning angle ϕ_k, where

$$\begin{aligned} \phi_k &= \cos^{-1}\left(\boldsymbol{t}^{k-1} \cdot \boldsymbol{t}^k\right), \\ \kappa_k &= \frac{2\sin(\phi_k)}{1+\cos(\phi_k)} = 2\tan\left(\frac{\phi_k}{2}\right), \end{aligned} \tag{14.3}$$

for $k = 1,\ldots,N-2$, the internal nodes of the rod.

The intrinsic shape of a PDER is defined by the set of intrinsic lengths $\bar{l}^i$ of each edge and intrinsic curvatures $\bar{\kappa}_i$ at each node. This represents the resting or undeformed shape of the rod. Following Bergou et al. [1], the mass m_k associated with the kth node is the average mass of the edges meeting at this node, i.e.,

$$m_1 = \frac{1}{2} m^1 \tag{14.4}$$

$$m_k = \frac{1}{2}\left(m^k + m^{k-1}\right), \quad k = 2, \ldots, N-1 \tag{14.5}$$

$$m_N = \frac{1}{2} m^{N-1}. \tag{14.6}$$

The mass matrix M takes the form

$$M = \begin{bmatrix} m_1 & 0 & \ldots & 0 & 0 \\ 0 & m_1 & \ldots & 0 & 0 \\ \vdots & \vdots & \ddots & \vdots & \vdots \\ 0 & 0 & \ldots & m_N & 0 \\ 0 & 0 & \ldots & 0 & m_N \end{bmatrix}. \tag{14.7}$$

Conservative elastic forces arising in the deformed rod are derived via scalar potential functions, where bending energy E_b is a function of the deviation between intrinsic and actual node curvatures and stretching energy E_s is a function of the deviation in length. The total elastic energy E_e is [7]

$$E_e = E_s + E_b, \tag{14.8}$$

where

$$E_s = \frac{1}{2}\sum_{k=0}^{N-2} EA^k \left(\frac{\|\boldsymbol{e}^k\|}{\bar{l}^k} - 1\right)^2 \bar{l}^k \tag{14.9}$$

$$E_b = \frac{1}{2}\sum_{j=1}^{N-2} \frac{EI_j}{\bar{l}_j}\left(\kappa_j - \bar{\kappa}_j\right)^2 \tag{14.10}$$

and E is the elastic modulus of the rod, A^k is the cross-sectional area of the kth edge, I_j is the area moment of inertia of the jth node, and $\bar{l}_j = \frac{1}{2}(\|\boldsymbol{e}^j\| + \|\boldsymbol{e}^{j-1}\|)$ is the length of the Voronoi region of the jth node. The area moment of inertia in the bending energy is based on the average area moment of inertia of neighboring edges, $I_j = \frac{1}{2}\left(I^j + I^{j-1}\right)$.

The conservative elastic force acting on node j is $F_{\text{elastic},j} = -\partial E_e/\partial \mathbf{r}_j$. Stretching forces are tangent to the edges, whereas bending forces are normal to the edges. The force due to stretching at a node is a function of its own position, along with the positions of its nearest neighbor to either side, and the force due to bending at a node is a function of its two nearest neighbors to either side. The localized nature of the force computation gives the Jacobian matrix of the system a banded structure, which allows for the use of efficient numerical solvers [1]. Figure 14.1 illustrates the directions that forces act on the nodes due to stretching and bending deformation. For a full derivation of the force expressions, see [7].

To write the dynamics in state-space form, let $\boldsymbol{q} = (x_1, y_1, \ldots, x_N, y_N)^T \in \mathbb{R}^{2N}$ represent the Cartesian coordinates of all of the nodes and $\boldsymbol{v} = (\dot{x}_1, \dot{y}_1, \ldots, \dot{x}_N, \dot{y}_N)^T \in \mathbb{R}^{2N}$ denote the node velocities. Let $\bar{\boldsymbol{\kappa}} \in \mathbb{R}^{N-2}$ be the vector of intrinsic curvatures of the interior nodes and $\bar{\boldsymbol{l}} \in \mathbb{R}^{N-1}$ denote the vector of intrinsic lengths of the edges. The state-space model of the PDER is

$$\begin{aligned}
\dot{\boldsymbol{q}} &= \boldsymbol{v} \\
\dot{\boldsymbol{v}} &= -M^{-1}\left(\frac{\partial E_t}{\partial \boldsymbol{q}} + \boldsymbol{F}_{ext}\right) \\
\dot{\bar{\boldsymbol{\kappa}}} &= \boldsymbol{B}_\kappa \boldsymbol{u}_\kappa \\
\dot{\bar{\boldsymbol{l}}} &= B_l \boldsymbol{u}_l,
\end{aligned} \tag{14.11}$$

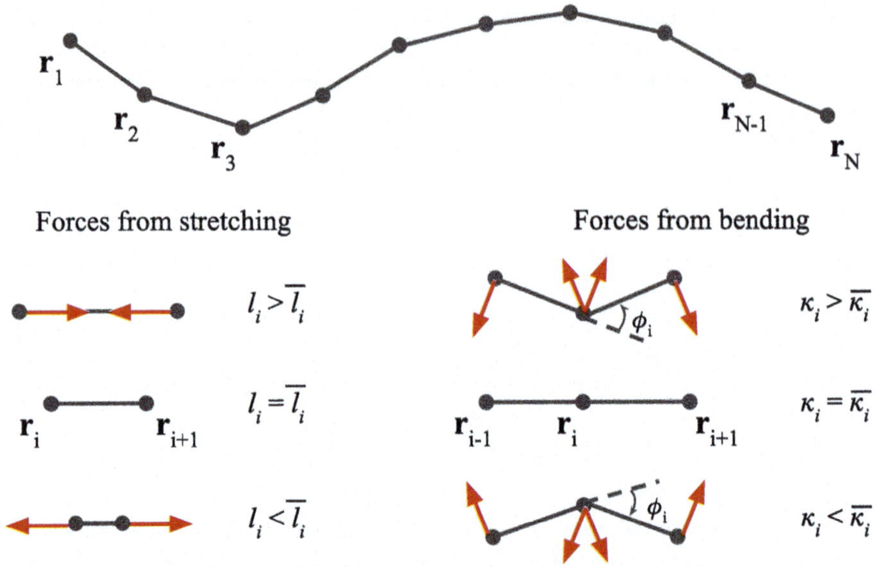

Fig. 14.1 Diagram of the states of the PDER rod (top), and illustration of forces arising due to stretching and bending deformation (bottom)

where $\boldsymbol{F}_{ext}$ is the column matrix containing the components of all external forces acting on the nodes, including the fluid and/or non-conservative damping forces discussed in later sections. Consider two types of input to the system, corresponding separately to the stretching and bending motions. Input $\boldsymbol{u}_\kappa$ controls the rate of change of the intrinsic curvature, with matrix B_κ describing how the inputs are mapped to individual nodes, such that a single input may control curvature at many nodes, and a single node may be affected by multiple inputs. Similarly input $\boldsymbol{u}_l$ controls the rate of change of the intrinsic length of the edges, with matrix B_l describing which input maps to which edge. In the remainder of this chapter, assume B_k to be the $N-2$ by $N-2$ identity matrix, such that each internal node ($i = 2, \ldots, N-1$) has a single unique bending input associated with it. This assumption idealizes the representation of multiple soft rod-like actuators connected in series.

The elastic stretching and bending forces at each node are conservative and arise from the partial derivative with respect to the state $\boldsymbol{q}$ of the corresponding potential energy. (For an additional description of these forces refer to [10].)

14.2.2 Internal Damping Forces

Previous work on PDER used a simple viscous damping model with damping force proportional to individual node velocities [7]. Here we present a new formulation for internal damping forces due to stretching and bending. These forces sum to zero, ensuring they do not contribute to locomotion directly through a net force on the robot.

Suppose that the elastic rod dissipates energy through internal damping at a rate proportional to the rate of change of curvature and the rate of change of edge length. We derive the appropriate node forces through the use of the following dissipation functions for bending $\mathcal{F}_b$ and stretching $\mathcal{F}_s$:

$$\mathcal{F}_b = \frac{d_b}{2}\sum_{i=2}^{N-1} \dot{\kappa}_i^2, \qquad \mathcal{F}_s = \frac{d_s}{2}\sum_{i=1}^{N-1} (\dot{l}^i)^2, \tag{14.12}$$

where d_b and d_a are the bending and stretching damping coefficients, respectively. The force at a particular node j due to bending damping is

$$\boldsymbol{F}_j^{bd} = -\frac{\partial \mathcal{F}_b}{\partial \boldsymbol{v}_j} = -d_b \sum_{i=1}^{N-2} \dot{\kappa}_i \left(\frac{\partial \dot{\kappa}_i}{\partial \boldsymbol{v}_j}\right). \tag{14.13}$$

The rate of change of curvature at a node is

$$\dot{\kappa}_i = \sum_{k=1}^{N-2} \frac{\partial \kappa_i}{\partial \boldsymbol{r}_k}\boldsymbol{v}_k \tag{14.14}$$

$$= \frac{\partial \kappa_i}{\partial \boldsymbol{r}_{i-1}}\boldsymbol{v}_{i-1} + \frac{\partial \kappa_i}{\partial \boldsymbol{r}_i}\boldsymbol{v}_i + \frac{\partial \kappa_i}{\partial \boldsymbol{r}_{i+1}}\boldsymbol{v}_{i+1} \tag{14.15}$$

$$= \frac{\partial \kappa_i}{\partial \phi_i}\left(\frac{\partial \phi_i}{\partial \boldsymbol{r}_{i-1}}\boldsymbol{v}_{i-1} + \frac{\partial \phi_i}{\partial \boldsymbol{r}_i}\boldsymbol{v}_i + \frac{\partial \phi_i}{\partial \boldsymbol{r}_{i+1}}\boldsymbol{v}_{i+1}\right) \tag{14.16}$$

$$-\frac{2}{1+\cos\phi_i}\left(\frac{1}{l^{i-1}}\boldsymbol{n}^{i-1}\cdot\boldsymbol{v}_{i-1} - \left(\frac{1}{l^{i-1}}\boldsymbol{n}^{i-1} + \frac{1}{l^i}\boldsymbol{n}^i\right)\cdot\boldsymbol{v}_i + \frac{1}{l^i}\boldsymbol{n}^i\cdot\boldsymbol{v}_{i+1}\right). \tag{14.17}$$

Then

$$\boldsymbol{F}_1^{bd} = d_b\left(-\frac{2\dot{\kappa}_2}{1+\cos\phi_2}\right)\frac{\boldsymbol{n}^1}{l^1},$$

$$\boldsymbol{F}_2^{bd} = d_b\left(\frac{2\dot{\kappa}_2}{1+\cos\phi_2}\right)\frac{\boldsymbol{n}^1}{l^1} + d_b\left(\frac{2\dot{\kappa}_2}{1+\cos\phi_2} - \frac{2\dot{\kappa}_3}{1+\cos\phi_3}\right)\frac{\boldsymbol{n}^2}{l^2},$$

$$\boldsymbol{F}_j^{bd} = d_b\left(\frac{2\dot{\kappa}_{j-1}}{1+\cos\phi_{j-1}} - \frac{2\dot{\kappa}_j}{1+\cos\phi_j}\right)\frac{\boldsymbol{n}^{i-j}}{l^{j-1}}$$

$$+ d_b\left(\frac{2\dot{\kappa}_j}{1+\cos\phi_j} - \frac{2\dot{\kappa}_{j+1}}{1+\cos\phi_{j+1}}\right)\frac{\boldsymbol{n}^j}{l^j}, \quad j = 3, \ldots, N-2,$$

$$\begin{aligned}
\boldsymbol{F}^{bd}_{N-1} =& d_b \left(- \frac{2\dot{\kappa}_{N-2}}{1+\cos\phi_{N-2}} + \frac{2\dot{\kappa}_{N-1}}{1+\cos\phi_{N-1}} \right) \frac{\boldsymbol{n}^{N-2}}{l^{N-2}} \\
&+ d_b \left(\frac{2\dot{\kappa}_{N-1}}{1+\cos\phi_{N-1}} \right) \frac{\boldsymbol{n}^{N-1}}{l^{N-1}}, \\
\boldsymbol{F}^{bd}_{N} =& d_b \left(- \frac{2\dot{\kappa}_{N-1}}{1+\cos\phi_{N-1}} \right) \frac{\boldsymbol{n}^{N-1}}{l^{N-1}}
\end{aligned} \tag{14.18}$$

The force at node j due to stretching damping is obtained similarly:

$$\boldsymbol{F}^{sd}_j = -\frac{\partial \mathcal{F}_s}{\partial \boldsymbol{v}_j} = -d_s \sum_{i=1}^{N-1} \dot{l}^i \left(\frac{\partial \dot{l}^i}{\partial \boldsymbol{v}_j} \right). \tag{14.19}$$

Noting that the rate of change of an edge length is $\dot{l}^i = \boldsymbol{t}^i \cdot (\boldsymbol{v}_{i+1} - \boldsymbol{v}_i)$, we have

$$\begin{aligned}
\boldsymbol{F}^{sd}_1 &= d_s \boldsymbol{t}^1 \cdot (\boldsymbol{v}_2 - \boldsymbol{v}_1), \\
\boldsymbol{F}^{sd}_i &= d_s \boldsymbol{t}^i \cdot (\boldsymbol{v}_{i+1} - \boldsymbol{v}_i) - d_s \boldsymbol{t}^{i-1} \cdot (\boldsymbol{v}_i - \boldsymbol{v}_{i-1}), \quad i = 2, \ldots, N-1, \\
\boldsymbol{F}^{sd}_N &= -d_s \boldsymbol{t}^{N-1} \cdot (\boldsymbol{v}_N - \boldsymbol{v}_{N-1}).
\end{aligned} \tag{14.20}$$

Like conservative elastic forces, damping stretching forces are tangent to edges, whereas damping bending forces are normal to edges. We verify that the damping forces act internally to the rod by noting that the sum of the damping forces is zero.

14.2.3 Hydrodynamic Forces

This section provides the framework for modeling hydrodynamic forces acting on a multi-link swimmer. Hydrodynamic forces induced by the motion of a body in an underwater environment are complex and nonlinear. For control design purposes, a simplified hydrodynamic model in closed form is preferred. Here, we use the model applied by Kelasidi et al. [11] to a rigid eel-like robot to describe hydrodynamic forces in terms of the instantaneous velocities and accelerations of the edges in the PDER model. We choose our system parameters (see Table 14.1) to be similar to a rigid eel-like swimming robot, which operates with Reynolds number in the range 10^4–10^5 [11].

The proposed hydrodynamic model is based on Morison's equations [16] and models the forces between the fluid and the cylindrical links of an underwater eel-like robot. The assumptions underlying the model are as follows [11]:

1. The fluid is viscid, incompressible, and irrotational in the inertial frame.
2. The robot is neutrally buoyant, such that it operates in a horizontal plane within a volume of water.

Table 14.1 Simulation parameter values

Name	Parameter	Value	Units
Number of nodes	N	10	
Number of bending inputs	N_u	8	
Time step	h	0.01	[s]
Edge length	$\bar{l}$	0.18	[m]
Edge mass	m	0.8	[kg]
Rod height	a	0.055	[m]
Rod width	b	0.05	[m]
Elastic modulus	E	1.875×10^6	[Pa]
Damping coefficient	d	0	
Water density	ρ	1000	$\left[\frac{\text{kg}}{\text{m}^3}\right]$
Drag coefficient	C_f	0.3	
Drag coefficient	C_D	1.75	
Added mass coefficient	C_A	1.5	
Inter-node phase difference	δ	$\pi/4$	[rad]
Proportional gain	K_p	10	
Derivative gain	K_d	1	
Integral gain	K_i	5	

No units are provided for non-dimensional quantities

3. The fluid is at rest in the inertial frame, such that there are no time-varying currents.

Fluid forces are calculated separately for each cylindrical link in the robot model [11], which here we associate with an edge in the PDER model. Let $\boldsymbol{v}^i = (\boldsymbol{v}_i + \boldsymbol{v}_{i+1})/2$ represent the velocity of edge i in the inertial frame. The scalar tangential and normal components of the ith edge velocity are $v_t^i = \boldsymbol{v}^i \cdot \boldsymbol{t}^i$ and $v_n^i = \boldsymbol{v}^i \cdot \boldsymbol{n}^i$, respectively.

Fluid drag forces are calculated using separate drag coefficients, denoted by c_t for tangential and c_n for normal directions:

$$c_t = \frac{1}{4}\rho\pi C_f(a+b)l, \tag{14.21}$$

$$c_n = \rho C_D a l, \tag{14.22}$$

where a is the half height and b is the half-width of the elliptical cross section of the robot and l is the length of the edge. Drag coefficients $C_f = 0.3$ and $C_d = 1.75$ come from studies of drag on a cylinder [11].

The fluid drag force acting on an edge $\boldsymbol{F}_D^i$ contains a linear term and a nonlinear quadratic term in each of the tangential and normal directions, i.e.,

$$\boldsymbol{F}_D^i = -c_t\left(v_t^i + v_t^i|v_t^i|\right)\boldsymbol{t}^i - c_n\left(v_n^i + v_n^i|v_n^i|\right)\boldsymbol{n}^i. \tag{14.23}$$

Additionally, the fluid force model includes an added mass term acting only in the normal direction,

$$\boldsymbol{F}^i_A = -\mu_n a^i_n = -\mu_n \dot{v}^i_n, \tag{14.24}$$

with coefficient $\mu_n = \rho\pi C_A a^2 l$, and added mass coefficient $C_A = 1.5$.

For the PDER model, assume that the fluid forces acting on an edge are evenly distributed to the two nodes associated with the edge, such that the fluid force at each node is

$$\begin{aligned}
\boldsymbol{F}^{\text{fluid}}_1 &= \frac{1}{2}\left(\boldsymbol{F}^1_D + \boldsymbol{F}^1_A\right),\\
\boldsymbol{F}^{\text{fluid}}_i &= \frac{1}{2}\left(\boldsymbol{F}^{i-1}_D + \boldsymbol{F}^{i-1}_A + \boldsymbol{F}^i_D + \boldsymbol{F}^i_A\right), \quad i = 2, \ldots, N-1,\\
\boldsymbol{F}^{\text{fluid}}_N &= \frac{1}{2}\left(\boldsymbol{F}^{N-1}_D + \boldsymbol{F}^{N-1}_A\right).
\end{aligned} \tag{14.25}$$

Although the fluid force model is idealized, the difference in drag coefficients for tangential and normal directions is sufficient to induce locomotion under a traveling-wave gait, thus serving as a useful starting point for feedback control design in an underwater soft robot.

14.2.4 Numerical Integration

To simulate the PDER model of the eel robot system, we utilize an implicit backwards Euler integration scheme as described in [1] and [7]. At each time step t_k, numerically solve the following equations of motion for the velocity at the next time step $\mathbf{v}(t_{k+1})$:

$$\begin{aligned}
0 = &-\boldsymbol{M}\mathbf{a}(t_{k+1}) + F_{\text{elastic}}(\mathbf{q}(t_{k+1}), \bar{\boldsymbol{\kappa}}(t_k), \bar{\boldsymbol{l}}(t_k)) + F_{\text{damp}}(\mathbf{q}(t_{k+1}), \mathbf{v}(t_{k+1}))\\
&+ F_{\text{fluid}}(\mathbf{q}(t_{k+1}), \mathbf{v}(t_{k+1}), \mathbf{a}(t_{k+1})),
\end{aligned} \tag{14.26}$$

where

$$\mathbf{q}(t_{k+1}) = \mathbf{q}(t_k) + h\mathbf{v}(t_{k+1}), \tag{14.27}$$

$$\mathbf{a}(t_{k+1}) = (\mathbf{v}(t_{k+1}) - \mathbf{v}(t_k))/h \tag{14.28}$$

are the updated node positions and accelerations, respectively. The Jacobian matrix of (14.26) has a banded structure, due to the fact that forces at a node depend on the states of nodes only one or two steps away in either direction, which enables the use of efficient numerical solution algorithms for fast computation [1].

The intrinsic curvature and length are updated outside of the implicit integration step in an alternating manner, with inputs $\boldsymbol{u}$ based on the control inputs:

$$\bar{\boldsymbol{\kappa}}(t_{k+1}) = \bar{\boldsymbol{\kappa}}(t_k) + hB_\kappa \boldsymbol{u}_\kappa(t_k)$$
$$\bar{\boldsymbol{l}}(t_{k+1}) = \bar{\boldsymbol{l}}(t_k) + hB_l \boldsymbol{u}_l(t_k). \quad (14.29)$$

14.3 Traveling-Wave Reference Trajectory

The serpenoid curve has been proposed to describe the shape of a snake or snake-like segmented robot in motion [9, 18], such that the angle between neighboring segments varies sinusoidally in time and in space (along the length of the body). The equation for the desired turning angle at node i includes amplitude α and turning bias γ parameters, along with a phase variable ψ_i associated with the node:

$$\phi_{i,des}(t) = \alpha \sin(\psi_i(t)) + \gamma. \quad (14.30)$$

Traveling-wave motion occurs when the phase angles increase linearly with time with frequency ω, while keeping the phase difference between neighboring nodes a constant value δ, i.e.,

$$\dot{\psi}_i = \omega, \quad (14.31)$$
$$\psi_i = \psi_{i-1} + \delta, \quad (14.32)$$

at every internal node $i = 2, \ldots, N-1$. Examples of serpenoid curves for a range of values of δ and α are illustrated in Fig. 14.2. For a given value of δ, there is a maximum amplitude $\alpha_{\max}$ above which the curve will self-intersect.

To achieve a traveling-wave serpenoid motion given the phases ψ_i, the desired curvature and rate of change of curvature at each node are

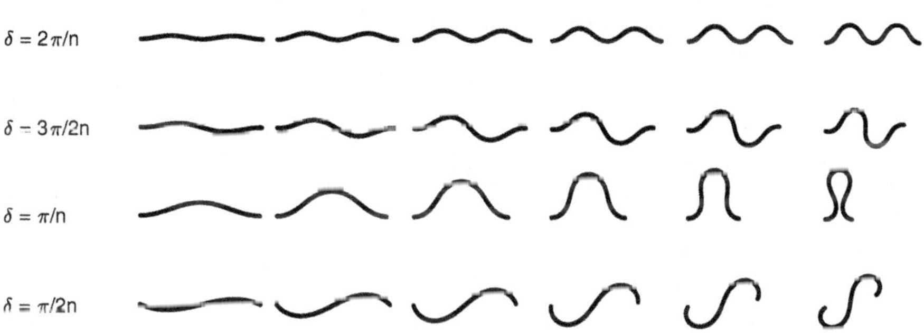

Fig. 14.2 Illustrations of serpenoid curves described by (14.30) with $N = 100$ nodes. Each row uses a different value of the phase offset δ, with amplitude α increasing from left to right

$$\kappa_{i,des} = 2\tan\left(\frac{\alpha \sin \psi_i + \gamma}{2}\right),$$

$$\dot{\kappa}_{i,des} = \alpha\dot{\psi}_i \cos\psi_i \sec^2\left(\frac{\alpha \sin \psi_i + \gamma}{2}\right). \tag{14.33}$$

We consider several types of system architectures for determining the value of $\kappa_{i,des}(t)$. With a centralized approach, a single pattern generator provides phase information to each node. Distributed approaches can achieve similar performance with fewer requirements on information transfer between nodes. We present a distributed controller wherein each node only needs to measure the curvature of its direct neighbor in order to estimate its proper phase over time. With that method, either the first node acts as the central pattern generator, or the first node is virtually connected to the last node to form a circulant topology to be fully decentralized.

The traveling-wave phase constraints (14.31) can be achieved through a clock signal t, such that

$$\psi_i(t) = \omega t + \delta i + \psi_0, \tag{14.34}$$

for some arbitrary initial phase ψ_0. Note, the centralized approach corresponds to a hub and spoke communication architecture, where a central pattern generator sends the timing signal to each of the nodes. The next section describes curvature tracking control strategies to achieve the traveling-wave gait using feedback.

14.4 Curvature Feedback Control

A feedback controller ensures that the actual curvature κ_i tracks the desired curvature $\kappa_{i,des}$ over time, given that the nodes have coordinated their phases based on a controller from Sect. 14.3. Assume that each node is able to measure its local curvature, rate of change of local curvature, and intrinsic curvature, and optionally that nodes can share this information with one or more neighbors. Recall that the input is the rate of change of intrinsic curvature so feedback is required to track the desired curvature.

14.4.1 Intrinsic Curvature Tracking Control

First consider a control law that does not require measurement of the actual curvature of the robot. Instead it uses a measurement of the intrinsic curvature, analogous to the fluid pressure in a fluidic bending actuator, in order to track the intrinsic curvature $\bar{\kappa}_i$ to the desired curvature $\kappa_{i,des}$ locally at each node.

Consider a proportional control law with positive gain K_p, i.e.,

$$u_i = \dot{\kappa}_{i,des} - K_p(\bar{\kappa}_i - \kappa_{i,des}). \tag{14.35}$$

This control law brings the intrinsic curvature to the desired curvature at each node. It does not take into account the actual curvature of the rod. When ψ_i and $\dot{\psi}_i$ are known, the controller (14.35) drives the error $e_{\kappa_i} = \bar{\kappa}_i - \kappa_{i,des}$ to zero exponentially quickly, as

$$\dot{e}_{\bar{\kappa}_i} = \dot{\bar{\kappa}}_i - \dot{\kappa}_{i,des} = u_i - \dot{\kappa}_{i,des} = -C(\bar{\kappa}_i - \kappa_{i,des}) = -Ce_{\bar{\kappa}_i}. \tag{14.36}$$

The control design (14.35) has acceptable performance when the rod has higher bending stiffness. However as the bending stiffness decreases, higher curvature differences $\Delta\kappa_i = \kappa_i - \bar{\kappa}_i$ are necessary to generate the same internal moment at a node, which cannot be achieved with the intrinsic curvature controller.

14.4.2 Actual Curvature Feedback Control

This section introduces a Proportional-Integral-Derivative (PID) control law that uses a measurement of the actual curvature κ_i at each node, as opposed to the controller of the previous section, which measured only the intrinsic curvatures $\bar{\kappa}_i$. PID controllers are widely used and simple to implement due to the fact that they do not require a model of the system being controlled. The expression for the PID control input at each node is

$$u_i = \dot{\kappa}_{i,des}(t) + K_p e_i(t) + K_d \dot{e}_i(t) + K_i \int_0^t e_i(\tau)d\tau, \tag{14.37}$$

where $e_i = \kappa_{i,des} - \kappa_i$ is the error in actual curvature at node i. For numerical simulations, we choose $K_p = 10$, $K_d = 1$, and $K_i = 5$.

The individual gains can be tuned to achieve the desired performance. Proportional gain K_p increases the speed of controller, but can lead to instability if taken too high. Derivative gain K_d serves to add damping to the system, which can remove unwanted oscillations at the cost of slowing down the dynamic response. The integral term with gain K_i serves to remove steady-state errors.

To compare performance of the intrinsic and actual curvature feedback controllers, simulations were run using a range of values of amplitude α and frequency ω, using $\delta = 2\pi/(N-2)$, which gives a single full wave along the body. Parameter values for the simulations are summarized in Table 14.1. For the physical dimensions of the robot and corresponding fluid force coefficients, we use values from the rigid eel-like robot described in [12]. The elastic modulus $E = 1.875$ MPa is chosen to match that of Smoothsil 950 silicone rubber (Smooth-On, Inc.), which is commonly used to make bellows-style fluidic actuators [5]. We use $N = 10$ nodes

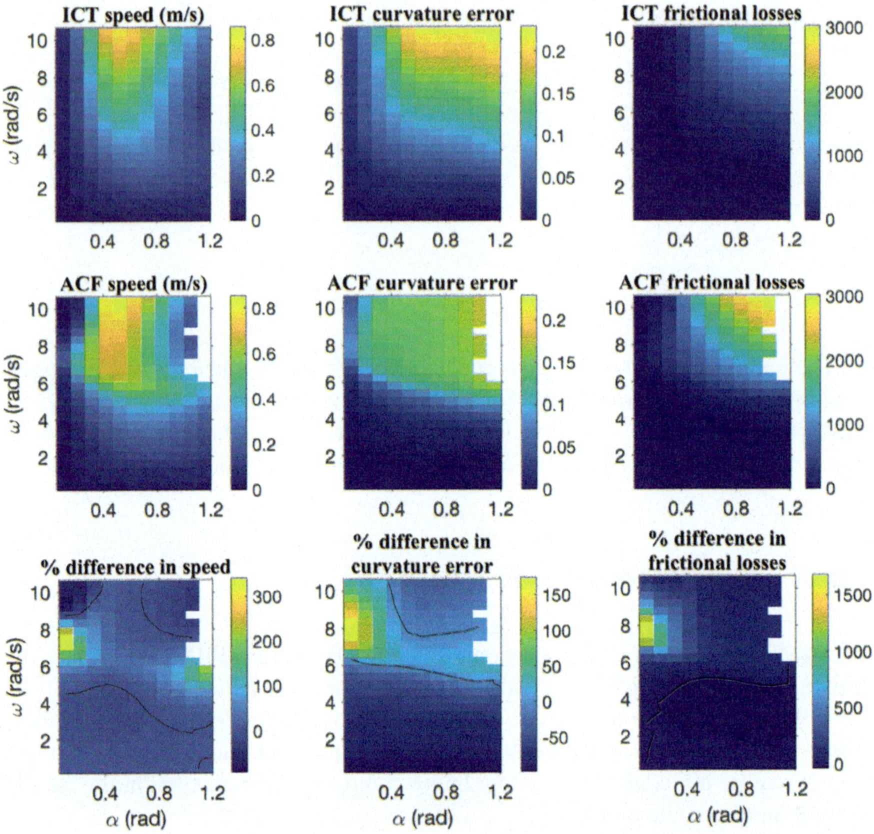

Fig. 14.3 Comparison of intrinsic curvature tracking (ICT) and actual curvature feedback (ACF) controllers over a range of frequencies ω and amplitudes α. Average speed (m/s), curvature error (unitless), and frictional losses (W) are calculated over a single cycle. Percent change is calculated with respect to ICT. Black lines in the bottom row of figures show the contours separating positive and negative values. White areas indicate simulations where self-intersection of the robot occurred

in the PDER model, with eight intrinsic curvature rate inputs each associated with a single interior node.

Figure 14.3 illustrates the performance of the two controllers, in terms of average speed, deviation from desired curvature, and energy loss due to fluid drag forces. In general, for both controllers, the speed increases with increasing frequency and, for a given frequency, speed is maximized at a certain optimal amplitude with decreased speeds at higher and lower amplitudes. Frictional losses due to fluid drag forces increase monotonically with both amplitude and frequency. Figure 14.4 shows snapshots of the shape of the simulated robot over a single gait cycle, using two different amplitudes at the same frequency with actual curvature feedback control. Although the shapes traced out by the nodes are quite different in the two cases, the average speeds are within 2% (12.8 and 12.9 cm/s, respectively), whereas frictional losses are more than doubled at higher amplitude from 4.65 to 10.33 W.

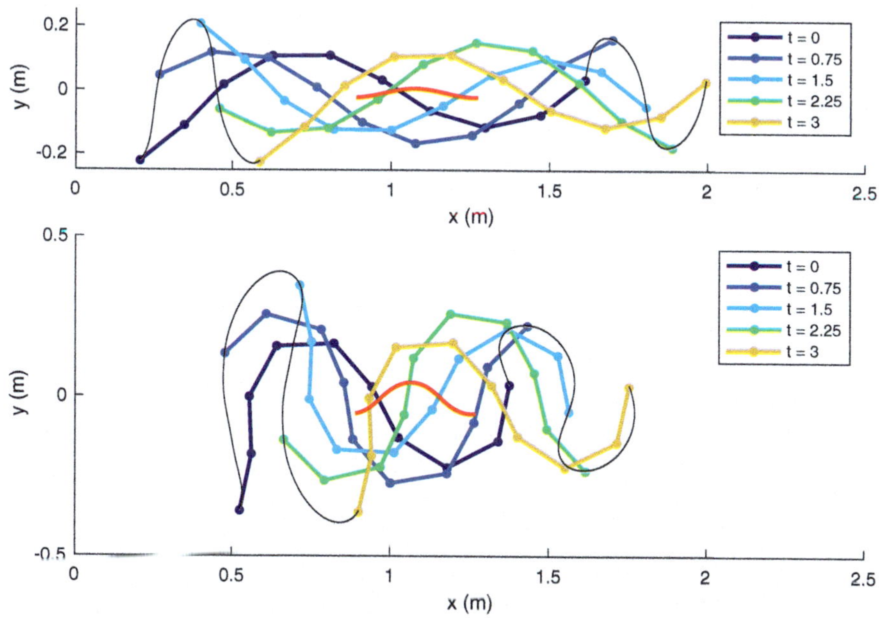

Fig. 14.4 Snapshots of the simulated robot over a single cycle using the actual curvature feedback law, colored by time. Top: $\alpha = \pi/6$, bottom: $\alpha = \pi/3$; both with $\omega = 2\pi/3$, and $\delta = \pi/4$. Black lines trace the trajectories of the first and last nodes, and the red line represents the motion of the center of mass

14.5 Decentralized Control Strategy for Undulatory Locomotion

This section presents a decentralized controller for traveling-wave motion where each node measures the curvature of its neighbor to estimate the current phase rather than relying on a central pattern generator. The estimated phase at each node is used to calculate the desired curvature and control input using the PID controller (14.37) from the previous section.

Suppose that each node can measure the curvature and rate of change of curvature of itself and its neighbors. If we assume that the curvature varies sinusoidally, we can estimate the current phase at node i with

$$\hat{\psi}_i = \operatorname{atan2}\left(\phi_i,\ \dot{\phi}_i/\omega\right), \tag{14.38}$$

where turning angles and rates are calculated from the measured curvature and curvature rate:

$$\psi_i = 2\tan^{-1}(\kappa_i/2)$$

$$\dot{\phi}_i = \frac{4\dot{\kappa}_i}{\kappa_i^2 + 4}. \tag{14.39}$$

We connect the nodes in a circulant topology, such that node i chooses its phase as an offset of the phase of its neighbor with index $i+1$, except for the last node $N-1$, which looks to node 2 to close the loop:

$$\psi_i = \hat{\psi}_{i+1} - \delta, \; i = 2, \ldots, N-2,$$
$$\psi_{N-1} = \hat{\psi}_2. \tag{14.40}$$

With this topology, information flows backward along the eel robot in the same direction as the traveling-wave motion. This architecture has the advantage of not requiring an additional state in the system to represent the phase. Stability may be further improved through additional measurement filtering, which is the subject of ongoing work.

Figure 14.5 demonstrates that the performance of the decentralized phase control nearly matches that of the centralized phase. For both simulations, the robot starts at rest at $t = 0$ with curvatures equal to the desired curvature under the centralized phase control, with gait parameters $\alpha = \pi/6$, $\omega = 2\pi/3$, and $\delta = \pi/4$ and PID gains $K_p = 10$, $K_d = 2$, $K_i = 1$. Observe that for both cases, transient errors die out within 1 s, though errors in the decentralized case remain slightly higher in steady state. Larger oscillations exist in the decentralized case, with control becoming saturated initially at the prescribed maximum value of $|u_i| \leq 1.5\alpha\omega$.

14.6 Conclusion

This chapter presents an abstract model of an eel-like underwater robot, where the body of the robot is represented as a flexible elastic rod with the capability to control the rate of change of the intrinsic curvature along the length of the rod. We introduce curvature feedback controllers acting at each node of the PDER to achieve traveling-wave motion in the robot, using either a central pattern generator or distributed control approach to determine the desired reference curvature. Numerical simulations using an idealized fluid force model illustrate the effects of traveling-wave frequency and amplitude on the robot locomotion performance. The decentralized feedback control achieves similar performance as the centralized control, with reduced communication requirements.

We continue to study this model, examining the effects of disturbances on the performance either from sensor noise, actuator noise, or unknown external forces. In the distributed controller, small changes to node positions can cause large variation in the estimated phase, so a filtering approach may be necessary to remove unwanted oscillations. We also look to optimal control theory to determine the inputs necessary to track the desired reference trajectory exactly, which could be

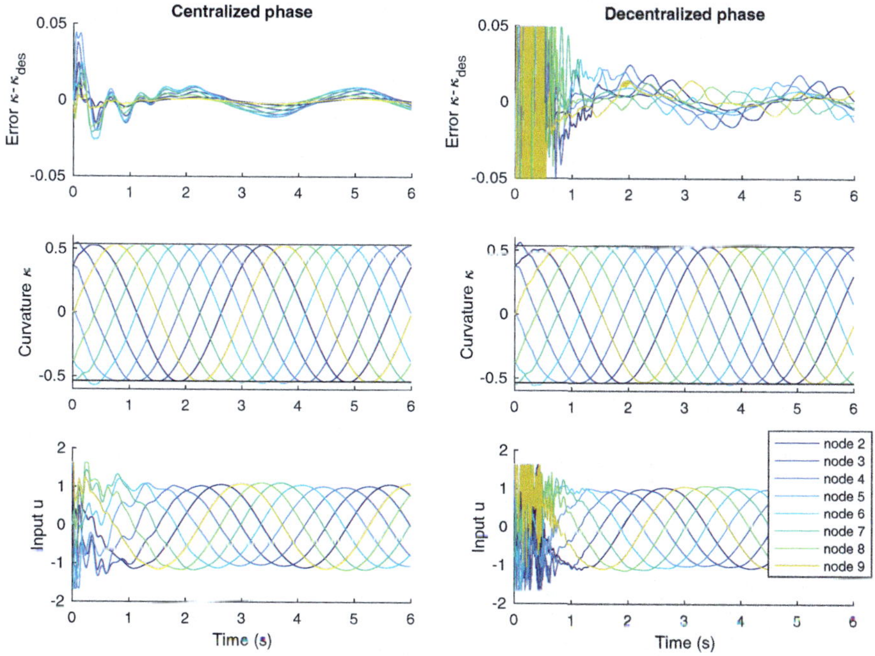

Fig. 14.5 Comparison of centralized (left) and decentralized phase control (right) under the actual curvature feedback law, colored by node. In both cases $\alpha = \pi/6$, $\omega = 2\pi/3$, and $\delta = \pi/4$ with PID gains $K_p = 10$, $K_d = 2$, $K_i = 1$. The top plots show the difference between actual curvature and desired curvature (as a function of phase) over time at each node. Middle plots show actual curvature, with black lines representing the nominal maximum value of $2\tan(\alpha/2)$. Bottom plots show the input u at each node, with saturation applied such that $|u_i| \leq 1.5\alpha\omega$

precomputed and used as a feedforward term in a new feedback controller design. The current work focuses on bending deformations relevant to snake- and eel-inspired motion, but soft actuators exist that can change their intrinsic length as well. Future work will examine whether simultaneous curvature and length control can offer advantages in traveling-wave locomotion.

The model described in this chapter places an independent control input at each node. In a hardware robot implementation, size and weight constraints may limit the number of inputs, in which case a single input may act on multiple nodes in series. Future work will study the effects of the number of nodes and number of inputs on locomotion performance and shape tracking accuracy.

Although this chapter considers an eel-like robot design, the PDER model with fluid forces may be applied to other robot morphologies that incorporate soft bending actuators, such as a fish robot with multiple soft bending fins. Additionally, the fluid force model presented here can be replaced with a frictional ground contact model to describe the motion of a terrestrial snake like or worm-inspired robot.

Fast numerical simulations of idealized soft robot models enable researchers to test creative new designs before committing to building hardware prototypes. On the other hand, experiments in hardware reveal limitations in models and control design, spurring new innovation. As soft actuator and soft shape sensor technology matures, we envision feedback control as an integral aspect of soft robotic development enabling increased performance and robustness for complex tasks in uncertain environments.

Acknowledgement This work was supported in part by ONR Grant No. N000141712063.

References

1. M. Bergou, M. Wardetzky, S. Robinson, B. Audoly, E. Grinspun, Discrete elastic rods. ACM Trans. Graph. **27**(3), 63 (2008). https://doi.org/10.1145/1360612.1360662
2. F. Boyer, M. Porez, W. Khalil, Macro-continuous computed torque algorithm for a three-dimensional eel-like robot. IEEE Trans. Robot. **22**(4), 763–775 (2006). https://doi.org/10.1109/TRO.2006.875492
3. C. Christianson, N.N. Goldberg, D.D. Deheyn, S. Cai, M.T. Tolley, Translucent soft robots driven by frameless fluid electrode dielectric elastomer actuators. Sci. Robot. **3**(17), eaat1893 (2018). https://doi.org/10.1126/scirobotics.aat1893
4. H. Feng, Y. Sun, P.A. Todd, H.P. Lee, Body wave generation for anguilliform locomotion using a fiber-reinforced soft fluidic elastomer actuator array toward the development of the eel-inspired underwater soft robot. Soft Robot. **7**(2), 233–250 (2020). https://doi.org/10.1089/soro.2019.0054
5. K.C. Galloway, K.P. Becker, B. Phillips, J. Kirby, S. Licht, D. Tchernov, R.J. Wood, D.F. Gruber, Soft robotic grippers for biological sampling on deep reefs. Soft Robot. **3**(1), 23–33 (2016). https://doi.org/10.1089/soro.2015.0019
6. G.B. Gillis, Undulatory locomotion in elongate aquatic vertebrates: anguilliform swimming since Sir James Gray. Am. Zool. **36**(6), 656–665 (1996). https://doi.org/10.1093/icb/36.6.656
7. N.N. Goldberg, X. Huang, C. Majidi, A. Novelia, O.M. O'Reilly, D.A. Paley, W.L. Scott, On planar discrete elastic rod models for the locomotion of soft robots. Soft Robot. **6**(5), 595–610 (2019). https://doi.org/10.1089/soro.2018.0104
8. J. Gray, Studies in animal locomotion: I. the movement of fish with special reference to the eel. J. Exp. Biol. **10**(1), 88–104 (1933). https://jeb.biologists.org/content/10/1/88
9. S. Hirose, *Biologically-Inspired Robots: Snake-Like Locomotors and Manipulators* (Oxford University Press, Oxford, 1993)
10. M.K. Jawed, A. Novelia, O.M. O'Reilly, *A Primer on the Kinematics of Discrete Elastic Rods* (Springer, Berlin, 2018)
11. E. Kelasidi, K.Y. Pettersen, J.T. Gravdahl, P. Liljebäck, Modeling of underwater snake robots, in *IEEE International Conference on Robotics and Automation (ICRA)* (2014), pp. 4540–4547. https://doi.org/10.1109/ICRA.2014.6907522
12. E. Kelasidi, P. Liljebäck, K.Y. Pettersen, J.T. Gravdahl, Experimental investigation of efficient locomotion of underwater snake robots for lateral undulation and eel-like motion patterns. Robot. Biomimetics **2**(1), 8 (2015). https://doi.org/10.1186/s40638-015-0029-4
13. E. Kelasidi, P. Liljebäck, K.Y. Pettersen, J.T. Gravdahl, Integral line-of-sight guidance for path following control of underwater snake robots: theory and experiments. IEEE Trans. Robot. **33**(3), 610–628 (2017). https://doi.org/10.1109/TRO.2017.2651119
14. C. McCaffrey, T. Umedachi, W. Jiang, T. Sasatani, Y. Narusue, R. Niiyama, Y. Kawahara, Continuum robotic caterpillar with wirelessly powered shape memory alloy actuators. Soft Robot. (2020, Available online ahead of print). https://doi.org/10.1089/soro.2019.0090

15. K.A. McIsaac, J.P. Ostrowski, Motion planning for anguilliform locomotion. IEEE Trans. Robot. Autom. **19**(4), 637–652 (2003). https://doi.org/10.1109/TRA.2003.814495
16. J. Morison, J. Johnson, S. Schaaf, et al., The force exerted by surface waves on piles. J. Pet. Technol. **2**(05), 149–154 (1950). https://doi.org/10.2118/950149-G
17. B. Mosadegh, P. Polygerinos, C. Keplinger, S. Wennstedt, R.F. Shepherd, U. Gupta, J. Shim, K. Bertoldi, C.J. Walsh, G.M. Whitesides, Pneumatic networks for soft robotics that actuate rapidly. Adv. Funct. Mater. **24**(15), 2163–2170 (2014). https://doi.org/10.1002/adfm.201303288
18. M. Sato, M. Fukaya, T. Iwasaki, Serpentine locomotion with robotic snakes. IEEE Contr. Syst. Mag. **22**(1), 64–81 (2002). https://doi.org/10.1109/37.980248
19. P.E. Schiebel, J.M. Rieser, A.M. Hubbard, L. Chen, D.Z. Rocklin, D.I. Goldman, Mechanical diffraction reveals the role of passive dynamics in a slithering snake. Proc. Natl. Acad. Sci. **116**(11), 4798–4803 (2019). https://doi.org/10.1073/pnas.1808675116
20. Z. Wang, K. Li, Q. He, S. Cai, A light-powered ultralight tensegrity robot with high deformability and load capacity. Adv. Mater. **31**(7), 1806849 (2019). https://doi.org/10.1002/adma.201806849
21. M. Wehner, R.L. Truby, D.J. Fitzgerald, B. Mosadegh, G.M. Whitesides, J.A. Lewis, R.J. Wood, An integrated design and fabrication strategy for entirely soft, autonomous robots. Nature **536**(7617), 451–455 (2016). https://doi.org/10.1038/nature19100

Chapter 15
Bioinspired Neural-Based Control of Flexible Fish-Like Propulsors

Gabriel N. Carryon and James L. Tangorra

15.1 Introduction

Biological evidence suggests that animal locomotion is the result of a dynamic interaction of neural control, sensing, and mechanics [1–3]. Neural circuits, known as *central pattern generators* (CPGs), govern the periodic motion of limbs during tasks like running and swimming [4, 5]. During locomotion, sensory feedback from the limbs adapts the drive signals of the CPG so that movement is appropriate for a given task [3]. Sensory feedback allows an animal to exploit the dynamics of the interaction between their limbs and the environment, making movement efficient [6]. Due to the capabilities of animal locomotion, engineers have looked to biology to control the motion of soft robots.

This chapter focuses on CPG-based control methods applied to flexible fish-like propulsors, which can be considered models of simple soft robotic systems that are controlled for motion and force (Fig. 15.1). Flexible fish-like foils are increasingly used as propulsors for underwater vehicles, and understanding how to control the motions of fins is fundamental to their use as propulsors [7]. Conventional model-based control schemes (i.e., pre-recorded trajectories, sine-based generators, feedback linearization) are not always appropriate to control the motions of flexible fins [8]. These methods require the exact design of a movement trajectory that can be difficult to achieve given the high numbers of degrees of freedom of a flexible fin and the nonlinear interaction of the fin with the fluidic environment. Additionally, these methods do not take advantage of the natural dynamics of the interaction between the fin and the fluidic environment. CPG-based control methods are well suited to address the shortfalls of conventional control methods of flexible foils.

G. N. Carryon · J. L. Tangorra (✉)
Drexel University, Philadelphia, PA, USA
e-mail: jlt66@drexel.edu

D. A. Paley, N. M. Wereley (eds.), *Bioinspired Sensing, Actuation, and Control in Underwater Soft Robotic Systems*, https://doi.org/10.1007/978-3-030-50476-2_15

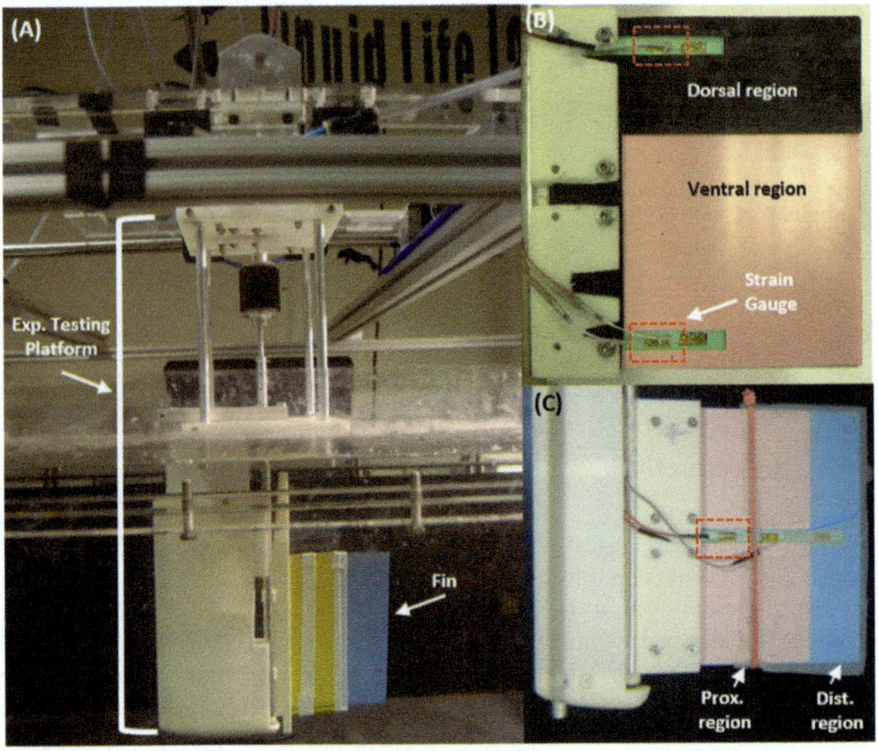

Fig. 15.1 (**a**) The experimental testing platform with fin attached. Two different types of fins were used in this study—*spanwise fins* (**b**) and *chordwise fins* (**c**). Both types of fins have spatially varying mechanical properties. The fins were designed with two distinct regions having different flexural rigidity values. The spanwise fins have a dorsal region and a ventral region and the chordwise fins have a proximal region and a distal region. For the spanwise fins the EI of the dorsal region is larger than the EI of the ventral region and for the chordwise fins the EI of the proximal region is larger than the EI of the distal region. The red dashed rectangles in (**b**) and (**c**) indicate where the strain gauges are attached to each of the fins. The fins have additional strain gauges; however they were not used to capture sensory information in this study

Despite knowledge of the benefits of CPG-based control, many questions remain about applying CPG-based control to drive the motions of soft robotic systems. In particular, there are questions regarding the modality of information that should be collected, such as bending, position, or force, and the way in which this information should be incorporated into the neural-based controller to produce desired performance. Addressing these questions could help the broader field of nonlinear oscillator-based control of robotic systems [9].

15.2 Mechanism of the Coupled System

CPG-based control can be implemented in both open-loop and closed-loop configurations. In either configuration, and in most studies of CPG-based control applied to robotic systems, the output of the CPG is used as a reference trajectory that drives the motion of a robotic limb. In the absence of sensory feedback (i.e., when $h = 0$), the CPG produces an endogenous frequency (f_{OSC}) and an endogenous amplitude (A_{OSC}) that depend upon the internal settings of the CPG. Under these conditions the fin oscillates with a frequency and amplitude equal to f_{OSC} and A_{OSC}, respectively; this configuration is referred to as an *uncoupled system*. A *coupled system* is formed when sensory feedback is applied to the CPG (i.e., when $h > 0$) and the output of the CPG is used to drive the motion of the fin (Fig. 15.2). The frequency and amplitude of the coupled system are different compared to f_{OSC} and A_{OSC}, and depend upon a variety of factors including the settings of the oscillator, the settings of the sensory feedback architecture, and the mechanical properties of the fin. The *entrainment state* refers to the frequency and amplitude of the coupled system. The frequency and amplitude of the coupled system refers to the frequency and amplitude of the base, or leading edge, of the fin. Understanding how sensory feedback affects the entrainment state is necessary in order to effectively apply CPG-based control to a robotic limb.

15.3 Experimental Apparatus and Methods

An experimental testing platform was developed in order to carry out experimentation (Fig. 15.1a). A flexible fin with embedded sensing was attached to the base of a

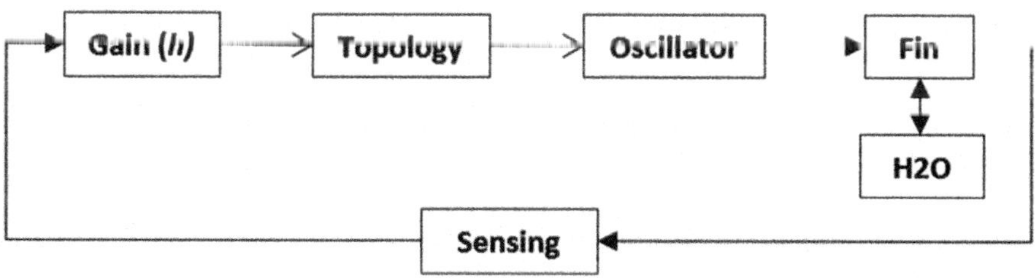

Fig. 15.2 The coupled system formed when a CPG (oscillator) is connected to a mechanical system (fin). The value of gain (h) can be set to vary the magnitude of the sensory feedback signal. The topology block represents the way in which the sensory feedback signal is applied to the oscillator and is explained in more detail in Sect. 7. The oscillator block represents the model of the CPG. In this study, the Matsuoka neural oscillator was used as the CPG [24]. The fin block represents the dynamics and the H_2O block represents the fluid-structure interaction that occurs as the fin flaps in the water. The sensing block represents the sensing modality that is acquired from the fin and applied to the oscillator

pitching rod and driven at a frequency, f_{SYS}, and amplitude, A_{SYS}, using a DC motor coupled to the rod via a belt. A potentiometer was attached to the top of the rod to measure the angular position of the base of the fin and determine the frequency and amplitude of the coupled system.

Experiments were conducted in a flow tank with static flow for the sensor location experiments, and a flow tank for the topology experiments. The test section of the flow tank was 0.25 m wide, 0.25 m deep, and 0.50 c long. The leading edge of the fin was attached to pitching rod that was housed in a NACA fairing.

In each set of experiments the oscillator frequency and sensory feedback gain were varied over a range of values. The oscillator frequency was set to values ranging from 30%f_{RES} to 140%f_{RES} in increments of 10%f_{RES}, where f_{RES} is the resonant frequency of each tested fin, which depends on the fin's geometry and material properties. At each oscillator frequency value, the sensory feedback gain was varied between 0.10 and 4.00 in increments of 0.10. For the experiments that examined the effect of sensor location on the entrainment characteristics of the coupled system, data were collected when feedback was acquired from the dorsal region of the fin and when feedback was acquired from the ventral region of the fin. For the topology experiments, data were collected under four different topology conditions. The details of the sensory feedback topology properties are explained in Sect. 7.

15.4 Design of Fins

Two types of fins were used for experimentation: *spanwise fins* and *chordwise fins*. The spanwise fins were used to investigate how the location of the bending sensor affects the entrainment state of the coupled system. The spanwise fins were designed with spatially varying mechanical properties in the spanwise direction (i.e., in the dorsal-ventral direction, Fig. 15.1b). The chordwise fins were used to investigate how sensory feedback topology affects the entrainment state of the coupled system. The chordwise fins were designed with spatially varying mechanical properties in the chordwise direction (i.e., leading edge to trailing edge, Fig. 15.1c).

The design of the fins used in this study was inspired by the morphology of fish fins. Fish fins have spatially varying mechanical properties and fish can actively change the spatial distribution of a fin's flexural rigidity [10]. Research has shown that fins with chordwise flexibility maximize propulsive performance [11–13]. Fins with spanwise flexibility can produce forces that are parallel to the span of the fin, which could aid in maneuvering an underwater vehicle without changing the orientation of the fin.

The spanwise fins were designed with two discrete regions, a dorsal region and a ventral region, each with different flexural rigidity values. The flexural rigidity (EI) of the ventral region was held constant at a value of 1.45 mPa·m^4, while the flexural rigidity of the dorsal region was set to three different values: 3.00, 11.75, and 17.89 mPa·m^4. The ratio of the flexural rigidity of the dorsal region to the flexural rigidity of the ventral region was used to identify the fins. The ratios were approximately 2:1, 8:1, and 12:1. The different spanwise flexural rigidity configurations are referred to as *FS2*, *FS8*, and *FS12*. Likewise, the chordwise fins were designed with two discrete regions of flexural rigidity, a proximal region and a distal region, each with different flexural rigidity values. The ratio of the flexural rigidity of proximal region to the flexural rigidity of the distal region was held constant at approximately 10:1, with a mean ratio of 9.73:1. In a previous study three different flexural rigidity values were used for the proximal and distal sections: 2.98:0.31, 15.20:1.4, and 30.28:3.45 (all flexural rigidity values are given in units of mPa·m^4). However, here we present only the results for the fin with ratio of 15.20:1.40, which is referred to as Fin *FC15*.

The fins were fabricated out of flexible polyethylene terephthalate (PET) (McMaster-Carr, Robbinsville, NJ). The stiffness of the fin could be set by the geometry and flexural rigidity of the shim-stock. The shim-stock was encased in a silicone rubber sheath in order to keep the shim-stock layers together while the fin oscillates in the water.

For both types of fins, strain gauges (Omega Engineering, Stamford, CT, USA) were used to acquire bending sensory information from the fins. For the spanwise fin, a strain gauge was placed in both the dorsal region and the ventral region of the fin (indicated by the dashed red rectangles in Fig. 15.1b). For the chordwise fins, a strain gauge was placed along the midline of the fin (indicated by the dashed red rectangle in Fig. 15.1c). For both fins, the strain gauges were placed 15 mm from the leading edge of the fin.

15.5 Sensory Feedback

A fundamental property of CPGs is that their output can be modified by sensory feedback [5]. Using sensory feedback to modify the entrainment state of the coupled system raises the question: what type of sensory feedback should be used? Biology can serve as blueprint with which to address this question. Biological evidence suggests that fish fins are innervated with mechanosensors that encode information about movement and the position of a fin [14]. Based on this evidence, strain gauges were used to acquire bending sensory information about the fin, specifically the bending magnitude and direction.

For the studies highlighted in this work, the signal from the strain gauge was used to estimate the angular position of the tip of the foil. This was done, in part, because studies have shown that driving a fin at its resonant frequency has propulsive benefits, such as maximizing propulsive efficiency [15] and maximizing thrust production [16]. Research into CPG-based control of robotic systems has shown that position feedback is capable of driving a robotic limb at the limb's resonant frequency, a behavior known as *resonance-entrainment* [17, 18]. For these reasons the bending sensor signal was used to estimate the angular position of the tip of the fin as the sensory feedback signal. A first-order model was used to map the bending sensor signal to an estimate of the angular position of the tip of the foil. The first-order model was derived by holding the base of the foil stationary while applying a point load to the tip of the foil.

15.6 Effect of Sensor Location on the Entrainment State of the Coupled System

The results show that the characteristics of the entrainment state are significantly influenced by the location of the bending sensor. When the EI-ratio of the spanwise fin is low, the entrainment frequency was similar regardless of whether the feedback signal was acquired from the stiff dorsal (stiff) region of the fin or the ventral (flexible) region of the fin (Fig. 15.3a, b, g). As the EI-ratio increased, the effect of the sensor location on the entrainment state became more significant. In general, a spanwise fin can be driven over a broader range of frequencies when feedback is acquired from its stiff lobe rather than the compliant lobe. However, there are instances when acquiring feedback from the compliant lobe of the fin could be beneficial. For example, at high values of f_{OSC}, the range of entrainment frequencies that are produced when feedback is acquired from the stiff lobe does not overlap with the range of entrainment frequencies produced when feedback is acquired from the compliant lobe (Fig. 15.3h, i). Under these conditions it may be useful to modulate the entrainment frequency by switching between sensor locations as this would broaden the range of entrainment frequencies over which a spanwise fin can be driven compared to acquiring feedback from a single sensor location.

15.7 Effect of Topology on the Entrainment State of the Coupled System

The sensory feedback topology has been shown to affect the entrainment state of the coupled system [17]. Biological evidence has shown that CPG circuits comprise connections that can be both *excitatory* and *inhibitory* [19]. Excitatory/inhibitory feedback causes a neuron to increase/decrease its firing rate. Additionally, the directionality of the sensory feedback signal can be either *ipsilateral* or *contralateral*. Ipsilateral/contralateral feedback is applied to the neuron of the CPG when the limb moves in the same/opposite direction that the neuron drives the movement of the limb. Four basic feedback topologies can be created by combining these two types of synapses and directionality: *ipsilateral inhibitory* (II), *contralateral inhibitory* (CI), *ipsilateral excitatory* (IE), and *contralateral excitatory* (CE) (Fig. 15.4).

To investigate how the topology effects the entrainment state, experiments were conducted using a chordwise fin swimming in a flow tank. The oscillator frequency (f_{OSC}) and sensory feedback gain were varied under each of the topologies and the

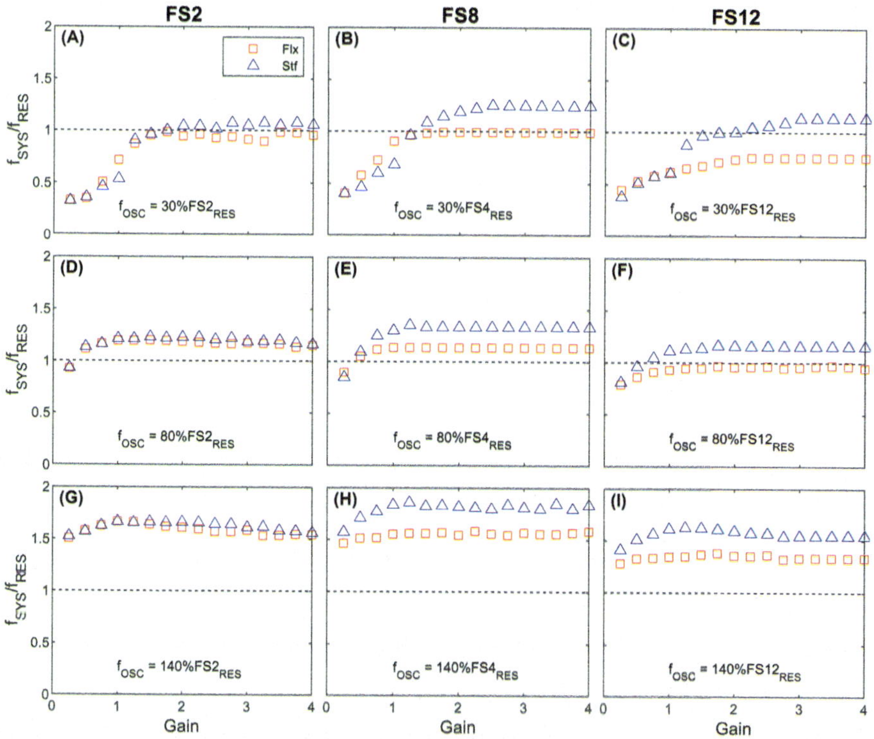

Fig. 15.3 The entrainment frequency as a function of the sensory feedback gain and sensory feedback location. The dashed line represents the resonant frequency of the fin. (**a**, **d**, **g**) represent the frequency of the coupled system when the oscillator frequency (f_{OSC}) is set to 30%Fres, 80%f_{RES}, and 140%f_{RES}, respectively. (**b**, **e**, **h**) represent the coupled system frequency at the same oscillator frequency settings for fin FS8. (**c**, **f**, **i**) represent the coupled system frequency at the same oscillator frequency settings for fin FS12. The frequency of the coupled system is normalized to the resonant frequency of the fin. The blue triangles and red squares represent the coupled system frequency when feedback is acquired from the stiff region of the fin or the flexible region, respectively. The dashed horizontal line represents the resonant frequency of the fin

entrainment frequency and amplitude were recorded. The results showed that the entrainment frequency can be modulated by varying f_{OSC} under all the sensory feedback topologies (Fig. 15.5a–h). Increasing f_{OSC} under any of the sensory feedback topologies caused the entrainment frequency to increase monotonically. But the size of the range of frequencies depended on the type of topology and the sensory feedback gain magnitude. Additionally, the entrainment frequency can be modulated by varying the sensory feedback gain under all the sensory feedback topologies. However, the effect that modulating the sensory feedback gain had on the entrainment frequency depended on the topology. Increasing the sensory feedback gain under the II or CE topologies caused the entrainment frequency to increase (indicated by the arrows in Fig. 15.5a, c), whereas under the CI and IE

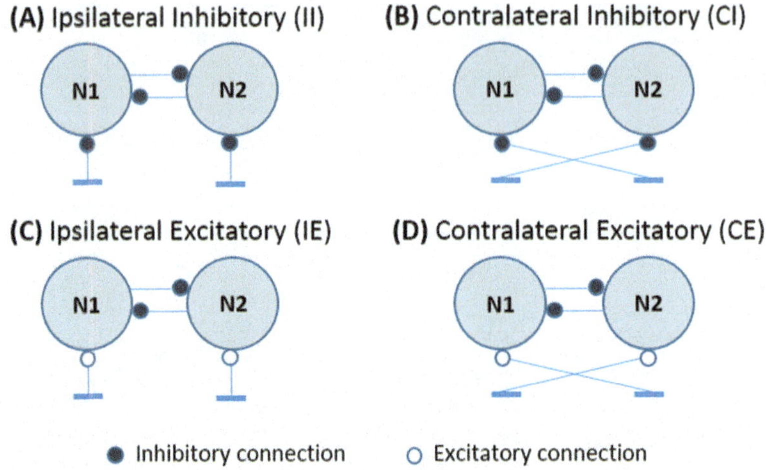

Fig. 15.4 Four types of feedback topologies. Ipsilateral/contralateral feedback is applied to the neuron of the CPG when the fin moves in the same/opposite direction that the neuron drives the movement of the fin

topologies increasing the sensory feedback gain caused the entrainment frequency to decrease (indicated by the arrows in Fig. 15.5b, d). Additionally, increasing the sensory feedback gain under the excitatory topologies also affected the entrainment amplitude—causing it to increase. The entrainment amplitude could be modulated by increasing the tonic drive signal under any of the topologies. The entrainment amplitude increased linearly as the tonic drive was increased (Fig. 15.5k).

15.8 Design of Coupled System for Desired Performance (Resonance-Entrainment)

A fundamental question in CPG-based robotics is how to design the coupled system to produce desired performance. For example, Bailey showed how to maximize running speed of a cockroach-like robot [20]. A focus for swimming robots has been understanding how to design the coupled system in order to swim efficiently. Research has shown that the propulsive efficiency of a fin is maximized when it is driven close to its resonant frequency [15]. The previous sections show how the coupled system parameters affect the characteristics of the entrainment state of the coupled system. This section shows how to design the parameters of the coupled system to produce the desired swimming performance (i.e., drive a fin at resonance).

Only certain combinations of sensory feedback topology and oscillator settings are capable of resonance-entrainment of a foil. Resonance-entrainment refers to an oscillator entraining to the resonant frequency and amplitude of a fin. For every feedback topology the oscillator's endogenous frequency (f_{OSC}), sensory feedback

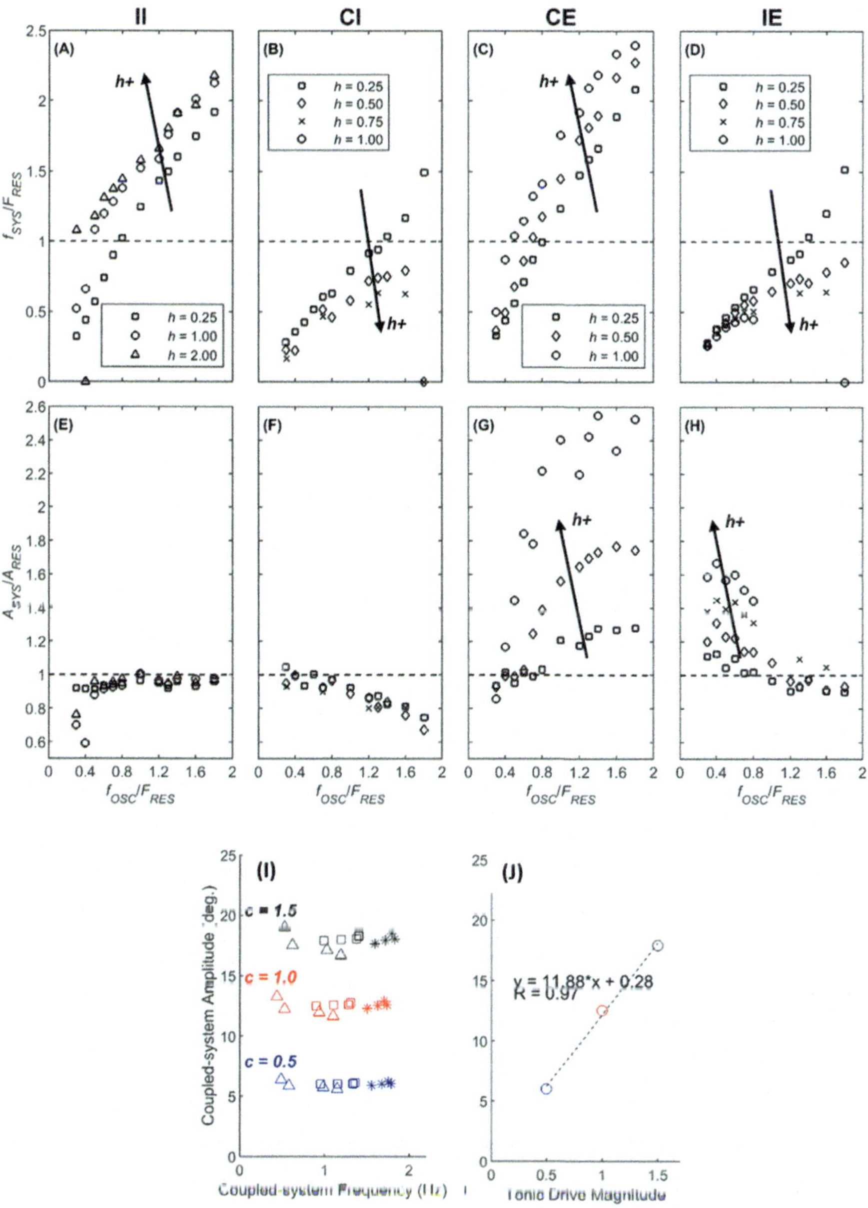

Fig. 15.5 (**a**) The frequency of the coupled system under the II topology. (**b**) The frequency of the coupled system under the CI topology. (**c**) The frequency of the coupled system under the CE topology. (**d**) The frequency of the coupled system under the IE topology. (**e**) The amplitude of the coupled system under the II topology. (**f**) The amplitude of the coupled system under the CI topology. (**g**) The amplitude of the coupled system under the CE topology. (**h**) The amplitude of the coupled system under the IE topology. The arrows in **a**, **b**, **c**, **d**, **g**, and **h** indicate the direction of increasing sensory feedback gain. (**i**) The coupled system frequency and amplitude as a function of the tonic drive. (**j**) The coupled system amplitude as a function of tonic drive

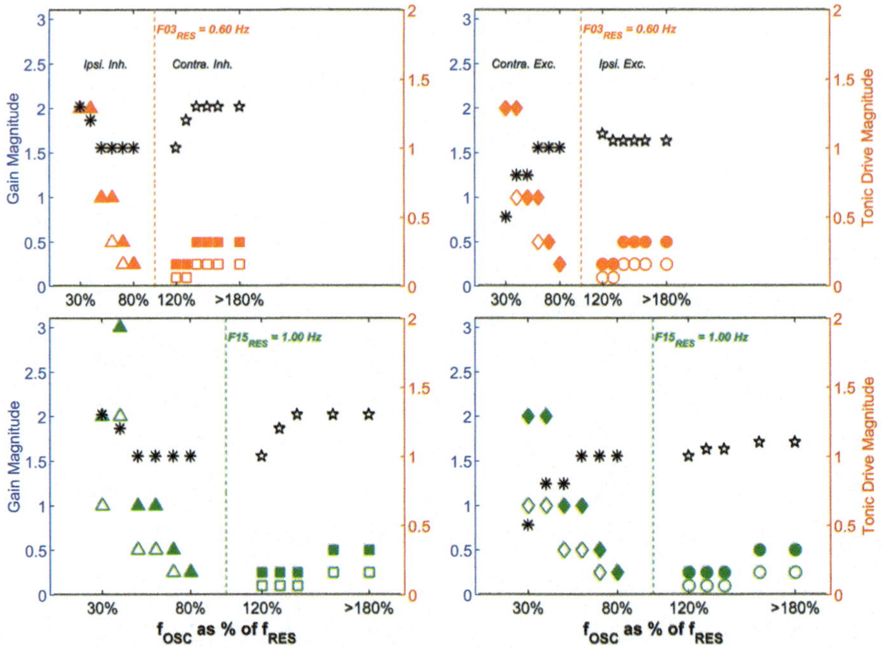

Fig. 15.6 The settings for f_{OSC}, gain, and tonic drive values under which resonance-entrainment occurred for fin *FC03* (**a**, **b**) and for fin *FC15* (**c**, **d**) under the different topologies. The triangle, square, diamond, and circle markers represent the gain values that caused resonance-entrainment under the II, CI, CE, and IE topologies, respectively. The unfilled markers represent the minimum sensory feedback gain value that causes resonance-entrainment and the filled markers represent the maximum sensory feedback gain value that causes resonance-entrainment. The asterisks and star markers represent the tonic drive value that was necessary to cause resonance-entrainment. The oscillator frequency is represented as a percentage of the resonant frequency of each fin

gain, and tonic drive must be set to specific values that depend upon the sensory feedback topology, in order for resonance-entrainment to occur. Under the II and CE topologies, the oscillator's endogenous frequency (f_{OSC}) must be set between 30% and 80% of the resonant frequency of the foil. The sensory feedback gain must be set in accordance with the setting of f_{OSC}: for every value of f_{OSC}, there is a range of gain values that will produce resonance-entrainment (the range of gain values that produce resonance-entrainment are identified by their minimum and maximum values in Fig. 15.6). If f_{OSC} is set to 30%f_{RES}, then the sensory feedback gain must be high; however, if f_{OSC} is set to 30%f_{RES}, then the sensory feedback gain must be set "low" (triangular markers in Fig. 15.5a, c; diamond markers in b, d). Additionally, the tonic drive must be set in accordance with the value of f_{OSC} value: high when f_{OSC} is set to close to 30%f_{RES}, and low when f_{OSC} is set closer to 80%f_{RES} (black asterisk markers in Fig. 15.6a, b). Under the CI and IE topologies f_{OSC} must be set between 120% and 180% of the resonant frequency of foil. The sensory feedback gain had to be low when f_{OSC} was set close to 120%f_{RES} and slightly higher as f_{OSC} was set closer to 180% f_{RES} (square markers in Fig. 15.6a, c; circular markers in b, d). The tonic drive had to be set increasingly higher as f_{OSC} was varied across its range of values that produced resonance-entrainment.

These results offer guidance on how to modulate the sensory feedback topology and neural oscillator settings to produce resonance-entrainment as the flexural rigidity of a fin is varied. For example, suppose that the flexural rigidity of a fin is varied over a range of values from a minimum value of f_{Min} to a maximum value of f_{Max}. Under the II and CE topologies, set f_{OSC} equal to $80\% f1_{RES}$ and set the gain appropriately for resonance-entrainment to occur. If the resonant frequency of the foil is changed such that $30\% f_{Max} \leq f_{OSC} = 80\% f_{Min} \leq 80\% f_{Max}$, then only the gain has to be increased in order for resonance-entrainment to be maintained. If the resonant frequency of the foil increases outside of this range, then f_{OSC} must be increased and reset. Under the CI and IE topologies, set $f_{OSC} = 180\% f_{Min}$ and set the gain appropriately for resonance-entrainment to occur. If the resonant frequency of the foil is changed such that $120\% f_{Max} \leq f_{OSC} = 180\% f_{Min} \leq 180\% f_{Max}$, then only the gain has to be increased in order for resonance-entrainment to be maintained. If the resonant frequency of the foil increases outside of this range, then f_{OSC} must be increased.

15.9 Discussion

The results presented in this chapter demonstrate how the location of the sensory feedback signal and sensory feedback topology can be used to modulate the performance of a compliant system. Results are shown for a compliant fin, but are applicable more generally to soft robotic systems where resonance characteristics affect locomotion or force production.

Sensory feedback can be acquired from a compliant robotic limb and used to modulate the entrainment state of a coupled system. By using positional sensory information obtained from strain gauges placed on the fins, the entrainment state of the coupled system could be modulated by varying the oscillator frequency and sensory feedback gain. These results demonstrate that angular position feedback is capable of successful entrainment of a compliant fin. However, other forms of feedback may also be capable of successful entrainment. Research has shown that raw bending feedback is capable of entrainment and affects the entrainment frequency and duty cycle of the coupled system [21]. The use of other forms of feedback will most likely enhance the capabilities of a CPG-driven compliant system.

The performance of compliant robotic system with spatially varying mechanical properties can be enhanced by acquiring sensory feedback from an array of sensors compared to the use of a single sensor. The range of frequencies over which entrainment occurred was maximized when feedback was acquired from multiple locations on the spanwise fin. When applying CPG-based control to these types of robotics systems, engineers should strive to place sensors in regions of the limb with different flexural rigidity properties.

Additionally, the performance of the system can be enhanced by modulation of the sensory feedback signal. The entrainment frequency can be altered solely by varying the location of the sensory feedback signal. This finding offers another parameter with which the entrainment state of the coupled system can be modulated. Another way that the entrainment state of the system can be modulated is by changing the sensory feedback topology. Mixing sensory feedback topologies maximizes the range of entrainment frequencies over which entrainment can occur. The range of frequencies over which entrainment occurs is similar for both inhibitory and excitatory topologies. However, the coupled system amplitude is significantly affected by variation of the sensory feedback gain magnitude. Therefore, the number of control parameters that are necessary to modulate the entrainment state of the coupled system with inhibitory topologies is reduced compared to excitatory topologies.

The entrainment state of the coupled system is sensitive to the mechanical properties of the fin. There are only a few parameter settings that produce resonance-entrainment. This result is in contrast to research into CPG-based control of second-order systems, which have been shown to have robust resonance-entrainment characteristics [17, 18]. Fins are highly damped systems that operate in a fluidic environment in which damping forces due to drag, and the inertial forces of the added mass of the fluid, are significant. A recently accepted study [22] showed, using describing function analysis, that the damping of a fin affects the frequency response of the mechanical properties of a fin, which in turn affect the entrainment state of the coupled system. These results imply that compliant robotic systems that operate in damped environments need to be more carefully tuned/designed in order for resonance-entrainment to occur.

15.10 Conclusion

A fundamental question in soft robotics is how to acquire and use sensory feedback from a compliant system in order to produce desired entrainment [23]. The results presented in this chapter offer one way in which to acquire sensory feedback from a compliant structure: a fin with spatially varying mechanical properties. We have shown that bending sensory feedback can be used to modulate the entrainment state of the coupled system. The entrainment state can be modulated by varying the sensory feedback location and the sensory feedback topology. Resonance-entrainment occurs only for certain combinations of the coupled system parameters. Taken together these results show the promise in using fin-based sensory feedback to modulate the entrainment state of a coupled system.

References

1. A.H. Cohen, D.L. Boothe, Sensorimotor interactions during locomotion: Principles derived from biological systems. **245**, 239–245 (1999)
2. T. Iwasaki, M. Zheng, Sensory feedback mechanism underlying entrainment of central pattern generator to mechanical resonance. Biol. Cybern. **94**(4), 245–261 (2006)
3. S. Rossignol, R. Dubuc, J. Gossard, Dynamic sensorimotor interactions in locomotion. Physiol. Rev. **86**(1), 89–154 (2006)
4. T.G. Brown, The intrinsic factors in the act of progression in the mammal. Proc R Soc Lond Ser B Biol Sci **84**(572), 308–319 (1911)
5. E. Marder, R.L. Calabrese, Principles of rhythmic motor pattern generation. Physiol. Rev. **76**(3), 687–717 (1996)
6. N.G. Hatsopoulos, W.H. Warren Jr., Resonance tuning in rhythmic arm movement. J. Mot. Behav. **28**(1), 3–14 (1996)
7. P.R. Bandyopadhyay, Trends in biorobotic autonomous undersea vehicles. IEEE J. Ocean. Eng. **30**(1), 109–139 (2005)
8. C. Laschi, B. Mazzolai, M. Cianchetti, Soft robotics: Technologies and systems pushing the boundaries of robot abilities. Sci. Rob. **1**(1) (2016)
9. S.H. Strogatz, Exploring complex networks. Nature **410**(6825), 268–276.5 (2001)
10. G.V. Lauder, Locomotion, in *The Physiology of Fishes*, ed. by D. H. Evans, J. B. Claiborne, 3rd edn., (CRC Press, Boca Raton, 2006), pp. 3–46
11. C.J. Esposito, J.L. Tangorra, B.E. Flammang, G.V. Lauder, A robotic fish caudal fin: Effects of stiffness and motor program on locomotor performance. J. Exp. Biol. **215**(Pt 1), 56–67 (2012).
12. A.K. Kancharala, M.K. Philen, Optimal chordwise stiffness profiles of self- propelled flapping fins. Bioinspir. Biomim. **11**, 056016 (2016)
13. K.N. Lucas, P.J.M. Thornycroft, B.J. Gemmell, S.P. Colin, J.H. Costello, G.V. Lauder, Effects of non-uniform stiffness on the swimming performance of a passively-flexing, fish-like foil model effects of non-uniform stiffness on the swimming performance of a passively-flexing, fish-like foil model. Bioinspir. Biomim. **10**, 056019 (2015)
14. B.R. Aiello, T.A. Stewart, M.E. Hale, T.A. Stewart, Mechanosensation in an adipose fin. Proc. R. Soc. B **283**, 1–5 (2016)
15. P.A. Dewey, B.M. Boschitsch, K.W. Moored, H.A. Stone, A.J. Smits, Scaling Laws for the thrust production of flexible pitching panels. J. Fluid Mech. **732**, 29–46 (2013)
16. Y. Park, S. Member, U. Jeong, J. Lee, S. Kwon, H. Kim, K. Cho, Kinematic condition for maximizing the thrust of a robotic fish using a compliant caudal fin. IEEE Trans. Robot. **28**(6), 1–12 (2012)
17. C.A. Williams, S.P. DeWeerth, A comparison of resonance tuning with positive versus negative sensory feedback. Biol. Cybern. **96**(6), 603–614 (2007)
18. M.M. Williamson, Neural control of rhythmic arm movements. Neural Netw. **11**(7–8), 1379–1394 (1998)
19. D.L. McLean, J. Fan, S. Higashijima, M.E. Hale, J.R. Fetcho, A topographic map of recruitment in spinal cord. Nature **446**(7131), 71–75 (2007)
20. S. Bailey, *Biomimetic Control with a Feedback Coupled Nonlinear Oscillator: Insect Experiments, Design Tools, and Hexapedal Robot Adaptation Results* (Stanford University, 2004)
21. C.L. Hamlet, K.A. Hoffman, E.D. Tytell, L.J. Fauci, The role of curvature feedback in the energetics and dynamics of lamprey swimming: A closed-loop model. Comput. Biol. **14**(8), 1–29 (2018)
22. G.N. Carryon, J.L. Tangorra, The effect of sensory feedback topology on the entrainment of a neural oscillator with a compliant foil for swimming systems. Bioinspir. Biomim (2020)
23. M. Ishige, T. Umedachi, T. Taniguchi, Y. Kawahara, Exploring behaviors of Caterpillar-like soft robots with a central pattern generator-based controller. Soft Robot. **6**(5), 579–594 (2019)
24. K. Matsuoka, Sustained oscillations generated by mutually inhibiting neurons with adaptation. Biol. Cybern. **376**, 367–376 (1985)

Index

D. A. Paley, N. M. Wereley (eds.), *Bioinspired Sensing, Actuation, and Control in Underwater Soft Robotic Systems*, https://doi.org/10.1007/978-3-030-50476-2

The manufacturer's authorised representative in the EU is Springer Nature Customer Service Centre GmbH, Europaplatz 3, 69115 Heidelberg, Germany. If you have any concerns regarding our products, please contact ProductSafety@springernature.com

Printed and bound by CPI Group (UK) Ltd, Croydon, CR0 4YY
28/09/2026
02224092-0003